11-30

A Short Course in Calculus with Applications

Under the Editorship of the Late Carl B. Allendoerfer

A Short Course in Calculus with Applications

Hugh G. Campbell & Robert E. Spencer

Virginia Polytechnic Institute and State University

Macmillan Publishing Co., Inc.

New York

Collier Macmillan Publishers

London

To Julia and Earl Spencer

PRINTED IN THE UNITED STATES OF AMERICA

MACMILLAN PUBLISHING CO., INC.
866 Third Avenue, New York, New York 10022

COLLIER-MACMILLAN CANADA, LTD.

LIBRARY OF CONGRESS CATALOGING IN PUBLICATION DATA

Campbell, Hugh G
A short course in calculus with applications.

Bibliography: p.
1. Calculus. I. Spencer, Robert E., joint author.
II. Title.
QA303.C19 515 73-22628
ISBN 0-02-318590-2

Printing: 1 2 3 4 5 6 7 8 Year 5 6 7 8 9 0

Preface

One distinguishing feature of this book is its preoccupation with motivation—a vital part of the learning process. The authors have devoted considerable energy toward the goal of furnishing the reader with reasons for studying the mathematics included in this book. Almost every section includes an optional subsection entitled "Applications"; these subsections, however, are not designed as comprehensive lists of real world applications, nor is any claim made concerning the realism of many examples. Rather, these "indications of usefulness" are designed to point out in a limited and elementary way that the material is relevant to a broad spectrum of disciplines, and also that it has considerable potential for future applications.

A second distinguishing feature of this book is its organizational form. A conscious effort was made to organize the book in such a way that the instructor can alter the emphasis or length of a course to suit the needs of the class. The book is organized into 7 chapters with a total of 43 sections; each section is designed to constitute 1 lesson, but 2 lessons will be needed in some cases if the optional applications are studied in detail. Prerequisites for each chapter are stated at the beginning of the chapter. These statements of prerequisites make it clear what previous sections in the book are needed before studying each new section; in a few cases, however, additional previous sections are assumed in the optional subsections on applications.

The book is intended primarily for use by students of business, the social sciences, or the life sciences. The only prerequisites are high school algebra and geometry, although the maturity gained in a college course in either algebra or finite mathematics may be desirable. The book can be used for a period of one quarter, one semester, or two quarters. Some possibilities for a one-quarter course are (1) a course in differential calculus from Chapters 1–3; (2) an introduction to differential and integral calculus from Chapter 1, Sections 2.1–2.6, Sections 3.1–3.2, the first part of Section 3.6, Sections 4.1–4.2, and Sections 5.1–5.4; and (3) a course in differential calculus (including exponential functions, logarithmic functions, and functions of more than one variable) from Chapter 1, Sections 2.1–2.5, Sections 3.1–3.3, the first part of Section 3.6, Sections 6.1–6.3, and Sections 7.1–7.3. For a one-semester course it is suggested that one include Chapter 1, Sections 2.1–2.6, Sections 3.1–3.4, the first part of Section 3.6, Sections 4.1–4.2, Sections 5.1–5.4, and Sections 6.1–6.4. Several variations may be obtained by deciding upon a few more sections, such as 7.1–7.4, to add to this foundation. An emphasis on preparation for probability and statistics can be made by including Sections 5.5–5.6 and 6.7; *examples related to probability and statistics in the optional subsections on applications are labeled clearly and occur in Sections 1.4, 1.6, 1.8, 2.2, 3.4, 5.2, 5.3, 5.5, 5.6, and 6.7.* A two-quarter course can cover all sections in the book, although a few sections may be omitted to allow extra time for emphasis on applications.

Each set of exercises is preceded by a carefully chosen "suggested minimum assignment," which can assist the teacher and the student in determining a minimal set of exercises that need to be solved in order to gain an understanding of the material of the section.

This book has the same format as the text *Finite Mathematics* by the same authors. Together the two books present a carefully planned, well-motivated mathematics program for a two- or three-semester introductory course in those topics with particular usefulness in the business, social, and life sciences.

Above all, the guiding principle in the construction of this book has been consideration for the needs of the student. The sincerity of the authors in this regard is illustrated by the attempt to motivate through the subsections on applications, the list of new vocabulary at the end of each chapter, the numerous exercises, and the many examples illustrating the concepts.

The authors express sincere appreciation to the many people who have assisted us in the production of this book, especially the late Dr. Carl Allendoerfer. Special gratitude is also expressed to Everett Smethurst for his time and efforts in behalf of this project. Several others, especially William F. Tyndall and H. Earl Spencer, have made many valuable suggestions. Our respective wives, Allen and Carole, also deserve considerable thanks for their encouragement and typing during the development of the book.

H. G. C.
R. E. S.

Blacksburg, Virginia

Contents

1 Preliminary Topics

Before one can begin a successful study of calculus, certain preliminary topics must be mastered. The first chapter is a collection of such topics.

Much of the first part of this chapter is classified as analytic geometry. Analytic geometry has been described as "a subject generated by the marriage of algebra and geometry"; its development in the first half of the 17th century by René Descartes (1596–1650) and Pierre de Fermat (1601–1665) provided a necessary foundation for the development of much of modern science and mathematics, including calculus.

Much of the latter part of this chapter is devoted to a brief introduction to the concept of the limit of a function. This is necessary because two of the basic concepts of the calculus—the derivative and the definite integral—are defined as limits. Historically, the attainment of precision and clarity in the presentation of the basic ideas concerning limits proved to be very elusive. More than one hundred years elapsed after the discovery of calculus before the underlying notion of a limit was refined to the extent that the calculus was placed on a rigorous foundation; many of these refinements were made by Augustin Louis Cauchy (1789–1857) in the early part of the 19th century.

Some immediate applications of these preliminary topics will be given; moreover, the reader should recognize that several applications which occur in later chapters also depend to a large extent upon the topics in this chapter.

Prerequisites: high school algebra and geometry
Suggested Sections for Moderate Emphasis: 1.1–1.9 should all be covered; 1.1 and 1.2 can be combined into one lesson.
Suggested Sections for Minimum Emphasis: 1.1–1.9 should all be covered; 1.1 and 1.2 can be combined into one lesson. For a short course with emphasis on techniques, 1.7–1.9 can be skimmed in one or two lessons.

1.1 An Orientation

Over the years, the development of mathematics has closely paralleled the need for quantification of natural and other phenomena. From the time man invented symbols that correspond to our numerals 1, 2, . . . , there has been a relentless march toward the use of mathematics in ever more diverse ways. Certainly, the rate of this advance has not been constant, and often it has been related to the intellectual and political climates of the age as well as the maturity and influence of sister disciplines. Early interest in astronomy, for example, provided considerable impetus for the development of related mathematical concepts and techniques. Over a period of years, many of these concepts and techniques have led to a comparatively sophisticated development of the engineering sciences; the development of the computer and space travel give ample evidence of this fact. The computer and other engineering developments, in turn, are now having a tremendous effect upon modern mathematics, as well as having the effect of opening vast new areas for quantification in the social, management, and life sciences.

In this book we shall introduce certain topics of calculus and illustrate their use in the social, management, and life sciences. Extensive quantification in these areas is a relatively new thing. Consequently, the art of application of the topics of this book to the behavioral, management, and life sciences is in its infancy. There is considerable reason to believe that calculus can contribute significantly to our understanding of these areas of knowledge, but much work remains to be done.

As participants in a physical, biological, economic, political, and social environment characterized by continuing change, we find it helpful to be able to study these changes and associated rates of change through the construction of mathematical models and other quantifying techniques. Calculus is invaluable in this endeavor. There are two major classifications of problems that are particularly susceptible to the techniques of calculus.

1. A relationship between two variables is known; find a second relationship that involves the original two variables and the *rate of change* of one variable with respect to the other.
2. The converse of classification 1, where the second relationship is known and the first relationship is sought.

As examples of these two types of problems consider a population of objects, such as people, plants, voters, dollars, or blood cells. Let y represent the number of objects at a given time t. As time t passes, y may vary. In fact, it may be possible to establish an equation that will model the relationship between y and t. Suppose that this is the case and that this relationship is expressed by an equation,

$$y = f(t),$$

which is read "y is a function of t." Now if t_1 and t_2 are two different measures of time, then $f(t_1)$ and $f(t_2)$ are measures of the corresponding populations. It is customary to represent the change $t_2 - t_1$ in time by the symbol Δt (read "delta t"), and likewise, the change in population by

$$\Delta y = f(t_2) - f(t_1).$$

If the change in population is divided by the change in time, then the result is referred to as the *average rate of change* of population with respect to time over the time interval from t_1 to t_2. This average rate of change of y with respect to t can be expressed as

$$\frac{\Delta y}{\Delta t} = \frac{f(t_2) - f(t_1)}{t_2 - t_1}.$$

Under certain conditions, if t_1 remains unchanged, and smaller and smaller changes $t_2 - t_1$ in time are considered, then the average rate of change

$$\frac{f(t_2) - f(t_1)}{t_2 - t_1}$$

will approximate what might be called the *immediate* or *instantaneous rate of change* of population with respect to time at time t_1. In other words, we are raising the following question: What value does $\Delta y/\Delta t$ approach as Δt approaches zero? The answer to this question is so important that we give it a special name; we say that it is the value of the **derivative** of y with respect to t at time t_1. The question of finding the immediate or instantaneous rate of change of one variable with respect to another variable, when the relationship between them is known, occurs so frequently in applied problems that it is difficult to overemphasize its importance.

Other very important applications arise from the inverse question of finding the relationship between two variables when the instantaneous rate of change is known. For example, if we reason that the rate of change of a population y with respect to time t is proportional to the current population, then $\Delta y/\Delta t$ approaches ky as Δt approaches zero, where k is the constant of proportionality. From this assumption concerning the rate of change, the techniques of calculus can be used to find an equation representing the relationship between the population y and the time t; this relationship can be used as a model to make projections into the future.

In a serious approach to the solution of the many applied problems similar to those just mentioned, we must gain some precision in our understanding of the terms and concepts that have been mentioned so far. Hence in the first chapter we shall deal with those topics that are essential to the main concept of the derivative that is to be presented in Chapter 2.

Primary credit for the discovery of calculus is usually given to Isaac Newton (1642–1727) and Gottfried Leibniz (1646–1716). The student who wishes to read an interesting historical article about calculus before pursuing the study of the subject in detail is referred to Newman's book [17], pp. 53–62.

EXERCISES

Suggested minimum assignment: Exercises 1, 2, and 4.

1. Suppose that the number n of bacteria in a culture after t hours is given by Table 1.1. By guessing a formula relating n and t, determine a mathematical model that expresses the number of bacteria at any given time. Be sure that your formula checks for the data in Table 1.1.

Table 1.1

n	100	150	200	250	300
t	0	1	2	3	4

2. Suppose that a child attempts to measure the area of a circle. Let A represent the area and r represent the radius of the circle. The child uses the formula $A = 3r^2$ as the mathematical model of the relationship between the area and radius of a circle. What are the advantages and disadvantages of this model over the formula $A = \pi r^2$?
3. Let R_0 be the sensation evoked by a stimulus of intensity S_0. Suppose that ΔR is the increase in sensation caused by an increase of ΔS in the stimulus. It has been shown that

$$\frac{\Delta S}{S_0} = \Delta R.$$

Therefore, for a given increase of ΔS in the stimulus, the increase in sensation is greater for a relatively small original stimulus S_0 than for a relatively large original stimulus. Verify this fact for the case that $\Delta S = 2$ by computing ΔR under the assumption that $S_0 = 50$ and also under the assumption that $S_0 = 100$. For further discussion see pp. 23–29 of *Quantification in Psychology* by Hays [12].
4. An automobile manufacturer wants to test a new car to find certain information about its performance. Let y be the distance in feet which the car travels from the starting point in t seconds. Suppose that $y = 10$ when $t = 1$ and $y = 100$ when $t = 3$. What is the change Δy in distance corresponding to the change in time of $\Delta t = 3 - 1 = 2$ seconds? Also find $\Delta y/\Delta t$ and interpret its meaning.
5. Find in your school library an article related to the history of calculus, and write a short summary of what you find.

1.2 Real Numbers

Before the subject known as calculus can be studied successfully, it is necessary to have some basic knowledge about the number system within which we shall operate. Thus we begin with a discussion of some of the terminology and important facts about real numbers. Rather than attempt to state all the definitions and postulates needed for a rigorous development of the properties of the real number system, we assume some familiarity with real numbers from high school algebra and proceed with the facts that the reader will need.

The real numbers can be represented geometrically by the **real line,** shown in Figure 1.1. We assume that for each real number there is a corresponding point on the real line, and vice versa. The point $\boldsymbol{O}$ is called the **origin,** and the number corresponding to any point P is called the

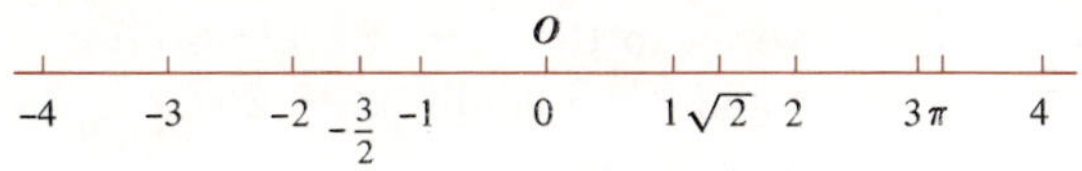

Figure 1.1. *Real line.*

coordinate of P. Included among the real numbers are the **positive integers** or **natural numbers,**

$$1, 2, 3, 4, \ldots,$$

the **negative integers,**

$$-1, -2, -3, -4, \ldots,$$

and the integer 0 (**zero**), which is neither positive nor negative. A **rational number** is a real number that can be expressed as the ratio a/b of two integers a and b, where $b \neq 0$. The rational numbers include the integers (any integer a can be expressed as $a/1$) and fractions such as $-\frac{3}{2}$, $\frac{1}{3}$, and $\frac{22}{7}$. Any real number that is not a rational number is called an **irrational number.** It can be shown that numbers such as $\sqrt{2}$, $\sqrt[3]{5}$, and π are irrational numbers. Every real number is known to have a decimal representation, and it may be helpful to know that terminating or repeating decimal representations always correspond to rational numbers. For example,

$$3.125 = \tfrac{25}{8}$$

and

$$.124242424\ldots = \tfrac{41}{330}.$$

Long division of 41 by 330 will verify the last equality. An irrational number has a decimal representation that is nonterminating and nonrepeating.

We say that real number a is less than real number b if the point corresponding to a lies to the left of the point corresponding to b on the real line, and we denote this relationship by $a < b$. For example,

$$\sqrt{2} < 3 \qquad \text{and} \qquad -7 < -6.$$

Another way to explain the meaning of $a < b$ is to say that $b - a$ is positive. To indicate that a is either equal to or less than b, we write $a \leq b$. For example,

$$4 \leq 5 \qquad \text{and} \qquad 5 \leq 5$$

are both correct. "Real number b is greater than real number a" is written $b > a$ and means exactly the same thing as $a < b$. Similarly, $b \geq a$ denotes that b is equal to or greater than a and means the same thing as $a \leq b$. The symbols $<$, $>$, $\leq$, and $\geq$ are called **inequality symbols,** and any mathematical statement involving any of these symbols is called an **inequality.** Notice that $a > 0$ means that a is positive, $a < 0$ means that a is negative, $a \geq 0$ means that a is nonnegative, and $a \leq 0$ means that a is nonpositive. If we write

$$2 < a < 3,$$

we mean that $2 < a$ *and* also $a < 3$. Therefore, to say that $\frac{5}{2}$ is between 2 and 3 we could write $2 < \frac{5}{2} < 3$.

Example 1

A broker recommends to a client that a particular common stock be purchased if the price–earnings ratio of the stock is less than 10. To express this statement as an inequality, let E be the earnings per share for the latest 12 months and let P be the current price per share. The broker's recommendation (assume that $E > 0$) is to buy the stock, provided that

$$\frac{P}{E} < 10. \blacksquare$$

We shall need to be able to refer to subsets of real numbers, and hence it shall be convenient to use set notation. For example,

$$\{x:2 < x < 5\}$$

denotes the set of numbers that are greater than two and also less than five. If a and b are real numbers and $a < b$, then the set

$$\{x:a < x < b\}$$

is called an **open interval** and is also designated by (a, b). The set

$$\{x:a \leq x \leq b\}$$

is called a **closed interval** and often is denoted by $[a, b]$. The sets

$$\{x:a \leq x < b\} \qquad \text{and} \qquad \{x:a < x \leq b\},$$

where $a < b$, are called **half-open intervals** and are denoted by $[a, b)$ and $(a, b]$, respectively. In each case a and b are called the **endpoints** of the interval, and each number in (a, b) is called an **interior point** of the interval. See Figure 1.2 for a geometric illustration of the types of intervals discussed above.

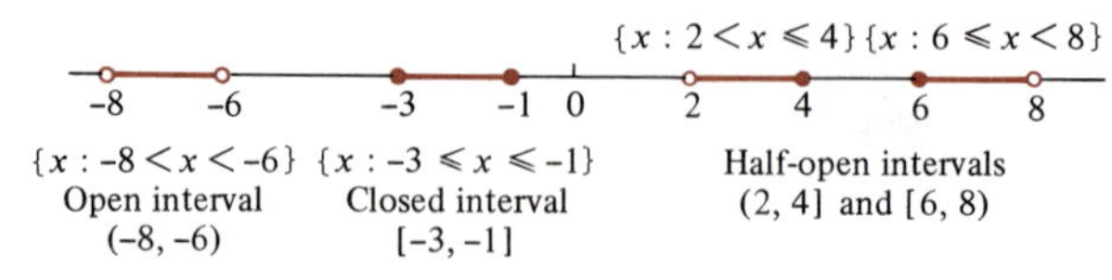

Figure 1.2. *Types of intervals.*

We now turn to the final new concept of this section. If the price of a stock changes from \$120 to \$130 over a period of time, we say that the price changed by \$10 in a positive direction. If, over the next time interval, the stock price changes from \$130 to \$120, we say that the change was −\$10. In some situations, only the amount of change, and not the direction of change, is desired. In such cases the concept of **absolute value** of a real number is important.

Definition 1

The **absolute value** of a real number a, denoted by $|a|$, is given by

$$|a| = \begin{cases} a & \text{if } a \geq 0, \\ -a & \text{if } a < 0. \end{cases}$$

Example 2

$$|5| = 5,$$
$$|0| = 0,$$
$$|-5| = -(-5) = 5.$$

Notice that the absolute value of a negative number is the negative of that negative number, which gives a positive result. ▮

The number $|a - b|$ is the number of units between a and b on the real line and is called the **distance** between a and b. If $a = -4$ and $b = 8$, then the distance between a and b (see Figure 1.3) is given by

$$|a - b| = |-4 - 8| = |-12| = 12.$$

Observe also that

$$|b - a| = |8 - (-4)| = |12| = 12.$$

In general

$$|a - b| = |b - a|,$$

which can be interpreted as meaning the distance between a and b without regard to direction.

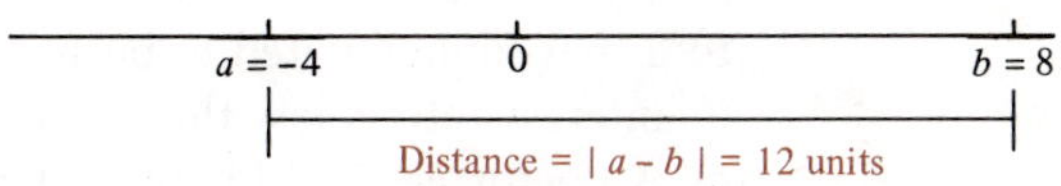

Figure 1.3

For any real number a the rules

$$|a|^2 = a^2$$

and

$$\sqrt{a^2} = |a|$$

are valid.

Example 3

If $a = -5$, then

$$|a|^2 = |(-5)|^2 = 5^2 = 25$$

and

$$a^2 = (-5)^2 = 25,$$

which illustrates that $|a|^2 = a^2$ when $a = -5$. Also

$$\sqrt{a^2} = \sqrt{(-5)^2} = \sqrt{25} = 5$$

and

$$|a| = |-5| = 5,$$

which illustrates that $\sqrt{a^2} = |a|$ when $a = -5$. Notice that $\sqrt{25} = 5$; it is *not* correct to write $\sqrt{25} = \pm 5$. We may think of the equation $\sqrt{a^2} = |a|$ as defining $\sqrt{25}$ to equal 5. ▮

In this section we have introduced real numbers, discussed inequality relationships among real numbers, defined various types of intervals on the real line, and presented the concept of the absolute value of a real number.

APPLICATIONS

The applications given at the end of each section are certainly not meant to be an exhaustive listing of the applications of that particular section, nor is any claim made concerning the realism of many of the examples. Rather, these "evidences of usefulness" are supplied primarily in the interest of motivating the reader and with the hope that occasionally one will provide a spark that may arouse a deeper, more productive study in some areas of interest. *All applications are optional in that they are not prerequisite to later material in the main part of the text.*

Example 4

Many important decisions are based upon statements that can be expressed mathematically as inequalities. In Exercise 14 the reader is asked to translate each of the following into an inequality:

(*a*) The plane can take off as soon as the visibility is at least 1 mile.
(*b*) Town council voted to borrow money to build the new school if the interest rate is less than $5\frac{1}{2}$ per cent.
(*c*) A doctor advised his patient to come back whenever the patient's normal pulse rate exceeds 85 beats per minute.
(*d*) A client instructs his stockbroker to buy 10 shares of stock A if the price of the stock reaches the amount necessary to make the annual dividend exceed 5 per cent of the price of the stock.
(*e*) The client in part (*d*) tells the stockbroker to place a stop-loss order for stock B at \$50 per share.
(*f*) A basketball coach tells his team to freeze the ball whenever the team has at least a 10-point lead.
(*g*) The trailing presidential candidate will campaign in a particular state if the latest opinion poll in that state shows that he is within 12 percentage points of the leading candidate. ▮

EXERCISES

Suggested minimum assignment: Exercises 1, 3, 5, 7, 8, 10, 11(*a*), and 13.

1. Tell whether each of the following real numbers is rational or irrational:

$$\tfrac{2}{3},\quad 0,\quad \pi,\quad 2\pi,\quad 5,\quad \sqrt{2},\quad .44444\ldots,\quad -\tfrac{9}{8}.$$

2. Tell whether each of the following real numbers is rational or irrational:

$$\sqrt{9},\quad \sqrt[3]{9},\quad -3,\quad 1+\pi,\quad \tfrac{3}{4},\quad .7239,\quad \tfrac{22}{7},\quad -\tfrac{3}{2}.$$

3. State whether each of the following is true or false.
(*a*) $5 > \pi$. (*b*) $-8 > -7$. (*c*) $-5 \leq 0$. (*d*) $0 \leq 0$.
4. State whether each of the following is true or false.
(*a*) $-9 < -8$. (*b*) $-8 > -9$. (*c*) $-3 \leq 2$. (*d*) $4 \geq 4$.
5. Change from set notation to interval notation and illustrate the set geometrically. Also tell in each case whether the interval is open, closed, or half-open.
(*a*) $\{x: -1 < x < 6\}$.
(*b*) $\{x: -2 \leq x < 1\}$.
(*c*) $\{x: -8 \leq x \leq -6\}$.
(*d*) $\{x: 4 < x \leq 7\}$.
6. Repeat Exercise 5 for each of the following.
(*a*) $\{x: 9 \leq x \leq 10\}$.
(*b*) $\{x: 6 \leq x < 7\}$.
(*c*) $\{x: -5 < x \leq 0\}$.
(*d*) $\{x: -1 < x < 2\}$.
7. Change each of the following intervals to set notation.
(*a*) $(-3, 8)$. (*b*) $[\pi, 5]$. (*c*) $[-19, -5]$. (*d*) $(-3, 3]$.
8. Evaluate each of the following.
(*a*) $|-8|$. (*b*) $|8|$. (*c*) $|5-5|$. (*d*) $|3-9|$.
(*e*) $|5-7-1|$. (*f*) $|-4|^2$. (*g*) $|(-3)(4)|$.
9. Evaluate each of the following.
(*a*) $|-4|$. (*b*) $|0|$. (*c*) $|5-3|$. (*d*) $\left|\dfrac{-4}{5}\right|$.
(*e*) $\dfrac{|7|}{|3-4|}$. (*f*) $|-5| + |3|$. (*g*) $|-5+3|$.
10. If $a = -3$ and $b = 7$, illustrate that
(*a*) $|a-b| = |b-a|$. (*b*) $|a|^2 = a^2$.
(*c*) $\sqrt{a^2} = |a|$. (*d*) $|ab| = |a||b|$ (this property is valid for any real numbers a and b).
11. Find the distance between a and b. (In each case your answer should be a positive real number.)
(*a*) $a = 5$, $b = -2$. (*b*) $a = -10$, $b = -3$.
(*c*) $a = 0$, $b = 7$. (*d*) $a = 8$, $b = -8$.
12. (*a*) For what values of x does $|2x+7| = 2x+7$?
(*b*) For what values of x does $|2x+7| = -2x-7$? *Hint:* The equation $|a| = -a$ is valid for $a = 0$ as well as for $a < 0$.
13. In order to be considered, suppose that a petition must have at least 100 signatures. Suppose, furthermore, that at least 30 signatures must be women's signatures. Express these conditions as inequalities.
14. Translate each of the statements of Example 4 into an inequality.

1.3 An Introduction to Analytic Geometry

Analytic geometry was first developed in the 17th century as a way of relating geometric and algebraic concepts. This was an important accomplishment because, through the use of analytic geometry, many well-known geometric figures can be represented by means of algebraic equations, and vice versa. Such representations are fundamental in the development of calculus. Necessary to analytic geometry is the concept of a coordinate system. In this section we shall introduce the **rectangular coordinate system,** which is also called the **Cartesian coordinate system** in recognition of René Descartes (1596–1650). Descartes is credited by many as being the founder of analytic geometry.

Let two real lines, one horizontal and one vertical, be drawn in a plane so that the point of intersection is the origin of each real line. The horizontal line is called the ***x*-axis** and its positive direction is chosen to the right; the vertical line is called the ***y*-axis** and its positive direction is chosen upward. The two real lines are called **coordinate axes,** and their point of intersection is called the **origin.** These lines divide a plane into four **quadrants,** which are numbered as shown in Figure 1.4.

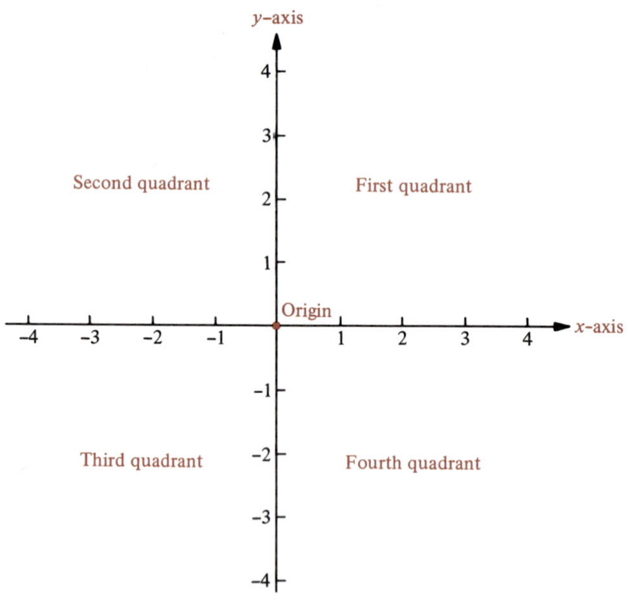

Figure 1.4

Select any point P that is *not* on either coordinate axis and draw perpendiculars from P to each coordinate axis. Let a and b be, respectively, the coordinates on the x and y axes of the points where these perpendiculars meet the axes. See Figure 1.5. The point P will be designated by (a, b), and although the same notation is used for an open interval, it should be clear in context whether (a, b) designates a point or an open interval.

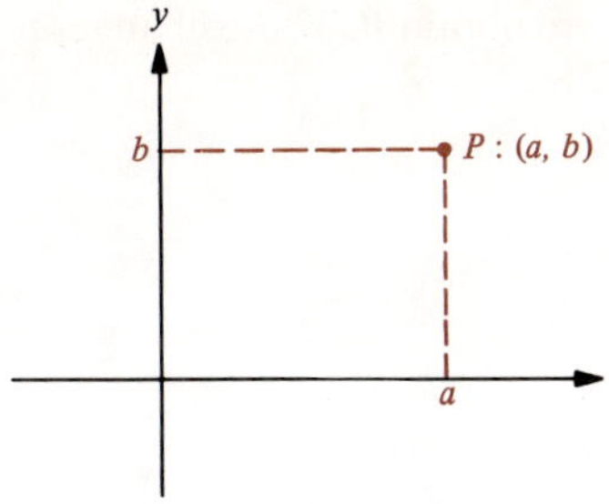

Figure 1.5

A point on the x-axis with coordinate a on the horizontal real line will be denoted by $(a, 0)$, and a point on the y-axis with coordinate b on the vertical real line will be denoted by $(0, b)$. The number a is called the **x-coordinate** or **abscissa** of the point (a, b), and b is called the **y-coordinate** or **ordinate.** When we designate a point by (a, b), we agree that the first number a is the one that corresponds to the horizontal axis. A point in any quadrant or on either coordinate axis has associated with it a unique designation (a, b), and conversely, if (a, b) is given, then a unique point in the plane is determined. In Figure 1.6 several points have been plotted and their coordinates are shown.

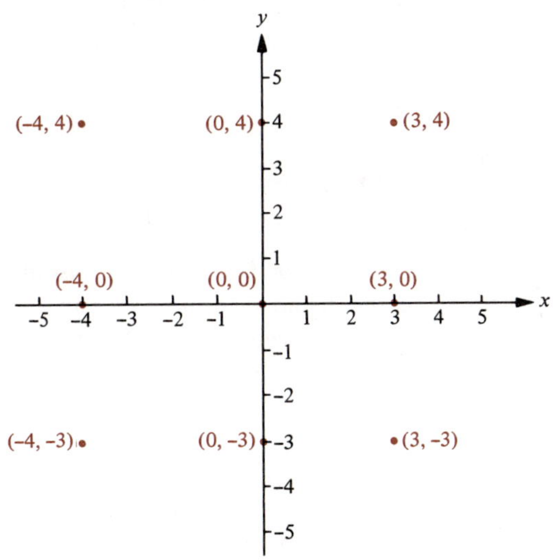

Figure 1.6

Next consider the problem of finding the length d of the line segment between two given points A and B in a plane. Assume that $A\colon(x_1, y_1)$ and $B\colon(x_2, y_2)$ are not on the same vertical or horizontal line (see Figure 1.7). Let us think of starting at A and changing the x-coordinate by $x_2 - x_1$ to get $C\colon(x_2, y_1)$, and then changing the y-coordinate of C by $y_2 - y_1$ in order to obtain B. These changes in x and y are called **increments** and are denoted by

$$\Delta x = x_2 - x_1$$

and

$$\Delta y = y_2 - y_1.$$

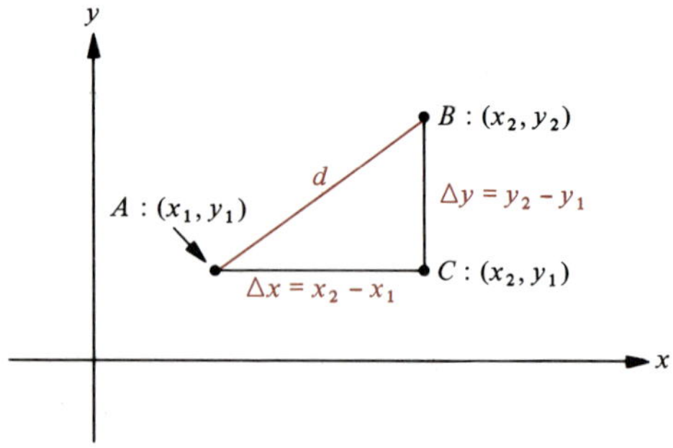

Figure 1.7

(The increments are read "delta x" and "delta y.") In Figure 1.7 the increments Δx and Δy are positive, but one or both of them can be *negative* so that, in general, the legs of right triangle ABC have lengths $|\Delta x|$ and $|\Delta y|$. The desired length d is found by the Pythagorean theorem:

$$\begin{aligned} d^2 &= |\Delta x|^2 + |\Delta y|^2 \\ &= (\Delta x)^2 + (\Delta y)^2 \\ &= (x_2 - x_1)^2 + (y_2 - y_1)^2, \end{aligned}$$

and therefore

$$d = \sqrt{(x_2 - x_1)^2 + (y_2 - y_1)^2}. \tag{1.1}$$

The number d is called the **distance** between A and B, and (1.1) is known as the **distance formula.** It can also be verified that (1.1) gives the correct length between A and B even if these points are on the same vertical or horizontal line (Exercise 14).

Example 1

A map of a region that contains valuable timber has been marked off using a rectangular coordinate system in such a way that one unit on each axis represents 1 mile. By sighting smoke from two different fire towers, it is determined that a fire is breaking out at approximately the location $(-6, -5)$. If a helicopter located at $(-2, 3)$ is ordered to fly directly to the fire, how far must the helicopter travel?

The answer to the problem is the distance d between $(-6, -5)$ and $(-2, 3)$. Let $x_1 = -6$, $y_1 = -5$, $x_2 = -2$, and $y_2 = 3$; then, by (1.1),

$$\begin{aligned} d &= \sqrt{(-2-(-6))^2 + (3-(-5))^2} = \sqrt{16 + 64} = \sqrt{80} \\ &= 4\sqrt{5} \text{ miles.} \end{aligned}$$ ▮

In a problem such as the one in Example 1, it does not matter which of the two points is chosen to be labeled (x_1, y_1); the answer is the same either way because $(x_2 - x_1)^2 = (x_1 - x_2)^2$ and $(y_2 - y_1)^2 = (y_1 - y_2)^2$.

An important problem in analytic geometry is that of sketching graphs (or geometric representations) of algebraic equations. This can be done after plotting a sufficient number of points whose coordinates satisfy a given equation.

Example 2

Sketch the graph of the equation

$$y = 4 - x^2.$$

By assigning various real values to x and then calculating the corresponding values of y, each of the following points can be shown to belong to the graph of the given equation: $(-4, -12)$, $(-3, -5)$, $(-2, 0)$, $(-1, 3)$, $(0, 4)$, $(1, 3)$, $(2, 0)$, $(3, -5)$, and $(4, -12)$. Then these points are joined together, as shown in Figure 1.8. ▮

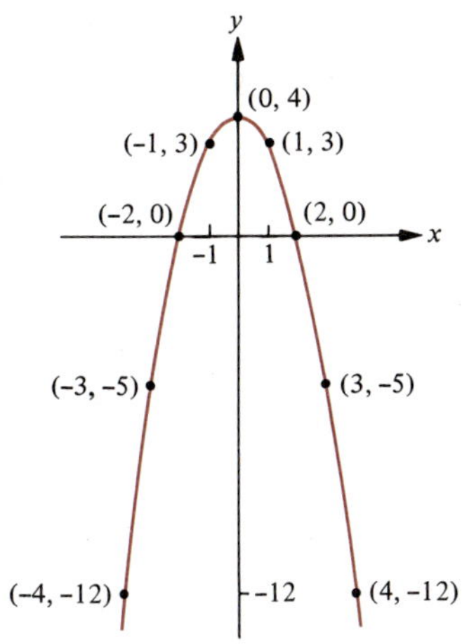

Figure 1.8. $y = 4 - x^2$.

The concept of the graph of an equation, as illustrated in the previous example, can be formalized by the following definition.

▶ Definition 2

The **graph of an equation** that involves x and y is the set of all points (x, y) whose coordinates satisfy the equation.

Later we shall frequently use the word **curve** when we are referring to the graph of a set of points (x, y); in this general usage, even a straight line may be referred to as a curve.

Perhaps the most important points on a graph are those, if any, which are on the coordinate axes. In Example 2, if we set $x = 0$ and solve to obtain $y = 4$, we find the y-coordinate of the only point that the graph has in common with the y-axis—in this case the number 4 is called a **y-intercept.** In general the y-intercepts of the graph of an equation are the ordinates of the points that are on the graph *and* on the y-axis. In Example 2, if we set $y = 0$, we find that $x = 2$ and $x = -2$; these are abscissas of the points at which the graph intersects the x-axis and are called the **x-intercepts** of the graph of the equation.

Another important problem in analytic geometry is that of finding algebraic equations which represent geometric figures.

Example 3

Find an equation for the circle with center (2, 3) and radius 5.

The desired curve consists of that set of points (x, y) which are at a distance of 5 units from the point (2, 3). Hence by (1.1) a point (x, y) is on the circle (see Figure 1.9) if and only if

$$\sqrt{(x-2)^2 + (y-3)^2} = 5$$

or

$$(x-2)^2 + (y-3)^2 = 25,$$

which is the desired equation. ▮

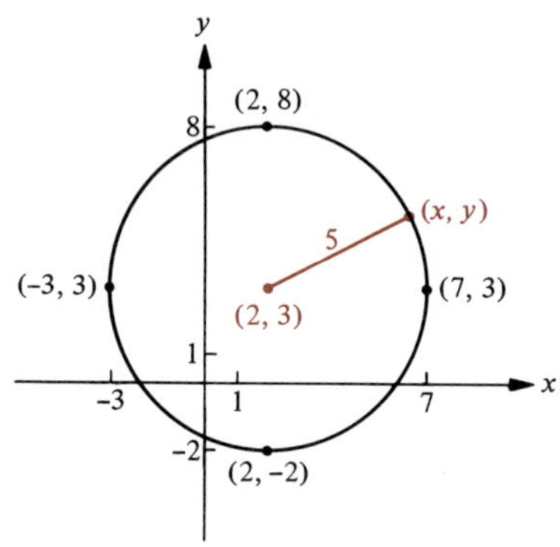

Figure 1.9

Example 3 can be generalized to show that an equation for a circle with center (h, k) and radius r is

$$(x-h)^2 + (y-k)^2 = r^2.$$

In this section we have introduced the rectangular coordinate system, increments, and the distance formula. In Example 2 we illustrated how to graph a geometric figure from an algebraic equation, and in Example 3 we illustrated the converse problem of writing an algebraic equation of a geometric figure.

APPLICATIONS

Example 4

Consider the hypothetical situation of three councilmen who allocate money for two items. The allocations advocated by each councilman can be thought of as a point (x_i, y_i) in our two-dimensional Cartesian coordinate system; that is, x_i represents the amount that councilman i would allocate to item 1 in the budget and y_i represents the amount that councilman i would allocate to item 2 in the budget. Once the issue is decided, the amounts actually budgeted for the two items can be represented by the point (a, b). The distances between the points (x_i, y_i) and the point (a, b) will provide a measure of the compromise made by the councilmen. Moreover, the distance formula can be generalized to apply to dimensions greater than two (see Exercise 11 of Section 7.1 for the three-dimensional case); this will permit an analysis of the allocations for a large number of items by a large group of arbitrators. Such an analysis can be used to rank the degree of compromise by the arbitrators and thereby, over a period of time, to study such things as leadership, influence, and middle-of-the-road positions. ▮

EXERCISES

Suggested minimum assignment: Exercises 1, 3(*b*), 5, 7, 8(*b*) (*c*), 10(*b*) (*d*), 12, and 13.

1. Plot each of the following points and tell in which quadrant each point is located: $(-3, 5)$, $(4, -1)$, $(2, 7)$, $(-5, -4)$.
2. Plot each of the following points and tell in which quadrant each point is located: $(4, -7)$, $(0, 5)$, $(-3, -1)$, $(-5, \pi)$.
3. Find the distance between each of the following pairs of points:
   (*a*) $(1, 4)$, $(5, 7)$. (*b*) $(-1, -5)$, $(6, -7)$. (*c*) $(-5, -3)$, $(4, -3)$.
4. Find the distance between each of the following pairs of points (let the first point be the initial point and identify Δx and Δy in each part):
   (*a*) $(3, -6)$, $(2, -5)$. (*b*) $(\sqrt{2}, 0)$, $(2\sqrt{2}, -1)$. (*c*) $(\frac{1}{2}, 5)$, $(\frac{-7}{2}, 5)$.
5. A circle passes through $(-4, 7)$. Its center is $(3, -1)$. What is the radius of the circle?
6. Use the distance formula and the Pythagorean theorem to determine whether or not the triangle with vertices $(-5, -1)$, $(-1, 4)$, and $(5, -9)$ is a right triangle. Sketch.
7. Given points $A\!:(-3, 6)$ and $B\!:(5, -1)$.
   (*a*) Starting at A and ending at B, find the increments Δx and Δy.
   (*b*) Starting at B and ending at A, find the increments Δx and Δy.
8. Sketch the graph of each of the following equations:
   (*a*) $y = x^2 - 9$. (*b*) $x = y^2$.
   (*c*) $y = |x|$. (*d*) $x^2 + y^2 = 4$.
   (*e*) $x^2 + 16y^2 = 16$. (*f*) $2x + 3y = 12$.

9. Is the point (1, 2) on the graph of the equation $x^4 + y^3 - 8x - 1 = 0$? Why?.

10. Find all x-intercepts and all y-intercepts.

    (a) $y = x^4 - x^2$. (b) $y = \frac{x^2 - 4}{x^2 - 9}$. (c) $y^2 = \frac{x^2 - 4}{x^2 - 9}$.

    (d) $x^2 - y^2 = 16$. (e) $x^2 + y^2 = 16$.

    (f) $x = (y - 1)(y - 2)(y - 3)$.

11. Find an equation of the set of points (x, y) that are at a distance of 3 from the point $(-1, -2)$.

12. Find an equation of the set of points (x, y) that are equidistant (the same distance) from (4, 3) and $(-3, -2)$.

13. State the center and radius of each of the following circles.

    (a) $(x - 7)^2 + (y + 10)^2 = 4$.

    (b) $x^2 + y^2 = 29$.

14. Show that the distance formula gives the correct length between two points if the points are on either the same vertical line or the same horizontal line.

15. A walkie-talkie has a maximum transmitting range of approximately 14 miles. Can the transmission be received by a person, appropriately equipped, who has walked 5 miles east and then 12 miles south of the sender?

1.4 Functions

Expressions of relationships between two or more entities are very common and important in the real world. These expressions may vary from simple statements such as "the demand for a commodity increases as the price decreases" to very complex mathematical formulas. The purpose of this section is to gain some precision in our understanding and expression of special types of relationships known as **functions.**

Example 1

Let $D = \{a, b, c, d\}$ be a set of four small countries. Table 1.2 defines a function f by associating with each country of set D, the gross national product (GNP) of that country in 1973 (to the nearest billion dollars). We

Table 1.2. THE GROSS NATIONAL PRODUCT OF FOUR COUNTRIES.

Country	Gross National Product
a	55,000,000,000
b	37,000,000,000
c	34,000,000,000
d	55,000,000,000

say that the **image** of each country is the corresponding GNP. The set D of countries is called the **domain** of the function f, and the set R of images that appear in the second column of Table 1.2 is called the **range** of f. Hence

$$R = \{34{,}000{,}000{,}000, \quad 37{,}000{,}000{,}000, \quad 55{,}000{,}000{,}000\}.$$

The fact that the GNP of b is 37,000,000,000 is indicated by writing

$$f(b) = 37{,}000{,}000{,}000,$$

which is read "the value of the function f at b is 37,000,000,000." ∎

Figure 1.10 is presented to help the reader understand the concept of a function. An element of the domain can have only one image in the range; however, it is possible for an element of the range to be the image of more than one element of the domain (consider the element 55 billion in Example 1). The sets D and R are different in general, but they may be the same. Next we give a definition of the terminology introduced above.

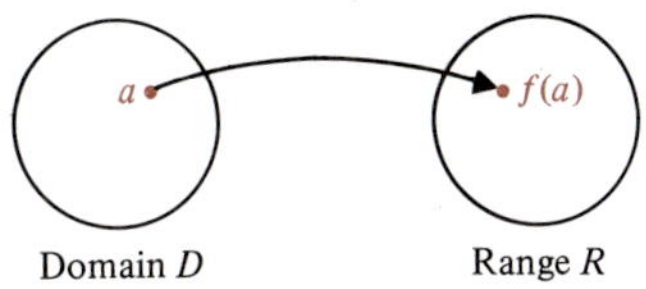

Figure 1.10. *Function concept.*

Definition 3

Let D be a given set. A **function** f is a rule that associates with each element a of D exactly one element of a set R. The element associated with a is denoted by $f(a)$ and is called the **image** of a. [The symbol $f(a)$ is read "f of a" or "the value of f at a."] Set D is called the **domain** of the function f, and the set R of all images of elements of D is called the **range** of f.*

In the next example a function will be defined by an equation rather than by a table. Observe also in Example 2 that the domain and range each contain an infinite number of elements rather than a finite number of elements.

Example 2

Corresponding to any real number x, let exactly one real number y be determined by the equation

$$y = 4 - x^2$$

of Example 2 of Section 1.3. This defines a function f whose domain is

* As an alternative definition, a **function** f is a set of ordered pairs (x, y), no two of which have the same first term. The **domain** of f is the set of all possible values of x, and the **range** of f is the set of all possible values of y.

the set of all real numbers. Figure 1.8 reveals that y can be any real number not greater than 4, and hence the range R of f is

$$R = \{y : y \leq 4\}. \blacksquare$$

Since x and y in the last example may vary throughout our discussion, they are called **variables.** When y is written in terms of x as in Example 2, it is customary to call x the **independent variable** and y the **dependent variable.** If x is replaced by a number in the domain of f, then the associated value of y can be computed. In this manner y is dependent upon x. We shall say that "y is a function of x" and write

$$y = f(x) = 4 - x^2.$$

Suppose that a function f is said to be defined by $y = f(x)$ and that no mention is made of the domain. *If the domain of f is not specified, we shall assume in this text that it consists of all real numbers which can be substituted for x in order to determine a unique real number y.* For example, if

$$y = f(x) = \frac{x}{x-2},$$

then the domain D of f, unless otherwise indicated, is implied to be

$$D = \{x : x \text{ is a real number and } x \neq 2\}.$$

Since division by zero is undefined in mathematics, the real number 2 is excluded from the domain. The range of the function in this example is not easy to identify from the original equation; one way to find the range is to first solve for x in terms of y as follows:

$$xy - 2y = x,$$
$$x(y - 1) = 2y,$$
$$x = \frac{2y}{y-1}.$$

Then it is possible to observe that y can take on any real value except 1, and hence the range R is

$$R = \{y : y \text{ is a real number and } y \neq 1\}.$$

If a function is said to be defined by

$$y = g(x) = \sqrt{x - 4},$$

then, in order to assure only real values for y, the domain of g is understood to be

$$\{x : x \geq 4\}.$$

If, however, a certain set of values of the independent variable is specified, such as

$$y = h(x) = \sqrt{x - 4}, \qquad \text{where} \qquad 7 \leq x \leq 8,$$

then the domain of h is the closed interval [7, 8]. Functions g and h above are *not* the same because the domains are different.

Example 3

Consider the function f defined by $f(x) = x^2 + 3x + 1$. The domain of f is assumed to be (as explained above) the set of all real numbers, so any real number can be substituted for x; in other words

$$f(\ ) = (\ )^2 + 3(\ ) + 1,$$

where any one real number can be inserted in all three parentheses. For example,

(*a*) $f(-2) = (-2)^2 + 3(-2) + 1 = -1.$

(*b*) $f(0) + f(3) = [(0)^2 + 3(0) + 1] + [(3)^2 + 3(3) + 1] = 1 + 19 = 20.$

(*c*) $f(a) = a^2 + 3a + 1.$

(*d*) $$\begin{aligned} f(2 + \Delta x) - f(2) &= [(2 + \Delta x)^2 + 3(2 + \Delta x) + 1] - [(2)^2 + 3(2) + 1] \\ &= 4 + 4\,\Delta x + (\Delta x)^2 + 6 + 3\,\Delta x + 1 - 11 \\ &= (\Delta x)^2 + 7\,\Delta x. \end{aligned}$$

(*e*) $$\begin{aligned} f(2x) - 2f(x) &= [(2x)^2 + 3(2x) + 1] - [2(x^2 + 3x + 1)] \\ &= 4x^2 + 6x + 1 - 2x^2 - 6x - 2 \\ &= 2x^2 - 1. \ ▮ \end{aligned}$$

An understanding of the notation used, and the calculations made, in Example 3 will be very helpful in preparing for a study of calculus in this text. A geometric understanding of the concept of a function is important also. The graph of the function f of Example 3 is shown in Figure 1.11

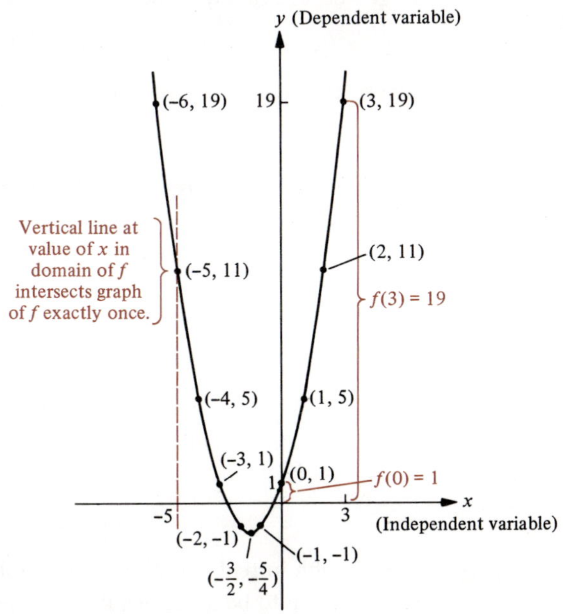

Figure 1.11. $y = f(x) = x^2 + 3x + 1$.

[by the **graph of f** we mean the graph of the equation $y = f(x)$]. Graphically, $f(3)$ is the y-coordinate or "height" of the curve at $x = 3$. The y-intercept is $f(0)$. A line drawn parallel to the axis of the dependent variable (the y-axis) cannot intersect the graph of f at more than one point; this is because the basic idea of a function is that each element of the domain has exactly *one* image in the range.

Example 4

Revenue functions, cost functions, and profit functions are the names of some important functions in economics. These functions are introduced here and will be used in future examples.

The revenue (perhaps in dollars) received by a manufacturer for the sale of a number x of a certain item might be given by

$$R(x) = 3x + \tfrac{1}{50}x^2,$$

where R is called a **revenue function** and expresses a relationship between the number of items sold and the amount of money received for these items. Some sort of revenue function may be assumed for planning purposes; often it is possible from experience to make a rather accurate determination of such a function. We can calculate

$$R(20) = 3(20) + \tfrac{1}{50}(20)^2 = 68,$$

which means that the company expects to receive \$68 if it can sell 20 items of this particular type.

The cost (perhaps in dollars) of producing a number x of a certain item might be estimated to be

$$C(x) = 10 + 2x,$$

where C is called a **cost function.** Observe that

$$C(0) = 10,$$

which means that there is a **fixed cost** of \$10 even if no items are produced—this may represent overhead. The cost of 20 items is given by

$$C(20) = 10 + 2(20) = 50.$$

If the revenue and cost functions above are valid for the same item, we can calculate

$$R(20) - C(20) = 68 - 50 = 18,$$

which gives the profit on the sale of 20 items. In general the **profit function** P is given by

$$P(x) = R(x) - C(x).$$

Although in real situations, the number of items sold or produced would usually have to be a nonnegative integer, it is customary to let the domain of a revenue, cost, or profit function (which serves as a model of the real situation) be an *interval* of nonnegative real numbers.

Cost accountants often plot both the revenue and cost functions on the same graph. They call this graph a **break-even chart.** The point, if it exists, at which the curves intersect is called the **break-even point.** For the revenue and cost functions of this example the break-even point occurs between $x = 8$ and $x = 9$ (see Figure 1.12). Hence, if 8 or fewer items are sold, there is a loss, but if at least 9 items are sold there is a profit. ▮

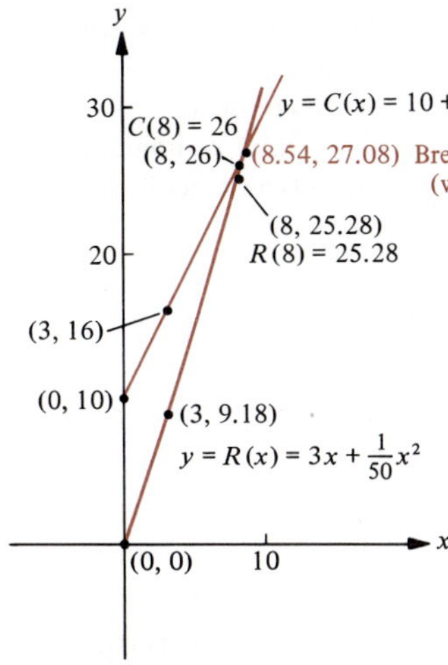

Figure 1.12. *Break-even chart.*

APPLICATIONS

★Example 5 Weighting Functions and Probability

This example is the first of a sequence of examples that apply the calculus and preliminary topics to a study of probability. The probability examples, which occur throughout the remainder of this book, often depend upon the content of previous examples in the sequence and will be designated by a ★.

In a study of probability one often begins with the idea of an experiment in which any one of a set of possible outcomes can occur. For instance, consider an experiment of choosing one card from an ordinary deck of 52 cards. There are 52 possible outcomes, and the experiment consists of the selection of one of these outcomes. The nonempty set U of possible outcomes of an experiment is known as a **sample space,** and any subset of U is called an **event.** An event is said to occur if one or more of its outcomes occur.* Several illustrations of sample spaces and events are given in Table 1.3.

* It is customary to say that the event which contains no outcomes does not occur, and it is called the impossible event.

Table 1.3. ILLUSTRATIONS OF SAMPLE SPACES AND EVENTS.

Experiment	Sample space	An event
Flip a coin on a flat surface and observe whether a head or a tail appears.	{head, tail}	{head}
Select a sample product from an assembly line and observe whether the product is passable, repairable, or nonrepairable.	{passable, repairable, nonrepairable}	{passable, nonrepairable}
Count the number of days it takes to breed a female offspring in a set of laboratory animals.	{1, 2, 3, . . .}	{3, 4, 5}
Measure the time in hours, or portion thereof, between accidents on a given stretch of highway.	The set of nonnegative real numbers	The set of nonnegative real numbers less than or equal to 4

In the remainder of this example we shall be concerned only with sample spaces having a finite number of outcomes. Prior to the occurrence of an experiment it is common to speculate about the likelihood or probability of achieving a single outcome, or a set of outcomes, of the sample space. Precision may be gained by thinking of the likelihood or probability of a single outcome in terms of a function w whose domain is the sample space $U = \{u_1, u_2, \ldots, u_n\}$; such a function, for which each $w(u_i) \geq 0$ and for which $w(u_1) + \cdots + w(u_n) = 1$, is often called a **weighting function.** For example, suppose that prior to the experiment of casting a pair of dice, one speculates about the likelihood that a given total will appear. The sample space is the set $U = \{2, 3, 4, 5, 6, 7, 8, 9, 10, 11, 12\}$; the weighting function w for this illustration is given in Table 1.4. The weight of each outcome can be determined by observing that since there are six faces on a single die, there are 6^2 or 36 equally likely ways for a pair of dice to land on a flat surface, and by observing the number of ways each number of the sample space can occur. Notice that the sum of the weights of all possible outcomes is 1. The set of numbers in the first row of Table 1.4 is the domain of the function w, and the set of numbers in the second row is the range of w. The graph of the weighting function w in this example is the set of eleven points shown in Figure 1.13.

For sample spaces that have a finite number of outcomes, it is customary to define $P(E)$, the **probability of an event** $\boldsymbol{E}$, in terms of the weighting

Table 1.4. WEIGHTS FOR THE OUTCOMES OF A PAIR OF DICE.

u_i	2	3	4	5	6	7	8	9	10	11	12
$w(u_i)$	$\frac{1}{36}$	$\frac{2}{36}$	$\frac{3}{36}$	$\frac{4}{36}$	$\frac{5}{36}$	$\frac{6}{36}$	$\frac{5}{36}$	$\frac{4}{36}$	$\frac{3}{36}$	$\frac{2}{36}$	$\frac{1}{36}$

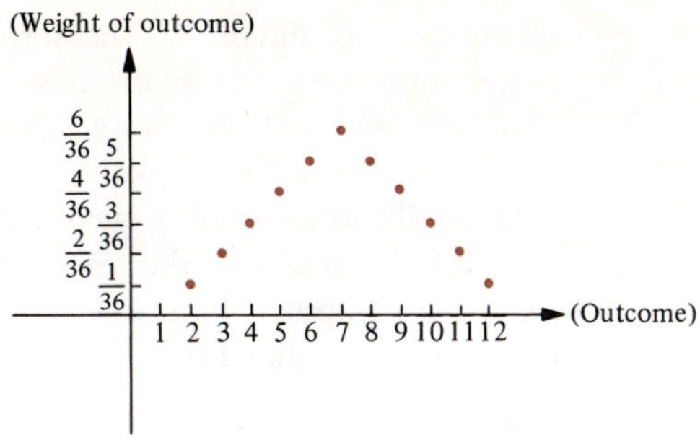

Figure 1.13. *Probabilities of various totals for a pair of dice.*

function as follows: If

$$E = \{e_1, e_2, \ldots, e_k\}$$

is a subset of a sample space U, then

$$P(E) = w(e_1) + w(e_2) + \cdots + w(e_k).$$

If E is the empty set, then $P(E)$ is defined to equal zero.

Suppose that it is desired to find the probability of achieving a sum of either seven or eleven for one throw of a pair of dice. In this case the event E is the subset $E = \{7, 11\}$ of the sample space U. Therefore,

$$P(E) = w(7) + w(11) = \tfrac{6}{36} + \tfrac{2}{36} = \tfrac{2}{9}. \blacksquare$$

*Example 6 Random Variable or Measurement

Consider an experiment in which a certain hospital determines a man's blood type in preparation for a possible transfusion. The set of outcomes in this particular test is $U = \{A+, A-, B+, B-, AB+, AB-, O+, O-\}$. The hospital has to buy blood for transfusions from a commercial blood bank. Associated with each of the outcomes of U is a cost in dollars of obtaining a pint of the particular type of blood. This information, given in Table 1.5, defines a function X. The domain and range of X are given in the respective rows of Table 1.5.

Table 1.5. Cost of pint of blood.

Outcomes	u	A+	A−	B+	B−	AB+	AB−	O+	O−
Cost	$X(u)$	55	55	60	60	60	65	50	55

A function X that associates a real number to each outcome u of a sample space U (finite or infinite) is called a **random variable.** In our illustration the values of the function X (the values of the random variable) are the real numbers 50, 55, 60, and 65.

Because a random variable simply assigns a real number to each outcome of a sample space, a random variable may be thought of as a **measurement** of the various outcomes. The term "measurement" seems to promote a better intuitive grasp of the basic concept than does the term "random variable," and for this reason "measurement" will be used more often in the remainder of this book.

Other illustrations of a measurement follow:

1. Let U consist of the set of grades {A, B, C, D, F}. Grade averages are often computed by the use of a random variable, or measurement, X as defined by Table 1.6.

Table 1.6

u	A	B	C	D	F
$X(u)$	4	3	2	1	0

2. Let $U = \{1, 2, 3, 4, 5, 6\}$ be the set of outcomes of a single throw of a single die. Consider a game in which the thrower will collect the amount in pennies shown on the die. Since the outcomes of U are real numbers, a measurement X can be defined that simply assigns to every outcome the same real number. In other words, $X(u) = u$. On the other hand, suppose that the thrower must *pay* the amount shown on the die if the number is *odd* and will continue to *collect* if the amount is *even*. In this case we have defined a different measurement Z, as shown in Table 1.7.

Table 1.7

u	1	2	3	4	5	6
$Z(u)$	-1	2	-3	4	-5	6

3. Consider an experiment in which a coin is flipped until a head appears. Here U has an infinite number of elements, and $U = \{\text{H, TH, TTH}, \ldots\}$. Let a random variable, or measurement, X be defined in such a way that $X(u)$ is the number of flips that it takes to obtain a head.

In all these illustrations it is important to note that a random variable, or measurement, is a *function* whose domain is a sample space U and whose range is a subset of the set of real numbers. ▮

EXERCISES

Suggested minimum assignment: Exercises 3, 5, 6, 9, 13, 15, and 17.

1. Suppose that the domain of a function f is $D = \{1, 3, 5, 7\}$ and that the image of each number of set D is 10 times the number. Find $f(1)$, $f(3)$, $f(5)$, $f(7)$, and the range R of f.
2. Make up an example of a function whose domain is the set of cities in the United States whose population is over 1,000,000.
3. A function f whose domain is the set of all real numbers is defined

by $y = f(x) = x^2 + 4$. What is the image of the number 3? Draw a graph of f and state the range of f.

4. Describe a function whose domain is the set of all stocks listed on the New York Stock Exchange.
5. Suppose that a function f is defined by $f(1) = 4$, $f(2) = 5$, and $f(3) = 6$. State the domain and range of f. If $f(2) = 7$ were added above, why would we no longer have a function defined?
6. Suppose that s is a function of t as defined by $s = f(t) = t^4 + 1$, provided that $t \geq 0$. Which variable is the dependent variable and which variable is the independent variable? Find $f(2)$.
7. Suppose that a function g is defined by $y = g(x) = \frac{1}{x}$, provided that $x \neq 0$. Which variable is the dependent variable and which variable is the independent variable? Find $g(5)$.
8. If $f(x) = 1 - x - 3x^2$, find $f(0)$ and $f(1)$.
9. State the implied domain of f if f is said to be defined by each of the following equations.

   (*a*) $f(x) = \dfrac{2x}{x+2}$. (*b*) $f(x) = 1 + 2x$.

   (*c*) $f(x) = \dfrac{1}{x^2 - 1}$. (*d*) $f(x) = 1 + \sqrt{x}$.

10. Let $y = f(x)$, where $f(x)$ is given in Exercise 9(*a*). Solve for x in terms of y and state the range of f.
11. If $f(x) = \dfrac{1}{2x+3}$, find $f(-3)$ and $f(a + b)$.
12. If $f(x) = 3 - \dfrac{x}{2}$, sketch the graph of f. Indicate $f(-4)$ and $f(10)$ on your sketch.
13. Suppose that $f(x) = x^2$. Calculate each of the following.

    (*a*) $f(3)$. (*b*) $f(3) - f(2)$. (*c*) $f(3 - 2)$. (*d*) $f(x + 3)$.

    (*e*) $f(x + 3) - f(x)$. (*f*) $\dfrac{f(x+h) - f(x)}{h}$, where $h \neq 0$.

14. If $f(x) = 3x - 2x^2$, calculate each of the following.

    (*a*) $f(4)$. (*b*) $f(2)$. (*c*) $f(x_1)$. (*d*) $f(x_1 + \Delta x)$.

    (*e*) $f(x_1 + \Delta x) - f(x_1)$. (*f*) $\dfrac{f(x + \Delta x) - f(x)}{\Delta x}$, where $\Delta x \neq 0$.

15. Let a cost function C be defined by $C(x) = 5 + 8x - \frac{1}{25}x^2$, where $0 \leq x \leq 10$. What is the domain of C? Find the cost of producing five items.
16. Solve algebraically the system of equations

$$\begin{cases} y = 3x + \frac{1}{50}x^2, \\ y = 10 + 2x, \end{cases}$$

in order to find the break-even point in Example 4.

17. The revenue and cost functions for an item are given by $R(x) = 10x$ and $C(x) = 40 + 8x$, provided that $0 \le x \le 100$. Find $P(x)$, and find the profit if 50 items are sold. What is the fixed cost? How many items must be sold to reach the break-even point?

18. An experiment consists of choosing one card from an ordinary deck of 52 cards and observing whether it is a spade, heart, diamond, or club.
    (*a*) What is the sample space U in this experiment?
    (*b*) Assign a weight to each outcome in a manner that satisfies the definition of a weighting function; let the outcomes that are equally likely have equal weights. State the domain and range of the weighting function w.
    (*c*) Find $P(E)$ if the event E is the subset {spade, club} of U.

19. (*a*) If U is a sample space and w is a weighting function with domain U, is it always correct to call w a random variable (or measurement)? Give reasons for your answer.
    (*b*) If U is a sample space and X is a measurement with domain U, is it always correct to call X a weighting function? Give reasons for your answer.

20. In illustration 3 of Example 6 a measurement X is defined. What is the image of the element TTH? What is the range of X?

1.5 Straight Lines

The owners of a business spend \$15,000 for equipment and decide for income tax purposes to depreciate the equipment over a 10-year period. If the straight line method of depreciation is used, the value of the equipment is assumed to decrease by \$1500 each year for 10 years. In Figure 1.14 the value y of the equipment in thousands of dollars is shown after x years, where $0 \le x \le 10$. The graph is a portion of a straight line. A

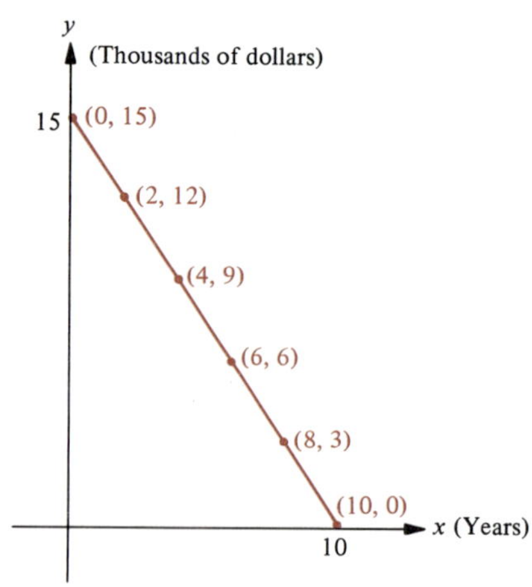

Figure 1.14. *Straight line depreciation.*

straight line is one of the simplest, and yet one of the most important, types of figures encountered in applications. In this section we shall define the useful concept of slope of a straight line; then we shall introduce various forms of the equation of a straight line.

First, let us agree not to distinguish between the terms "line" and "straight line." Also it is understood that if we refer to two points on a line, then we mean two different points. A vertical line is understood to be a line parallel to the y-axis, and a horizontal line is parallel to the x-axis. Assume that two points (x_1, y_1) and (x_2, y_2) determine a unique line l passing through the two points, and that any two points on a given line l can be used to determine l in this manner.

Definition 4

If (x_1, y_1) and (x_2, y_2) are points on a nonvertical line l, the **slope** m of l is given by

$$m = \frac{\Delta y}{\Delta x} = \frac{y_2 - y_1}{x_2 - x_1}.$$

The slope of a vertical line is not defined.

The slope of a nonvertical line is a real number that is a measure of the direction that the line has in relation to the coordinate axes. In Figure 1.15 a line with slope $\frac{7}{3}$ is shown. Start at any point on the line and change x by 3 units ($\Delta x = 3$); the change in y is 7 units ($\Delta y = 7$), and the slope is

$$m = \frac{\Delta y}{\Delta x} = \frac{7}{3}.$$

Any line that goes up as one traces it from left to right, as in Figure 1.15, has a positive slope. On the other hand, a line that goes down as one traces

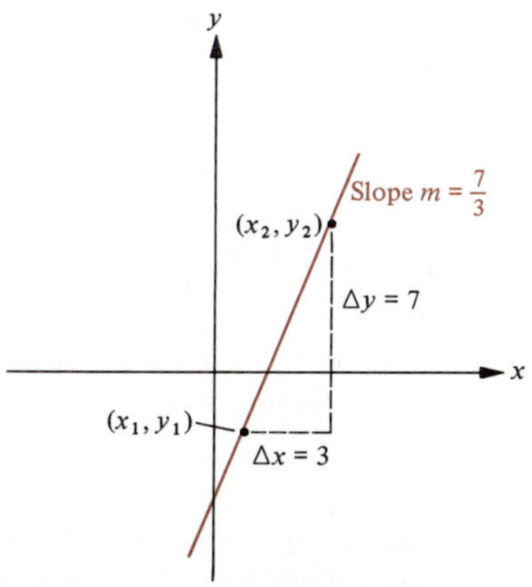

Figure 1.15. *Line with positive slope.*

it from left to right has a negative slope; for example, in Figure 1.16 each line goes down 3 units ($\Delta y = -3$) for every 4 units of movement to the right ($\Delta x = 4$), and the slope is

$$m = \frac{\Delta y}{\Delta x} = \frac{-3}{4} = -\frac{3}{4}.$$

As Figure 1.16 suggests, nonvertical parallel lines have the same slope and, conversely, lines with the same slope are parallel.* *The slope of a horizontal line is zero* because the change Δy between any two points on the line is 0, and hence $\Delta y/\Delta x = 0$.

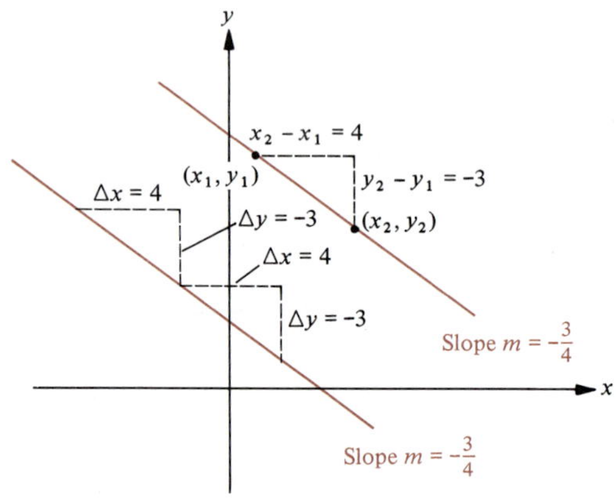

Figure 1.16. *Parallel lines with same negative slope.*

Example 1

Find the slope of the line that passes through $(-3, 8)$ and $(6, -4)$.

This problem can be solved without a graph, although the line is shown in Figure 1.18. Let $(x_1, y_1) = (-3, 8)$ and $(x_2, y_2) = (6, -4)$. Then

$$m = \frac{\Delta y}{\Delta x} = \frac{y_2 - y_1}{x_2 - x_1} = \frac{-4 - 8}{6 - (-3)} = \frac{-12}{9} = -\frac{4}{3}.$$

Since

$$\frac{y_2 - y_1}{x_2 - x_1} = \frac{y_1 - y_2}{x_1 - x_2},$$

it does not matter which of the two points above is labeled (x_1, y_1); the other is then (x_2, y_2). ∎

We now begin a discussion of *equations* of lines by considering vertical and horizontal lines. A vertical line has an equation of the form

$$x = a,$$

* The reader who has studied trigonometry may observe that the slope of a line equals $\tan \alpha$, where α is the smallest nonnegative angle whose initial side is the positive direction of the x-axis and whose terminal side is the given line.

and a horizontal line has an equation of the form

$$y = b.$$

For example, the equation $x = -5$ (which may be written $x + 0y = -5$) is satisfied by points such as $(-5, -2)$, $(-5, 0)$, $(-5, 2)$, and $(-5, y)$ for any real number y; its graph is a vertical line 5 units to the left of the y-axis. In Figure 1.17 some horizontal and vertical lines, together with their equations, are shown. Notice that the graph of $x = 0$ is the y-axis and the graph of $y = 0$ is the x-axis.

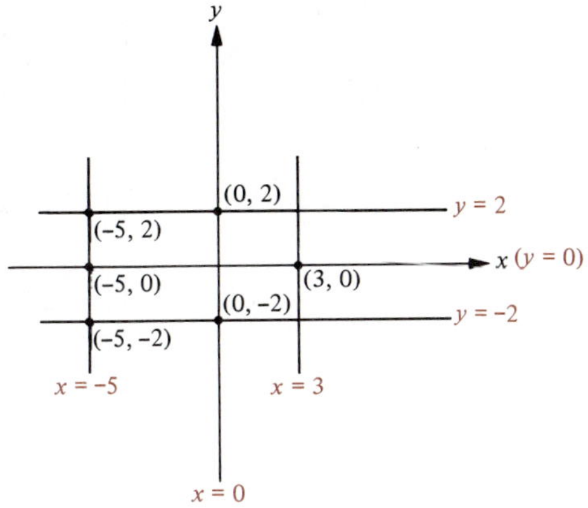

Figure 1.17. *Equations of vertical and horizontal lines.*

Next we examine the form of equations of other lines. A unique nonvertical line l is determined by a point (x_1, y_1) on the line and the slope m of the line. A point $(x, y) \neq (x_1, y_1)$ is on l if and only if

$$m = \frac{y - y_1}{x - x_1}$$

or

$$y - y_1 = m(x - x_1).$$

This last equation is also satisfied if $(x, y) = (x_1, y_1)$. Therefore, a point (x, y) is on l if and only if

$$y - y_1 = m(x - x_1),$$

which is called the **point-slope form** of the equation of a line.

Example 2

Find an equation of the line that passes through $(-3, 8)$ and $(6, -4)$.

In this type of problem, first find the slope of the line and then substitute in the point-slope form; in Example 1 we found that $m = -\frac{4}{3}$. Either of the given points can be substituted for (x_1, y_1) in the point-slope form; we select $(x_1, y_1) = (-3, 8)$ and substitute to obtain

$$y - 8 = -\tfrac{4}{3}(x - (-3))$$

or

$$y - 8 = -\tfrac{4}{3}(x + 3). \ \blacksquare$$

We shall use the letters a and b, respectively, for the x and y intercepts of a line. In Example 2 we can set $y = 0$ to find that the x-intercept is

$$a = 3,$$

and we can set $x = 0$ to find that the y-intercept is

$$b = 4.$$

See Figure 1.18.

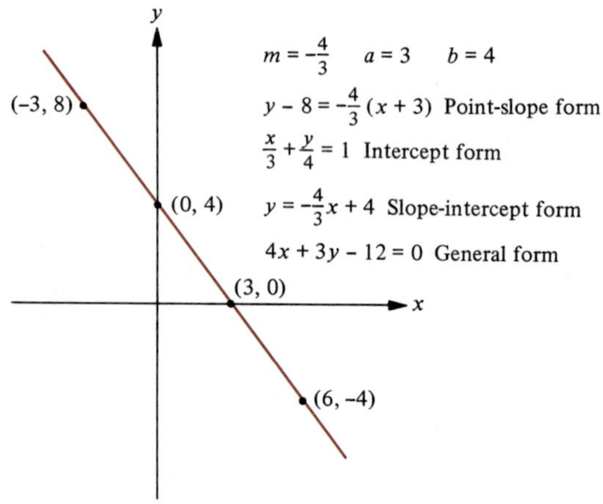

Figure 1.18

There are other useful forms of an equation of a line beside the point-slope form. Three other such forms are shown in Figure 1.18 for the line of Examples 1 and 2. These can be obtained from the point-slope form by simple algebra, and they may be compared with the information about the various forms of an equation of a line in Table 1.8.

Example 3

Suppose that the graph of an unknown cost function is known to be a line. Find an equation of the line if the fixed cost is \$10 and if each item costs an additional \$2 above the fixed cost.

The line has y-intercept $b = 10$ because the fixed cost is known to be \$10. The line has slope $m = 2$ because it is known that each item costs \$2 above the fixed cost. Since

$$b = 10$$

and

$$m = 2,$$

the slope-intercept form can be used immediately, and the equation is

$$y = 2x + 10.$$

(The cost function in this example is the same one that appeared in Example 4 of Section 1.4.) ▮

Table 1.8. FORMS OF EQUATION OF LINE.

Equation of line	Name	Comments
$y - y_1 = m(x - x_1)$	Point-slope form	Passes through (x_1, y_1); slope m.
$\frac{x}{a} + \frac{y}{b} = 1$	Intercept form	Passes through $(a, 0)$ and $(0, b)$. $(a \neq 0,\ b \neq 0.)$
$y = mx + b$	Slope-intercept form	Slope m; passes through $(0, b)$. Special case of point-slope form, obtained by setting $(x_1, y_1) = (0, b)$. May be obtained from any other form by simply solving for y.
$Ax + By + C = 0$	General form	Slope is $-A/B$ and y-intercept is $-C/B$, provided that $B \neq 0$. (A, B, and C are any real numbers, except that not both A and B are zero.)

Example 4

Find the slope of the line whose equation is

$$3x - 2y + 12 = 0$$

and draw its graph.

An excellent procedure for finding the slope is to solve the equation for y first:

$$2y = 3x + 12,$$
$$y = \tfrac{3}{2}x + 6.$$

This puts the equation in slope-intercept form $y = mx + b$, from which we can observe that the coefficient of x is $m = \frac{3}{2}$. (An alternative procedure is to observe that according to Table 1.8 the line is given in general form, and its slope is $m = -A/B = -(3/-2) = \frac{3}{2}$.) The line can be graphed quickly by plotting any two points, since two points determine a line. One way to find two points on the line is to put the equation in intercept form and read off the x- and y-intercepts:

$$3x - 2y = -12,$$

$$\frac{x}{-4} + \frac{y}{6} = 1.$$

Hence the x-intercept is -4, the y-intercept is 6, and the line passes through $(-4, 0)$ and $(0, 6)$. See Figure 1.19. By observation of the figure we can also verify that the slope of the line is $\frac{3}{2}$. ▮

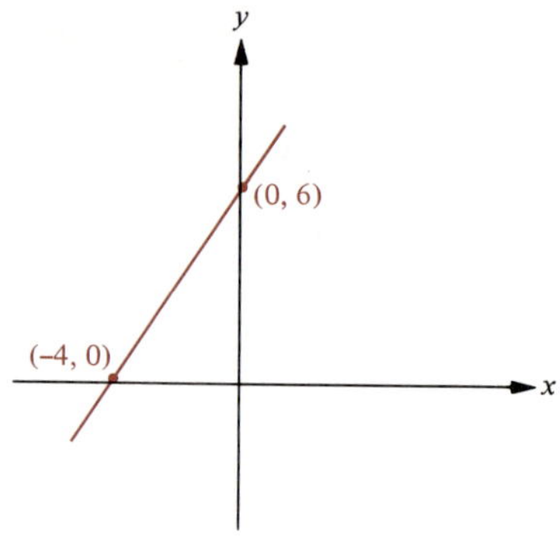

Figure 1.19. $3x - 2y + 12 = 0$.

APPLICATIONS

Example 5

Studies of some biological environments require knowledge of the two customary scales of measurement of temperature. The graph of the equation that relates the temperature y in degrees Celsius (centigrade) to the temperature x in degrees Fahrenheit is known to be a line. The line passes through $(32, 0)$ because water freezes at 32 degrees Fahrenheit and 0 degrees Celsius. It also passes through $(212, 100)$ because water boils at 212 degrees Fahrenheit (at standard pressure) and 100 degrees Celsius. Find the slope of the line, an equation of the line, and sketch. Also find the Celsius temperature that corresponds to 50 degrees Fahrenheit.

Let $(x_1, y_1) = (32, 0)$ and $(x_2, y_2) = (212, 100)$. Then

$$m = \frac{y_2 - y_1}{x_2 - x_1} = \frac{100 - 0}{212 - 32} = \frac{100}{180} = \frac{5}{9}.$$

The equation of the line in point-slope form is

$$y - 0 = \tfrac{5}{9}(x - 32)$$

or

$$y = \tfrac{5}{9}(x - 32).$$

The graph is shown in Figure 1.20. The slope $\frac{5}{9}$ may be interpreted to mean that for every 9-degree increase in the Fahrenheit temperature, there is an increase of 5 degrees Celsius. The Celsius temperature corresponding to 50 degrees Fahrenheit is found by substituting $x = 50$ in the equation to obtain

$$y = \tfrac{5}{9}(50 - 32) = \tfrac{5}{9}(18) = 10 \text{ degrees Celsius.} \blacksquare$$

EXERCISES

Suggested minimum assignment: Exercises 1, 2(*a*) (*b*), 4, 5(*b*), 8(*a*), 9(*a*), 12(*a*), and 13.

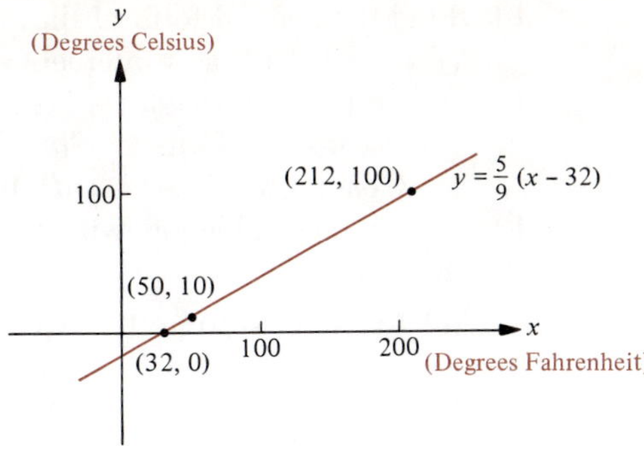

Figure 1.20

1. Find the slope, if possible, of the line passing through the given pair of points.
   (*a*) (4, 8) and (3, 1). (*b*) $(-5, -6)$ and $(-7, 4)$.
   (*c*) $(-4, 3)$ and (3, 3). (*d*) $(-1, 3)$ and $(-1, -4)$.
2. Sketch the line that passes through the given point and has given slope m.
   (*a*) (4, 3), $m = \frac{5}{4}$. (*b*) (4, 3), $m = -\frac{5}{4}$.
   (*c*) $(-1, -2)$, $m = 4$. (*d*) $(3, -4)$, $m = -3$.
   (*e*) (0, 4), $m = \frac{1}{3}$.
3. A line l passes through $(-5, -4)$ and (6, 3). What is the slope of any line parallel to l?
4. Graph each of the following lines.
   (*a*) $x = 6$. (*b*) $x = -7$. (*c*) $y = 8$. (*d*) $y = -9$.
5. Find the point-slope form of the equation of the line through each pair of points.
   (*a*) (5, 3) and (6, 5).
   (*b*) $(-2, -3)$ and (1, 2).
   (*c*) (3, 1) and $(-2, 8)$.
6. Find the general form of the equation of each line in Exercise 5.
7. Find the point-slope form of the equation of the line that passes through the given point and has given slope m.
   (*a*) $(6, -1)$, $m = \frac{2}{3}$.
   (*b*) $(-4, 2)$, $m = -5$.
8. Find the intercept form of the equation of the line whose x-intercept a and y-intercept b are given. Graph the line.
   (*a*) $a = 5$, $b = 2$.
   (*b*) $a = -6$, $b = \frac{1}{2}$.
9. Find the slope-intercept form of the equation of the line whose slope m and y-intercept b are given. Graph the line.
   (*a*) $m = 3$, $b = 2$.
   (*b*) $m = \frac{2}{3}$, $b = -1$.
   (*c*) $m = -7$, $b = 0$.

10. Find the general form of the equation of the line that passes through (7, 8) and that has x-intercept -2.
11. A line passes through $(-7, -5)$ and $(-3, 1)$. Find its equation in
(*a*) Point-slope form. (*b*) Slope-intercept form.
(*c*) General form. (*d*) Intercept form.
12. Sketch each of the following lines. Find the x-intercept, the y-intercept, and the slope.
(*a*) $5x - 2y + 10 = 0$. (*b*) $y = 4x - 7$.
(*c*) $x + y = 3$. (*d*) $\frac{x}{2} - \frac{y}{9} = 1$.
(*e*) $\frac{x}{2} - \frac{y}{9} = 2$. (*f*) $y - 4 = 2(x - 3)$.
13. Find an equation of the line that passes through (4, 2) and that is parallel to the line $5x + 2y + 7 = 0$. Leave your answer in general form.
14. Explain why the slope of a line, as defined in Definition 4, does not depend upon which two specific points on the line are used.
15. Let nonvertical lines l_1 and l_2 have slopes m_1 and m_2, respectively. It can be shown that l_1 and l_2 are perpendicular if and only if $m_2 = -\frac{1}{m_1}$. What is the slope of any line that is perpendicular to the line $2x - 3y + 5 = 0$?
16. A revenue function is given by $y = R(x) = 3x$, provided that $0 \leq x \leq 10$. Sketch its graph.
17. (*a*) Find the Celsius temperature corresponding to 140 degrees Fahrenheit.
(*b*) Find the Fahrenheit temperature corresponding to 30 degrees Celsius.
(*c*) Find the Fahrenheit temperature corresponding to -40 degrees Celsius.

1.6 Some Special Types of Functions

Suppose that a large corporation deposits \$10 million in a savings account for 1 year at 6 per cent interest compounded semiannually. After 6 months the savings account contains 103 per cent of the initial investment, which is

$$(10{,}000{,}000)(1.03) = \$10{,}300{,}000.$$

After 1 year the amount in the account is 103 per cent of the amount held after the first 6 months, namely,

$$(10{,}300{,}000)(1.03) = (10{,}000{,}000)(1.03)^2 = \$10{,}609{,}000.$$

Had the corporation deposited the money at 6 per cent at a place where the money was compounded quarterly, the amount after 1 year would have been

$$(10{,}000{,}000)(1.015)^4 = \$10{,}613{,}635.51.$$

If the money is compounded n times a year at 6 per cent, where n is a positive integer, the amount in the account after 1 year is a function of n as given by

$$f(n) = (10{,}000{,}000)\left(1 + \frac{.06}{n}\right)^n.$$

The domain of the function f above is the set of positive integers, since the number of times a year that interest is compounded is assumed to be a positive integer. The domain of many important functions is the set of positive integers 1, 2, 3, 4, . . . ; according to Definition 5, this special type of function is given a name.

Definition 5

A **sequence** is a function whose domain is the set of positive integers.

Example 1

Let f be the sequence defined by $f(n) = (n + 1)/(2n + 1)$. Since f is a sequence, we know that the domain of f is the set of positive integers 1, 2, 3, ... The positive integers (in order) can be substituted to obtain the elements of the range, which are called the **terms** of the sequence. Hence the **first term** is

$$f(1) = \frac{1+1}{2+1} = \frac{2}{3}.$$

The **second term** of the sequence is

$$f(2) = \frac{2+1}{4+1} = \frac{3}{5}.$$

$f(n) = (n + 1)/(2n + 1)$ is called the ***n*th term** or **general term** of the sequence because any term can be generated from it by substituting the appropriate positive integer. Since the domain of any sequence is the same, the sequence above can be accurately described by listing the terms of the sequence in order:

$$\frac{2}{3}, \frac{3}{5}, \frac{4}{7}, \frac{5}{9}, \ldots, \frac{n+1}{2n+1}, \ldots. \quad \blacksquare$$

Instead of using

$$f(1), f(2), \ldots, f(n), \ldots$$

as the general representation of the terms of a sequence, we find that it is customary to use

$$a_1, a_2, \ldots, a_n, \ldots,$$

where each subscript represents an element of the domain and each term represents an element of the range. In Example 1, $a_n = (n + 1)/(2n + 1)$. It is also customary to describe a sequence by just putting the nth term in braces—we can refer to the sequence in Example 1 by writing

$$\left\{\frac{n + 1}{2n + 1}\right\}.$$

We now consider another special type of function. Functions defined by

$$f(x) = \tfrac{3}{2}x^3 - 2x^2 + 5x + \sqrt{2}$$

and

$$g(x) = x^6 - 4x^2 + 7$$

are examples of polynomial functions; this type of function will be a favorite in this text because of its frequency of use in business, the social sciences, and the life sciences.

▶ Definition 6

A function f defined by

$$f(x) = b_n x^n + b_{n-1}x^{n-1} + \cdots + b_1 x + b_0,$$

where n is a nonnegative integer, $b_n \neq 0$, and $b_n, b_{n-1}, \ldots, b_1, b_0$ are real numbers, is called a **polynomial function of degree *n*.**

The two functions f and g immediately preceding Definition 6 are polynomial functions of degrees 3 and 6, respectively. The domain of any polynomial function, unless specified otherwise, is the set of all real numbers.

Three special cases of Definition 6 are now mentioned briefly:

1. A polynomial function of degree 0, as given by $f(x) = b_0$ with $b_0 \neq 0$, is called a **constant function.** Its graph is a horizontal line (Section 1.5). The function given by $f(x) = 0$ is also called a constant function.
2. A polynomial function of degree 1, as given by $f(x) = b_1 x + b_0$ with $b_1 \neq 0$, is called a **linear function.** Its graph is a line with slope b_1 and y-intercept b_0 (Section 1.5).
3. A polynomial function of degree 2, as given by $f(x) = b_2 x^2 + b_1 x + b_0$ with $b_2 \neq 0$, is called a **quadratic function.** Its graph is called a **parabola** and always has a characteristic appearance; if $b_2 > 0$, the parabola opens upward (see Figure 1.11), whereas if $b_2 < 0$, the parabola opens downward (see Figure 1.8).

A third special type of function is a **rational function,** which can be defined as a quotient of polynomial functions. For example, the function f defined by

$$f(x) = \frac{x^3 + 8x + 5}{2x^2 - 8}$$

is a rational function whose domain is the set of all real numbers except 2 and -2 (these numbers make the denominator zero).

A fourth special type of function is one that is defined by more than one equation; we illustrate such a function in Example 2.

Example 2

One item sold by a certain dealer in home and building supplies is storm windows. Most customers pay the dealer an extra fee for installing the windows instead of doing the installation themselves. The dealer charges y dollars per window for the installation, where y is a function of the number x of miles from the store to the customer's home—see Figure 1.21. A customer who lives less than 10 miles from the store is within the immediate city area, and the charge is \$6 per window for installation [the arrow at (10, 6) indicates that the \$6 fee applies up to but *not* including 10 miles]. A customer from a nearby town ($10 \leq x < 40$) pays \$8 per window, and a customer 40 or more miles away pays a little more ($\frac{1}{20}x + 6$), depending upon the mileage. The function f is graphed in Figure 1.21, and it can be defined by three equations as follows:

$$f(x) = \begin{cases} 6 & \text{if } 0 \leq x < 10, \\ 8 & \text{if } 10 \leq x < 40, \\ \frac{1}{20}x + 6 & \text{if } x \geq 40. \end{cases}$$ ∎

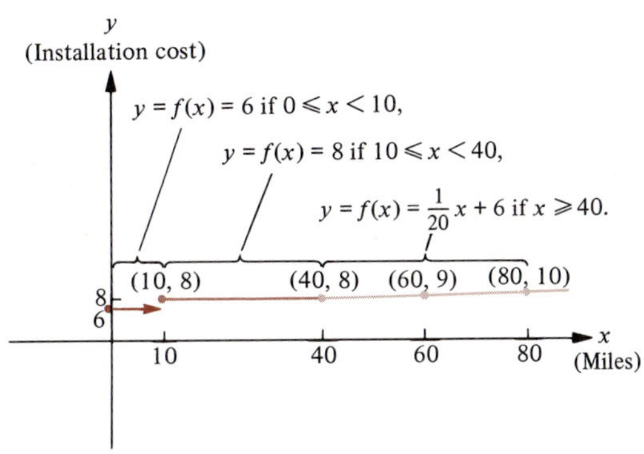

Figure 1.21. *Installation cost per window.*

It is important to realize in Example 2 that only *one* function f has been defined (not three functions). We are trying to point out that some functions can best be described by separate equations for different parts of the domain.

Example 3

Graph the function f defined by

$$f(x) = \begin{cases} 2 + |x| & \text{if } x < 2, \\ x^2 & \text{if } x \geq 2. \end{cases}$$

Table 1.9

x	-3	-1	0	1	1.9	2	3	4
$f(x)$	5	3	2	3	3.9	4	9	16

Use $f(x) = 2 + |x|$ if $x < 2$. (for $x = -3, -1, 0, 1, 1.9$)

Use $f(x) = x^2$ if $x \geq 2$. (for $x = 2, 3, 4$)

The domain of f is the set of all real numbers, but the manner in which $f(x)$ is calculated depends upon whether or not $x < 2$. Table 1.9 gives a table of values, and Figure 1.22 is a sketch of the curve that passes through the points $(x, y) = (x, f(x))$. Using the definition of absolute value (Definition 1), we can define the function f as follows:

$$f(x) = \begin{cases} 2 - x & \text{if } x < 0, \\ 2 + x & \text{if } 0 \leq x < 2, \\ x^2 & \text{if } x \geq 2. \end{cases}$$

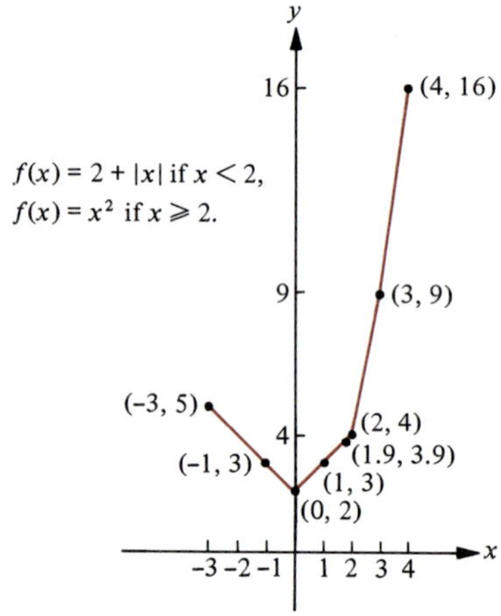

Figure 1.22

Thus the graph of f in Figure 1.22 consists of portions of the graphs of linear and quadratic functions. ∎

APPLICATIONS

Example 4

In the discussion at the beginning of this section, we showed that \$10 million deposited at 6 per cent compounded quarterly would yield after 1 year the amount \$10,613,635.51. Suppose, instead, that one wonders how much money would have to be deposited now at 6 per cent compounded quarterly in order to yield \$10 million 1 year from now. In other words, \$10 million 1 year from now has what **present value** at 6 per cent compounded quarterly?

Let P be the required present value. Then

$$P(1.015)^4 = 10{,}000{,}000$$

and

$$P = \frac{10{,}000{,}000}{(1.015)^4} = \frac{10{,}000{,}000}{1.061363550625} = \$9{,}421{,}842.30.$$

The type of calculations made in this example and at the beginning of this section are not restricted to problems in which the time period is 1 year. Suppose that a principal P is deposited at an annual interest rate r for t years. If interest is compounded n times a year, then the amount A after t years is given by

$$A = P\left(1 + \frac{r}{n}\right)^{nt}. \blacksquare$$

Example 5

Suppose that a person living in the United States wants to mail a domestic airmail letter, and suppose that the weight of the letter does not exceed 9 ounces ($\frac{9}{16}$ pound). The cost y, in cents, of mailing the letter is a function f of the weight x of the letter in ounces. As of this writing the cost of mailing a letter to Canada or Mexico is the same as the cost of a letter whose destination is within the United States, and the cost is given as follows:

$$f(x) = \begin{cases} 13 & \text{if } 0 < x \leq 1, \\ 26 & \text{if } 1 < x \leq 2, \\ 39 & \text{if } 2 < x \leq 3, \\ 52 & \text{if } 3 < x \leq 4, \\ 65 & \text{if } 4 < x \leq 5, \\ 78 & \text{if } 5 < x \leq 6, \\ 91 & \text{if } 6 < x \leq 7, \\ 104 & \text{if } 7 < x \leq 8, \\ 117 & \text{if } 8 < x \leq 9. \end{cases}$$

In Exercise 12 the reader is asked to graph $y = f(x)$. Incidentally, in the rare event that the letter weighs more than 9 ounces, the method of determining the cost is more complicated because it depends upon the distance the letter will travel as well as the weight of the letter. $\blacksquare$

★Example 6 Discrete and Continuous Measurements

The effect of defining a measurement (random variable) on a sample space U is to create a new, and corresponding, sample space S consisting of the set of values of the measurement. Such a set S has the advantage of having elements that are always real numbers. If the set S is finite or countably infinite,* then the measurement is said to be **discrete.** If, however, the set

* A set S is said to be **countably infinite** (or **countable**) if there is a one-to-one correspondence between S and the set of positive integers.

S consists of an interval of real numbers, then the measurement is said to be **continuous.**

The study of probability for discrete measurements presupposes information normally found in a course in finite mathematics, whereas the study of probability for continuous measurements requires some knowledge of calculus and will be pursued in later sections.

As an illustration of the discrete case, consider the experiment of flipping a coin until a head appears. The sample space U is

$$\{\text{H}, \text{TH}, \text{TTH}, \text{TTTH}, \ldots\}.$$

A measurement X is defined that associates with each outcome the number of times the coin is flipped until a head appears. The set of values of the measurement (the range of X) is the set $S = \{1, 2, 3, \ldots\}$ of all positive integers; since S is a countable set, X is a *discrete* measurement. A weighting function w, with domain S, will be defined below such that each weight $w(s) \geq 0$, and the sum of all weights is 1. Since, on one throw, the probability of throwing a head is $\frac{1}{2}$, on two throws the probability of throwing a tail followed by a head is $\frac{1}{4}$, and so on, then w is defined by

$$w(s) = \begin{cases} \dfrac{1}{2} & \text{if } s = 1, \\ \dfrac{1}{4} & \text{if } s = 2, \\ \vdots & \\ \dfrac{1}{2^n} & \text{if } s = n, \\ \vdots & \end{cases}$$

Notice that w is a sequence in this illustration and could be represented by $w(n) = 1/2^n$ or by $\{w_n\} = \{1/2^n\}$.

For the discrete case, the probability of an event $E = \{s_1, s_2, \ldots, s_k\}$, where E is a finite subset of S, can be defined as follows: If w, with domain S, is the weighting function, then

$$P(E) = w(s_1) + \cdots + w(s_k).$$

In the previous illustration let $E = \{1, 2, 3, 4\}$; hence E is the subset of measurement values that are less than or equal to 4. Here $P(E)$ represents the probability that a head will appear (at least once) if a coin is flipped four times. We calculate $P(E)$ as follows:

$$\begin{aligned} P(E) &= w(1) + w(2) + w(3) + w(4) \\ &= \tfrac{1}{2} + \tfrac{1}{4} + \tfrac{1}{8} + \tfrac{1}{16} \\ &= \tfrac{15}{16}. \end{aligned}$$

We assumed in the illustration of this example that the sum

$$\frac{1}{2} + \frac{1}{4} + \frac{1}{8} + \frac{1}{16} + \cdots + \frac{1}{2^n} + \cdots \qquad (1.2)$$

of weights was equal to 1. An expression, such as (1.2), in which addition of an infinite number of real numbers is indicated, will be discussed in Example 7 of Section 1.7; in that example we will prove that the sum (1.2) of weights is 1. The idea of the sum of an infinite number of real numbers also arises in defining $P(E)$ for an infinite subset $E = \{s_1, s_2, \ldots, s_n, \ldots\}$ of a countable sample space S corresponding to a discrete measurement X. (For example, if $E = \{2, 4, 6, \ldots, 2n, \ldots\}$ in our previous illustration, then $P(E)$ represents the probability that a head will appear first on an even-numbered flip of a coin.) If w, with domain S, is the weighting function, then $P(E)$ is defined by

$$P(E) = w(s_1) + w(s_2) + \cdots + w(s_n) + \cdots.$$

Example 7 of Section 1.7 should also aid in understanding this definition. ▮

EXERCISES

Suggested minimum assignment: Exercises 2(*c*), 3(*a*), 4, 5(*a*), 6, 8, and 11.

1. If \$8000 is deposited in a savings account at 5 per cent compounded semiannually, what is the amount in the account after 1 year?
2. Write the first five terms of each of the following sequences.

   (*a*) $\{4n - 1\}$. (*b*) $\left\{\frac{10}{n}\right\}$.

   (*c*) $\left\{\frac{n+2}{n+3}\right\}$. (*d*) $\left\{\frac{(-1)^n}{n+3}\right\}$.
3. Write the nth term of each of the following sequences.
   (*a*) 10, 20, 30, 40, 50, ... (*b*) $\frac{1}{2}, \frac{1}{5}, \frac{1}{8}, \frac{1}{11}, \frac{1}{14}, \ldots$.
   [In part (*a*) assume that the terms continue to increase by 10; in (*b*) assume that the numerators are always 1 and that the denominators continue to increase by 3.]
4. What is the twenty-fifth term of the sequence $\left\{\frac{n+7}{3n}\right\}$?
5. State the degree of each of the following polynomial functions. Also state the implied domain of each function.
   (*a*) $f(x) = 3 + 7x + x^5$. (*b*) $g(x) = (x^4)^2 + x^4 - 5$.
6. Tell whether each of the following polynomial functions is a constant function, a linear function, or a quadratic function.
   (*a*) $f(x) = 7x - 5$. (*b*) $g(x) = x^2 + 4x + 5$. (*c*) $h(x) = 8$.
7. Sketch the parabola defined by each of the following.
   (*a*) $y = f(x) = 16 - x^2$. (*b*) $y = g(x) = x^2 + 6x$.
8. State the implied domain of each of the following rational functions.

   (*a*) $f(x) = \frac{x}{(x-1)(x-2)(x-3)}$. (*b*) $g(x) = \frac{x+5}{x^2-9}$.
9. In Example 2 what is the cost of installation per window for a customer who lives 50 miles from the store?

10. Graph each of the following functions.

(a) $f(x) = \begin{cases} 2x & \text{if } x \leq 3, \\ 3 & \text{if } x > 3. \end{cases}$ (b) $g(x) = \begin{cases} |x| & \text{if } x < 4, \\ x - 4 & \text{if } x \geq 4. \end{cases}$

(c) $h(x) = \begin{cases} x & \text{if } 0 \leq x < 2, \\ 3 & \text{if } 2 \leq x \leq 3, \\ 4 & \text{if } x > 3. \end{cases}$

11. Graph the cost function C defined by

$$C(x) = \begin{cases} x^2 & \text{if } 0 \leq x \leq 2, \\ x + 2 & \text{if } x > 2. \end{cases}$$

12. Graph the function f defined in Example 5.
13. How much money would have to be deposited now at 8 per cent compounded semiannually in order to yield $2000 one year from now?
14. If $1000 is deposited at 5 per cent compounded quarterly, set up the formula (but do not calculate) for the amount in the account after 10 years.
15. In Theodore's book *Applied Mathematics: An Introduction* [22], several applications of linear and quadratic functions in business and economics are given on pp. 374–407. Solve the following problem, which is like one of Theodore's simpler examples. A retailer sells an item costing 60 cents for 78 cents and another item costing 70 cents for 91 cents. If these two examples represent the general markup policy for determining selling price, find the linear function that determines the selling price of an item.

1.7 Limits of Functions

The concept of the limit of a function is one of the most important concepts in mathematics and is fundamental in the development of calculus. We introduce the concept by means of two examples.

Example 1

Recall from the previous section that if $10 million is deposited at 6 per cent compounded n times a year, then the amount at the end of 1 year is

$$f(n) = (10{,}000{,}000)\left(1 + \frac{.06}{n}\right)^n.$$

In Table 1.10 the amount $f(n)$ is shown for several values of n (dollar signs omitted). It is not surprising that $f(n)$ gets larger as n gets larger, because if interest is compounded more frequently, one obtains more benefit from "interest on interest." What may be surprising is that the numbers near the bottom of the $f(n)$ column in Table 1.10 appear to show very little change; these numbers appear to be getting close to some **limit** that $f(n)$ cannot exceed, no matter how large n becomes. In Chapter 6 we shall be

Table 1.10

n	$f(n)$
1 (annually)	10,600,000.00
2 (semiannually)	10,609,000.00
4 (quarterly)	10,613,635.51
12 (monthly)	10,616,778.12
52 (weekly)	10,617,998.20
365 (daily)	10,618,313.11
1000	10,618,346.35
1,000,000	10,618,365.44
1,500,000	10,618,365.45

able to show that, as n gets very large, $f(n)$ approaches but never exceeds the limit $L = 10{,}618{,}365.47$. This is often expressed symbolically as

$$f(n) \to L \qquad \text{as } n \to \infty,$$

where $f(n) \to L$ is read "$f(n)$ approaches L" and $n \to \infty$ is understood to mean that n increases without bound and is read "n becomes infinite." The same idea is also expressed by the notation

$$\lim_{n\to\infty} f(n) = L,$$

which says that "the limit of $f(n)$ as n becomes infinite is L." Note that "lim" is an abbreviation of the word limit. (The value $L = 10{,}618{,}365.47$ corresponds to compounding interest "continuously," as more and more banks are doing—further discussion will appear in Section 6.1. Notice that compounding continuously gives only slightly more interest than compounding daily.) ∎

Example 2

Consider again the sequence

$$\frac{2}{3}, \frac{3}{5}, \frac{4}{7}, \frac{5}{9}, \ldots, \frac{n+1}{2n+1}, \ldots$$

of Example 1 of Section 1.6. The 100th term of the sequence is $\frac{101}{201}$, and we shall show later in this section that the terms of the sequence get very close to the limit $L = \frac{1}{2}$ for large values of n. We write

$$f(n) \to \frac{1}{2} \qquad \text{as } n \to \infty$$

or

$$\lim_{n\to\infty} f(n) = \lim_{n\to\infty} \frac{n+1}{2n+1} = \frac{1}{2}.$$

In Figure 1.23 the situation may be viewed geometrically. For the type of function known as a sequence, the independent variable n is called a **discrete variable;** on the graph there are gaps between the allowable values (positive integers) of the discrete independent variable. We also want to

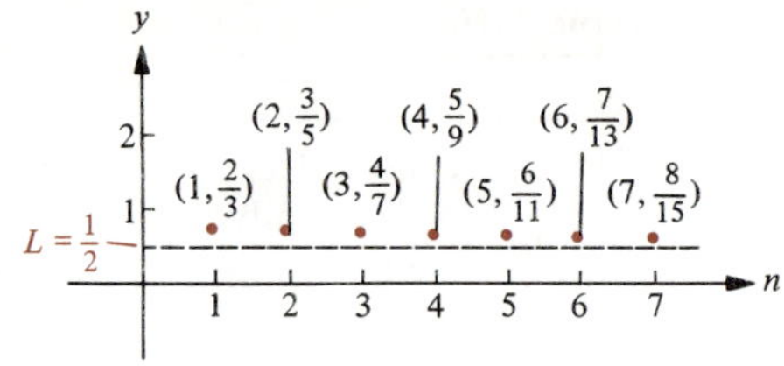

Figure 1.23. $y = f(n) = \dfrac{n+1}{2n+1}$.

study limits of functions of a **continuous variable,** in which case the independent variable takes on all real values in an interval. Consider

$$g(x) = \frac{x+1}{2x+1},$$

where the domain of g is the set of all real numbers except $-\frac{1}{2}$. A portion of the graph of g is shown in Figure 1.24. In this case also, $g(x)$ can be made as close to $\frac{1}{2}$ as desired by choosing x large enough, and we write

$$\lim_{x\to\infty} g(x) = \lim_{x\to\infty} \frac{x+1}{2x+1} = \frac{1}{2}. \quad \blacksquare$$

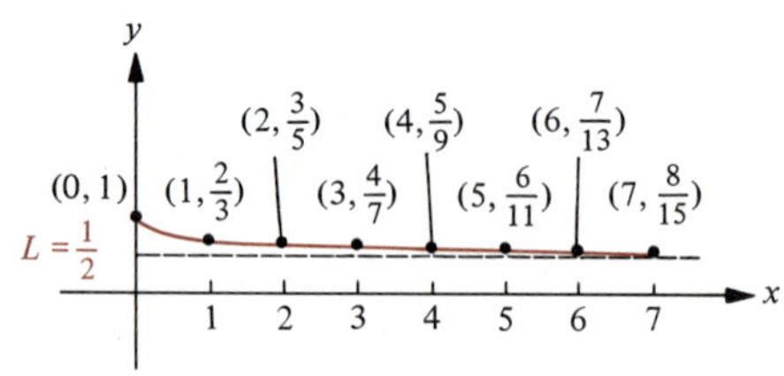

Figure 1.24. $y = g(x) = \dfrac{x+1}{2x+1}$.

In Definition 7 the meaning of $\lim_{x\to\infty} f(x) = L$ is given. It will suffice for the purposes of this text, but one should recognize that some precision can be gained by the introduction of more symbols, as shown in Exercise 6.

Definition 7

Suppose that the domain of f includes arbitrarily large numbers. The statement "**the limit of $f(x)$, as x becomes positively infinite, is the real number L** [written $\lim_{x\to\infty} f(x) = L$]" means that $|f(x) - L|$ can be made smaller than any given positive number by requiring that x (in the domain of f) be greater than a sufficiently large positive number.

Definition 7 applies whether x is a discrete variable or a continuous variable. In trying to understand Definition 7, the reader might review Figures 1.23 and 1.24 again, keeping in mind that the number L in Definition 7 is equal to $\frac{1}{2}$ in each case. Notice, for example, in Figure 1.24 that

$|g(x) - \frac{1}{2}| < \frac{1}{10}$ whenever $x > 2$ ($\frac{1}{10}$ is the difference between $g(2) = \frac{3}{5}$ and $L = \frac{1}{2}$). Moreover, $|g(x) - \frac{1}{2}|$ can be made even smaller by requiring that x be larger (see Exercise 11).

Not more than one number L can be the limit in Definition 7. Whenever the limit equals one real number (such as $L = \frac{1}{2}$), we say that the **limit exists.** Sometimes $\lim_{x\to\infty} f(x)$ does not exist—for example, $\lim_{x\to\infty} x^2$ does not exist because x^2 does not get close to any one number L as x gets large. Another example is the sequence

$$1, 2, 1, 2, 1, 2, \ldots, a_n, \ldots,$$

whose terms alternate between 1 and 2; the terms do not *all* get close to any one number L as one lets the term number n get large, and we say that the sequence does not have a limit (in other words, $\lim_{n\to\infty} a_n$ does not exist). Incidentally, a formula for a_n above is

$$a_n = \frac{3 + (-1)^n}{2}.$$

Theorem 1 is very useful in the calculation of limits. Its proof, however, is beyond the scope of this text.

▶ **Theorem 1**

If $\lim_{x\to\infty} f(x)$ and $\lim_{x\to\infty} g(x)$ exist, then

(*i*) $\lim_{x\to\infty} (f(x) + g(x)) = \lim_{x\to\infty} f(x) + \lim_{x\to\infty} g(x)$.

(*ii*) $\lim_{x\to\infty} (f(x)g(x)) = \left(\lim_{x\to\infty} f(x)\right)\left(\lim_{x\to\infty} g(x)\right)$.

(*iii*) $\lim_{x\to\infty} \frac{f(x)}{g(x)} = \frac{\lim_{x\to\infty} f(x)}{\lim_{x\to\infty} g(x)}$, provided that $\lim_{x\to\infty} g(x) \neq 0$.

(*iv*) $\lim_{x\to\infty} k = k$, where k is a real number.

(*v*) $\lim_{x\to\infty} (k\,f(x)) = k \lim_{x\to\infty} f(x)$, where k is a real number.

(*vi*) $\lim_{x\to\infty} \frac{1}{x^k} = 0$, provided that $k > 0$.

(*vii*) $\lim_{n\to\infty} k^n = 0$, provided that $-1 < k < 1$ and n takes only positive integral values. ◀

Example 3

Again consider $\lim_{x\to\infty} [(x + 1)/(2x + 1)]$ from Example 2. In order to evaluate this limit of a rational function, first divide numerator and denominator by the highest power of x in the fraction, and then apply various parts of

Theorem 1 as follows:

$$\lim_{x\to\infty}\frac{x+1}{2x+1}=\lim_{x\to\infty}\frac{1+(1/x)}{2+(1/x)}$$

$$=\frac{\lim\limits_{x\to\infty}[1+(1/x)]}{\lim\limits_{x\to\infty}[2+(1/x)]} \qquad \text{by Theorem 1}(iii)$$

$$=\frac{\lim\limits_{x\to\infty}1+\lim\limits_{x\to\infty}(1/x)}{\lim\limits_{x\to\infty}2+\lim\limits_{x\to\infty}(1/x)} \qquad \text{by Theorem 1}(i)$$

$$=\frac{1+\lim\limits_{x\to\infty}(1/x)}{2+\lim\limits_{x\to\infty}(1/x)} \qquad \text{by Theorem 1}(iv)$$

$$=\frac{1+0}{2+0} \qquad \text{by Theorem 1}(vi)$$

$$=\frac{1}{2}.$$

The above calculations are also valid for showing that

$$\lim_{n\to\infty}\frac{n+1}{2n+1}=\frac{1}{2},$$

where n takes only positive integral values (see the discussion in Example 2). ▮

Part (i) of Theorem 1 also holds for subtraction as well as for addition, and parts (i) and (ii) can be extended to more than two functions.

Example 4

$$\lim_{x\to\infty}\left[5-3\left(\frac{x+1}{2x+1}\right)+\frac{4}{x^2}\right]=5-3\left(\frac{1}{2}\right)+0=\frac{7}{2}$$

by Theorem 1(i), (iv), (v), (vi) and the result of Example 3. ▮

Example 5

If n takes only positive integral values,

$$\lim_{n\to\infty}(\tfrac{1}{2})^n=0 \qquad \text{and} \qquad \lim_{n\to\infty}(-\tfrac{1}{2})^n=0$$

by Theorem 1 (vii). ▮

One can also consider limits of functions in which the independent variable x becomes negatively infinite instead of positively infinite. This concept can be defined by making appropriate changes in Definition 7 (see Exercise 12). For example,

$$\lim_{x\to-\infty}\frac{1}{x}=0 \tag{1.3}$$

is correct because as x takes on values such as -10, -100, and -1000, the expression $1/x$ becomes closer and closer to 0. Parts (*i*) through (*v*) of Theorem 1 hold if $x \to -\infty$ instead of $x \to \infty$; part (*vi*) holds for values of k for which $1/x^k$ has meaning [see (1.3)]; but part (*vii*) is *not* valid for $n \to -\infty$. By following steps exactly like those in Example 3, it can be shown that

$$\lim_{x \to -\infty} \frac{x+1}{2x+1} = \frac{1}{2}.$$

APPLICATIONS

Example 6

Members of legislative bodies are often called on to respond with respect to a certain issue on repeated occasions. In time these legislators may alter their positions for a variety of reasons. On pp. 538–540 of [4] a formula is developed for the probability that a legislator will vote yes on the nth yes–no roll call pertaining to the same issue. The formula is

$$p_n = (1-2p)^n p_0 + \frac{1-(1-2p)^n}{2}$$

or

$$p_n = \frac{1}{2} + \left(p_0 - \frac{1}{2}\right)(1-2p)^n, \tag{1.4}$$

where n = number of roll calls taken;
p_n = probability that the legislator will vote yes on the nth roll call;
p = probability that the legislator will reverse his position on successive roll calls;
p_0 = probability that the legislator favors the issue upon initial consideration.

If $0 < p < 1$, then $-1 < (1-2p) < 1$; hence $(1-2p)^n$ gets closer to zero as n gets larger, by Theorem 1(*vii*). In other words,

$$\lim_{n \to \infty} (1-2p)^n = 0.$$

Using various parts of Theorem 1, the limit of p_n as $n \to \infty$ is found from (1.4) to be

$$\begin{aligned} \lim_{n \to \infty} p_n &= \lim_{n \to \infty} [\tfrac{1}{2} + (p_0 - \tfrac{1}{2})(1-2p)^n] \\ &= \left[\lim_{n \to \infty} \tfrac{1}{2}\right] + \left[\lim_{n \to \infty} (p_0 - \tfrac{1}{2})\right]\left[\lim_{n \to \infty} (1-2p)^n\right] \\ &= [\tfrac{1}{2}] + [(p_0 - \tfrac{1}{2})][(0)] \\ &= \tfrac{1}{2}. \end{aligned}$$

The result indicates that, subject to the assumptions made in the derivation of the formula, the probability of a legislator's voting yes on a given issue approaches $\frac{1}{2}$ as the number of roll calls increases. Notice that the limit is $\frac{1}{2}$ regardless of the values of p and p_0. This result helps explain the rationality of perseverance by a minority in favor of a certain issue. ▮

Example 7

An expression

$$a_1 + a_2 + a_3 + \cdots + a_n + \cdots, \tag{1.5}$$

in which addition of an infinite number of real numbers is indicated, is called an **infinite series.** For example,

$$\frac{1}{2} + \frac{1}{4} + \frac{1}{8} + \frac{1}{16} + \cdots + \frac{1}{2^n} + \cdots \tag{1.6}$$

is an infinite series. This particular infinite series was encountered in a previous application concerning probability (Example 6 of Section 1.6). In this example we shall develop a formula for finding what we shall call the **sum** of (1.6); in the process the reader will be introduced to the subject of infinite series, which has many applications.

In (1.5) let

$$s_1 = a_1,$$
$$s_2 = a_1 + a_2,$$
$$s_3 = a_1 + a_2 + a_3,$$

and, in general, for any positive integer n,

$$s_n = a_1 + a_2 + a_3 + \cdots + a_n.$$

The sequence $\{s_n\}$ is called the **sequence of partial sums** of the infinite series (1.5). If $\lim_{n\to\infty} s_n$ exists and equals S, then S is called the **sum** of the infinite series, and we write

$$S = a_1 + a_2 + a_3 + \cdots + a_n + \cdots.$$

An infinite series of the form

$$a + ar + ar^2 + ar^3 + \cdots + ar^{n-1} + \cdots \tag{1.7}$$

is called a **geometric series.** The number a is the first term of the infinite series, and the number r is called the **common ratio.** For the case that $-1 < r < 1$, let us try to find, if possible, the sum S of (1.7). Observe that

$$s_n = a + ar + ar^2 + ar^3 + \cdots + ar^{n-1}$$

and hence

$$rs_n = \qquad ar + ar^2 + ar^3 + \cdots + ar^{n-1} + ar^n.$$

Thus, by subtraction,

$$s_n - rs_n = a - ar^n.$$

Therefore,

$$s_n(1 - r) = a(1 - r^n)$$

and

$$s_n = \frac{a}{1-r}(1 - r^n).$$

Using $S = \lim_{n\to\infty} s_n$ and Theorem 1, we find that

$$\begin{aligned} S &= \lim_{n\to\infty} \frac{a}{1-r}(1 - r^n) \\ &= \frac{a}{1-r}\left(\lim_{n\to\infty} 1 - \lim_{n\to\infty} r^n\right) \\ &= \frac{a}{1-r}(1 - 0) \\ &= \frac{a}{1-r}. \end{aligned}$$

Notice especially that $\lim_{n\to\infty} r^n = 0$ by Theorem 1(*vii*), since we assumed that $-1 < r < 1$.

Let us demonstrate how the formula $S = a/(1 - r)$ can be used to find the sum of (1.6). The infinite series (1.6) is a geometric series with first term $a = \frac{1}{2}$ and common ratio $r = \frac{1}{2}$. Therefore,

$$S = \frac{a}{1-r} = \frac{\frac{1}{2}}{1 - \frac{1}{2}} = 1. \blacksquare$$

EXERCISES

Suggested minimum assignment: Exercises 1, 2, 5, and 9.

1. Plot the first six terms of the sequence $\left\{\frac{3n}{n+1}\right\}$ in a manner similar to that used in Figure 1.23. Estimate the value of $\lim_{n\to\infty} \frac{3n}{n+1}$.
2. If $g(x) = \frac{3x}{x+1}$, provided that $x \neq -1$, graph the portion of g for which $x > 0$ (see Exercise 1). Estimate the value of $\lim_{x\to\infty} \frac{3x}{x+1}$.
3. Repeat Exercise 1 for $\left\{\frac{n+3}{n+2}\right\}$.
4. If $g(x) = \frac{x+3}{x+2}$, provided that $x \neq -2$, graph the portion of g for which $x > 0$ (see Exercise 3). As $x \to \infty$, $g(x) \to L$. Estimate L.
5. Plot enough points to estimate the limit of each of the following sequences, provided that a limit exists.

   (*a*) $2, -2, 2, -2, 2, -2, \ldots, (-1)^{n+1}(2), \ldots.$

(*b*) $\left\{\frac{n^2 + 3}{n^2 + 2}\right\}$.

(*c*) $\left\{\frac{n}{n^2 + 2}\right\}$.

(*d*) $\frac{1}{2}, -\frac{1}{3}, \frac{1}{4}, -\frac{1}{5}, \frac{1}{6}, -\frac{1}{7}, \ldots, (-1)^{n+1}\frac{1}{n+1}, \ldots$.

(*e*) $5, 10, 15, 20, 25, \ldots, 5n, \ldots$.

6. The meaning of $\lim_{x\to\infty} f(x) = L$ in Definition 7 can be stated more precisely as follows: For any $\varepsilon > 0$ there exists a number $N > 0$ such that $|f(x) - L| < \varepsilon$ whenever x is in the domain of f and $x > N$. Draw a sketch to illustrate this definition and label ε, L, and N on your sketch.
7. Explain why $\lim_{x\to\infty} (f(x) - g(x)) = \lim_{x\to\infty} f(x) - \lim_{x\to\infty} g(x)$ follows from Theorem 1(*i*) and (*v*).
8. Use Theorem 1 to evaluate each of the following if possible.

   (*a*) $\lim_{x\to\infty} \frac{2x + 7}{x + 3}$. (*b*) $\lim_{x\to\infty} \frac{x^2 + 1}{x + 2x^2}$.

   (*c*) $\lim_{x\to\infty} (2x - 3)$. (*d*) $\lim_{x\to\infty} \frac{x^2}{x^3 - 3}$.
9. Use Theorem 1 (or its counterpart for $x \to -\infty$) to evaluate each of the following if possible.

   (*a*) $\lim_{x\to\infty} \frac{-3}{x^2}$. (*b*) $\lim_{x\to\infty} \frac{x^2}{x^2 + 2}$.

   (*c*) $\lim_{x\to-\infty} \frac{2x + 7}{x}$. (*d*) $\lim_{x\to-\infty} \frac{2x + 7}{x} \frac{x^3}{1975 + 2x^3}$.

   (*e*) $\lim_{n\to\infty} (-\frac{1}{3})^n$, where n takes only positive integral values.
10. If digital computer facilities are available, verify in Table 1.10 that $f(1000) = 10{,}618{,}346.35$ is correct.
11. In Figure 1.24 $|g(x) - \frac{1}{2}| < \frac{1}{30}$ whenever $x > N$. What is the smallest number that N can equal?
12. State the meaning of $\lim_{x\to-\infty} f(x) = L$ by making the appropriate changes in Definition 7.
13. Find the sum of each of the following geometric series.

    (*a*) $3 + \frac{9}{4} + \frac{27}{16} + \cdots$.

    (*b*) $\frac{1}{2} - \frac{1}{4} + \frac{1}{8} - \frac{1}{16} + \cdots$.
14. Change the repeating decimal .272727 . . . to a rational number by finding the sum of the geometric series

$$.27 + .0027 + .000027 + \cdots.$$

1.8 Limits of Functions (Continued)

In Section 1.7 we considered limits of functions as the independent variable became either positively or negatively infinite. Many applied problems involve the limit of a function as the independent variable approaches a finite number.

Example 1

In the population problem of Section 1.1 the average rate of change of population $y = f(t)$ with respect to time over a time interval $[t_1, t_2]$ was found to be

$$\frac{\Delta y}{\Delta t} = \frac{f(t_2) - f(t_1)}{t_2 - t_1}.$$

The immediate, or instantaneous, rate of change of population with respect to time at $t = t_1$ can be found by finding the value to which

$$\frac{f(t) - f(t_1)}{t - t_1}$$

gets close as t gets close to t_1. Suppose that $f(t) = t^2$, and we want to find the instantaneous rate of change of population with respect to time at $t = 3$.

$$\frac{f(t) - f(t_1)}{t - t_1} = \frac{t^2 - 3^2}{t - 3} = \frac{(t + 3)(t - 3)}{t - 3} = t + 3 \qquad \text{when } t \neq 3.$$

Intuitively, one expects $(t + 3)$ to get close to 6 as t gets close to 3. This can be stated

$$t + 3 \to 6 \qquad \text{as } t \to 3$$

or

$$\lim_{t \to 3} (t + 3) = 6. \blacksquare$$

Before giving a formal definition of the meaning of $\lim_{x \to c} f(x) = L$, where c and L are real numbers, we shall try to impart an intuitive understanding of its meaning by means of more examples.

Example 2

Suppose that a function f whose domain is the set of all real numbers is defined by

$$f(x) = \begin{cases} 2x + 1 & \text{if } x \neq 2, \\ \frac{5}{2} & \text{if } x = 2. \end{cases}$$

The graph of f is shown in Figure 1.25 (recall that $y = 2x + 1$ is a line with slope 2 and y-intercept 1). Despite the fact that $f(2) = \frac{5}{2}$, it is correct to write

$$\lim_{x \to 2} f(x) = 5$$

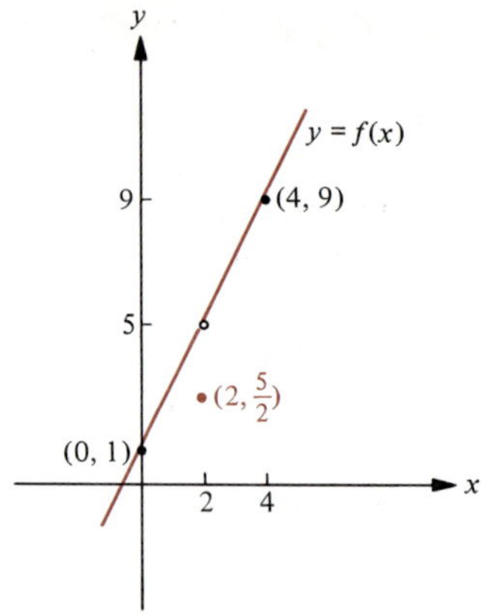

Figure 1.25

because the closer x is to 2 (but *not* equal to 2), the closer $f(x)$ is to 5. The value of f at 2 makes no difference at all in determining the limit of f as x approaches 2; even if $f(2)$ were undefined, $\lim_{x \to 2} f(x)$ would still equal 5. Also from Figure 1.25 it can be seen that if x is very close to 4, then $f(x)$ is very close to 9, and we write

$$f(x) \to 9 \qquad \text{as } x \to 4$$

or

$$\lim_{x \to 4} f(x) = 9. \blacksquare$$

Example 3

Referring back to Example 2 of Section 1.6 and to Figure 1.21, we can estimate from the figure that

$$\lim_{x \to 40} f(x) = 8.$$

However,

$$\lim_{x \to 10} f(x)$$

does not exist; this is because $f(x)$ is 6 when x is slightly less than 10, but $f(x)$ is 8 when x is slightly greater than 10. In order for the limit as x approaches 10 to exist, $f(x)$ must be near to some *one* value whenever x is *both* slightly less than 10 *and* slightly greater than 10. $\blacksquare$

In Definition 8 the meaning of $\lim_{x \to c} f(x) = L$ is given. It will suffice for the purposes of this text, but again one should recognize that precision can be gained by the introduction of more symbols, as shown in Exercise 4.

► **Definition 8**

Suppose that a function f is defined at every number in some open interval containing the number c, except possibly at the number c. The statement "**the limit of $f(x)$, as x approaches c, is the real number L** [written $\lim_{x\to c} f(x) = L$]" means that $|f(x) - L|$ can be made smaller than any given positive number by requiring that x be sufficiently close to c (but not equal to c).

As we did in Section 1.7, we present a theorem without proof that will be used in the calculation of limits.

Theorem 2

Let c be a real number and suppose that $\lim_{x\to c} f(x)$ and $\lim_{x\to c} g(x)$ exist. Then

(*i*) $\lim_{x\to c} (f(x) + g(x)) = \lim_{x\to c} f(x) + \lim_{x\to c} g(x)$.

(*ii*) $\lim_{x\to c} (f(x)\,g(x)) = \left(\lim_{x\to c} f(x)\right)\left(\lim_{x\to c} g(x)\right)$.

(*iii*) $\lim_{x\to c} \dfrac{f(x)}{g(x)} = \dfrac{\lim_{x\to c} f(x)}{\lim_{x\to c} g(x)}$, provided that $\lim_{x\to c} g(x) \neq 0$.

(*iv*) $\lim_{x\to c} k = k$, where k is a real number.

(*v*) $\lim_{x\to c} (k\,f(x)) = k \lim_{x\to c} f(x)$, where k is a real number.

(*vi*) $\lim_{x\to c} x = c$.

(*vii*) $\lim_{x\to c} \sqrt[n]{f(x)} = \sqrt[n]{\lim_{x\to c} f(x)}$, provided that either n is a positive odd integer, or $\lim_{x\to c} f(x) > 0$ and n is a positive even integer. ◀

Theorem 2(*i*) also holds for subtraction as well as for addition; parts (*i*) and (*ii*) can be extended to more than two functions.

Example 4

Evaluate $\lim_{x\to 2} \dfrac{3x^2 + 4}{x}$.

$$\lim_{x\to 2} \frac{3x^2 + 4}{x} = \frac{\lim_{x\to 2}(3x^2 + 4)}{\lim_{x\to 2} x} \qquad \text{by Theorem 2}(iii)$$

$$= \frac{\lim_{x\to 2}(3x^2 + 4)}{2} \qquad \text{by Theorem 2}(vi)$$

$$= \frac{\lim_{x\to 2} 3x^2 + \lim_{x\to 2} 4}{2} \qquad \text{by Theorem 2}(i)$$

$$= \frac{3\lim_{x\to 2} x^2 + \lim_{x\to 2} 4}{2} \qquad \text{by Theorem 2}(v)$$

$$= \frac{3\left(\lim_{x\to 2} x\right)\left(\lim_{x\to 2} x\right) + \lim_{x\to 2} 4}{2} \qquad \text{by Theorem } 2(ii)$$

$$= \frac{(3)(2)(2) + 4}{2} \qquad \text{by Theorem } 2(iv) \text{ and } (vi)$$

$$= 8. \blacksquare$$

Example 5

$$\lim_{x\to 9} \sqrt{x} = \sqrt{\lim_{x\to 9} x} \qquad \text{by Theorem } 2(vii)$$

$$= \sqrt{9} \qquad \text{by Theorem } 2(vi)$$

$$= 3. \blacksquare$$

Example 6

The limit $\lim_{x\to 0} 1/x^2$ does not exist. Rather than approach any one real number, the expression $1/x^2$ becomes larger and larger as x gets closer and closer to zero. ▮

In Example 1 we were concerned with $\lim_{t\to 3} [(t^2 - 9)/(t - 3)]$. If t is close to 3, both numerator and denominator of $(t^2 - 9)/(t - 3)$ are close to zero, and Theorem 2(*iii*) does not apply. When this happens try to obtain a cancellation—it is permissible to cancel $(t - 3)$ from numerator and denominator because in evaluating the limit, t approaches 3 but is not equal to 3, and hence we are not dividing by zero when the cancellation is made. Example 7 is similar to ones that we shall encounter in the next chapter. Notice that the numerator and denominator each approach zero again, so a cancellation is made to overcome this difficulty.

Example 7

$$\lim_{\Delta x\to 0} \frac{(2 + \Delta x)^2 - 4}{\Delta x}$$

$$= \lim_{\Delta x\to 0} \frac{4\,\Delta x + (\Delta x)^2}{\Delta x} \qquad \text{by expansion of } (2 + \Delta x)^2 \text{ and simplification}$$

$$= \lim_{\Delta x\to 0} (4 + \Delta x) \qquad \text{by cancellation of } \Delta x, \text{ since } \Delta x \neq 0$$

$$= \lim_{\Delta x\to 0} 4 + \lim_{\Delta x\to 0} \Delta x \qquad \text{by Theorem } 2(i)$$

$$= 4 + 0 \qquad \text{by Theorem } 2(iv) \text{ and } (vi)$$

$$= 4.$$

The reader should think of (Δx) as a single quantity in this example and realize that we evaluated a limit as $\Delta x \to 0$ of a function of Δx, whereas previous examples considered a limit as $x \to c$ of a function of x. ▮

APPLICATIONS

*Example 8 Cumulative Probability Distribution Function

In Example 6 of Section 1.6 an experiment was discussed in which a coin was flipped until a head appeared. A random variable or measurement was defined over the sample space U. The set of values (the range) of this discrete measurement constituted a new sample space $S = \{1, 2, 3, \ldots\}$. The probability of an event E, where E is a subset of S, was expressed in terms of the weighting function w; for example, if $E = \{1, 2, 3\}$, $P(E) = w(1) + w(2) + w(3)$. Recall that

$$w(s) = \begin{cases} \frac{1}{2} & \text{if } s = 1, \\ \frac{1}{4} & \text{if } s = 2, \\ \vdots & \\ \frac{1}{2^n} & \text{if } s = n, \\ \vdots & \end{cases}$$

In order to consider a second means of expressing the probability of an event of the type $E = \{1, 2, \ldots, k\}$, we shall introduce a function F whose domain is the set of *all* real numbers. Let F be related to w in such a way that $F(x)$ equals the sum of the weights of those measurement values that are less than or equal to x. In other words, the value of F for any real number x is the cumulative weight of those elements of S that are less than or equal to x; the function F is called the **cumulative probability distribution function.** In our illustration

$$F(x) = \begin{cases} 0 & \text{if } x < 1, \\ \frac{1}{2} & \text{if } 1 \leq x < 2, \\ \frac{3}{4} & \text{if } 2 \leq x < 3, \\ \vdots & \\ \frac{2^n - 1}{2^n} & \text{if } n \leq x < n + 1, \\ \vdots & \end{cases}$$

Portions of the graphs of w and F are given in Figures 1.26 and 1.27. Notice that the domain of F is the set of *all* real numbers, whereas the domain of w is the sample space S. The probability of the event $E = \{1, 2\}$, which is to say the probability that $s \leq 2$, can be expressed as

$$P(E) = F(2) = \tfrac{3}{4}.$$

In general, if $E = \{s : s \leq k\}$, then $P(E) = F(k)$.

A cumulative probability distribution function F may be defined independently of its relation to w. A **cumulative probability distribution**

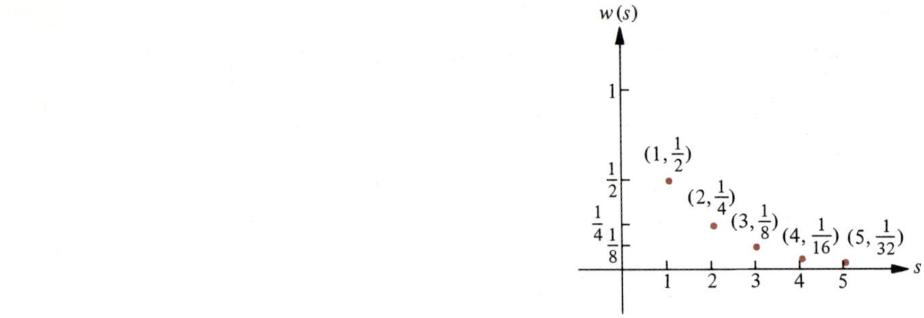

Figure 1.26. *A weighting function.*

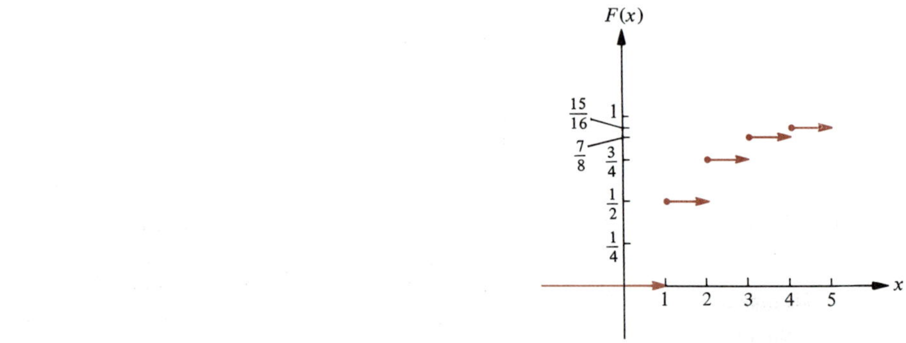

Figure 1.27. *A cumulative probability distribution function.*

function is any function F, whose domain is the set of all real numbers, possessing the following four properties:

1. $\lim_{x \to -\infty} F(x) = 0$.
2. $\lim_{x \to \infty} F(x) = 1$.
3. If $x_1 < x_2$, then $F(x_1) \leq F(x_2)$.
4. $\lim_{\Delta x \to 0} F(x + \Delta x) = F(x)$ for any real number x, where $\Delta x > 0$.

This abstract definition is consistent with our previous discussion of F, and the key point in understanding this consistency is to associate $F(k)$ with $P(E)$, where $E = \{s : s \leq k\}$.

Next we explain why the function F of the illustration used in this example possesses all four properties required of a cumulative probability distribution function.

1. $\lim_{x \to -\infty} F(x) = 0$ because $F(x) = 0$ for $x < 1$.
2. $\lim_{x \to \infty} F(x) = \lim_{n \to \infty} [(2^n - 1)/2^n] = \lim_{n \to \infty} 1 - \lim_{n \to \infty} (\frac{1}{2})^n = 1 - 0 = 1$.

3. The graph of F never "falls" as one moves from left to right along the curve. In other words, if $y = F(x)$ then y never decreases as x increases. This is clearly true from the definition of F as illustrated in Figure 1.27.

4. Notice that the graph of F consists of portions of *horizontal* lines, each of which is defined on an interval that does not include the right endpoint.

For example, $F(x) = \frac{1}{2}$ on the interval $[1, 2)$, and that interval does not include the right endpoint 2. Therefore, for any real number x, a sufficiently small positive change Δx in x will cause no change in the value of F. Hence we can say that

$$\lim_{\Delta x \to 0} F(x + \Delta x) = \lim_{\Delta x \to 0} F(x) = F(x), \qquad \text{where } \Delta x > 0.$$

If the intervals above had included the right endpoint, this property would not be satisfied when x is equal to a right endpoint.

In this example we have illustrated the concept of a cumulative probability distribution function for the case of a discrete measurement. An example for the case of a continuous measurement will be given in Example 7 of Section 2.2. ▮

EXERCISES

Suggested minimum assignment: Exercises 3, 5, 7, 8, and 9(*e*).

1. By the observation of Figure 1.22 on p. 38, estimate the values, if possible, of each of the following.
   (*a*) $\lim_{x\to 0} f(x)$. (*b*) $\lim_{x\to 2} f(x)$.
2. (*a*) Sketch the function f defined by

$$f(x) = \begin{cases} -x & \text{if } x \le -2, \\ 2x & \text{if } x > -2. \end{cases}$$

   (*b*) Estimate the value of $\lim_{x\to -2} f(x)$, if it exists.
   (*c*) Estimate the value of $\lim_{x\to 1} f(x)$, if it exists.
3. From Figure 1.28 estimate the value of each of the following or state that it does not exist.
   (*a*) $f(-5)$. (*b*) $\lim_{x\to -5} f(x)$. (*c*) $f(-3)$. (*d*) $\lim_{x\to -3} f(x)$.
   (*e*) $\lim_{x\to -2} f(x)$. (*f*) $\lim_{x\to 1} f(x)$. (*g*) $\lim_{x\to 3} f(x)$.

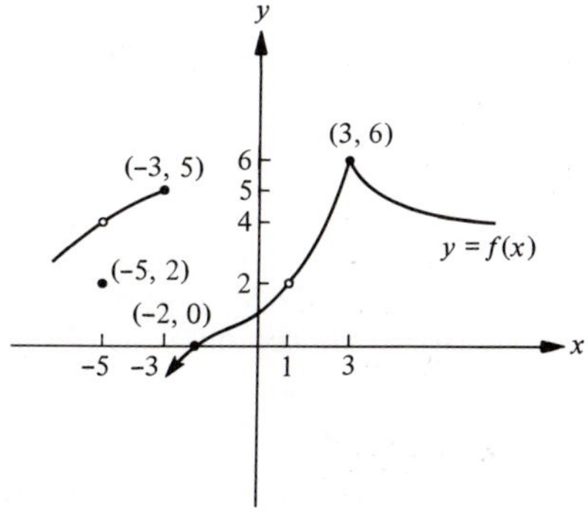

Figure 1.28

4. A more precise definition of the meaning of $\lim_{x \to c} f(x) = L$ than that given in Definition 8 is the following: For any $\varepsilon > 0$ there exists a $\delta > 0$ such that $|f(x) - L| < \varepsilon$ whenever $0 < |x - c| < \delta$. Draw a sketch illustrating the statement above, showing ε, δ, c, and L on your figure.

5. Use Theorem 2 to evaluate each of the following.

   (*a*) $\lim_{x \to 4} \frac{x^2 + 3}{5x + 2}$. (*b*) $\lim_{x \to 4} (2x - \sqrt[3]{16x})$.

6. Use Theorem 2 to evaluate each of the following.

   (*a*) $\lim_{x \to -2} \frac{x}{3x + 1}$. (*b*) $\lim_{x \to 3} \frac{x^2 + 3x}{x^3 + 1}$.

7. In Example 1 find the instantaneous rate of change of population with respect to time at $t = 4$ by evaluating $\lim_{t \to 4} \frac{t^2 - 16}{t - 4}$.

8. In Example 1 an alternative way of writing $\frac{f(t) - f(t_1)}{t - t_1}$ is to let $t = t_1 + \Delta t$; then the expression becomes $\frac{f(t_1 + \Delta t) - f(t_1)}{\Delta t}$. Use the latter expression to rework Exercise 7, after noting that $\Delta t \to 0$ as $t \to t_1$; that is, evaluate $\lim_{\Delta t \to 0} \frac{(4 + \Delta t)^2 - 4^2}{\Delta t}$.

9. Evaluate each of the following if possible.

   (*a*) $\lim_{x \to 5} \frac{x^2 - 25}{x^2 - 5x}$. (*b*) $\lim_{x \to 2} \frac{3x^2 - 7x + 2}{x - 2}$.

   (*c*) $\lim_{x \to 5} \frac{x - 5}{x + 5}$. (*d*) $\lim_{x \to 5} \frac{x + 5}{x - 5}$.

   (*e*) $\lim_{\Delta x \to 0} \frac{(1 + \Delta x)^2 - 1}{\Delta x}$.

10. Evaluate $\lim_{t \to -2} \frac{t^3 + 8}{t + 2}$.

11. Describe the cumulative probability distribution function F that corresponds to the weighting function given in Table 1.4.

1.9 Continuity

Many results in calculus are valid if a particular function in question possesses a property known as **continuity.** In this section we shall distinguish between continuous and discontinuous functions.

Example 1

Company A sells an item by the pound (or fraction thereof). In order to encourage large purchases, the company charges \$1.10 per pound if less than 8 pounds are ordered but charges only \$1.00 per pound on all orders of 8 or more pounds. Hence, if x is the number of pounds purchased, a cost function C is described by

$$C(x) = \begin{cases} 1.1x & \text{if } 0 < x < 8, \\ x & \text{if } x \geq 8. \end{cases}$$

The graph of C is shown in Figure 1.29. We say that the function C has a discontinuity at $x = 8$ because of the sudden change in cost. A customer would be foolish to buy 7.5 pounds of this item from Company A, since he can purchase 8 pounds at a lower cost. Observe from Figure 1.29 that $\lim_{x \to 8} C(x)$ does not exist, whereas $C(8) = 8$. Because $\lim_{x \to 8} C(x)$ does *not* equal $C(8)$, we say that C is discontinuous (not continuous) at $x = 8$. ▮

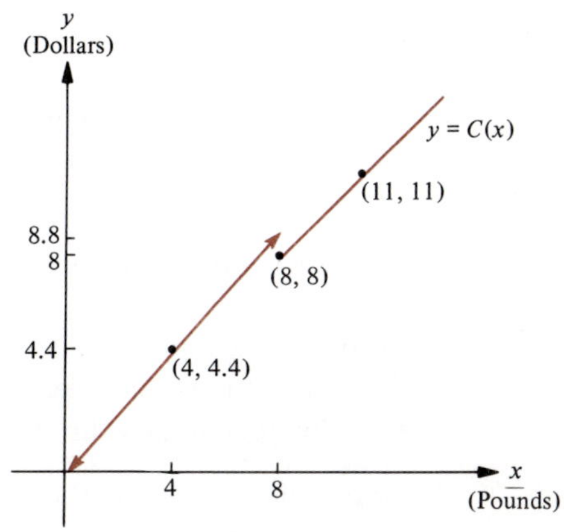

Figure 1.29

Example 2

Company B sells a competitive product by the pound (see Example 1). Its cost function f is given by

$$f(x) = \begin{cases} 1.2x & \text{if } 0 < x \leq 5, \\ 6 + .9(x - 5) & \text{if } x > 5. \end{cases}$$

In other words, Company B charges \$1.20 per pound for the first 5 pounds, and for purchases over 5 pounds it charges \$6.00 plus \$.90 for each pound over 5 pounds. See Figure 1.30. Even though Company B changes its method of computing the cost at 5 pounds, the function f is continuous at $x = 5$. This is because there is no sudden jump in the cost at $x = 5$. It can be shown that $\lim_{x \to 5} f(x) = 6$ and also that $f(5) = 6$; since

$$\lim_{x \to 5} f(x) = f(5),$$

we say that f is continuous at $x = 5$. ▮

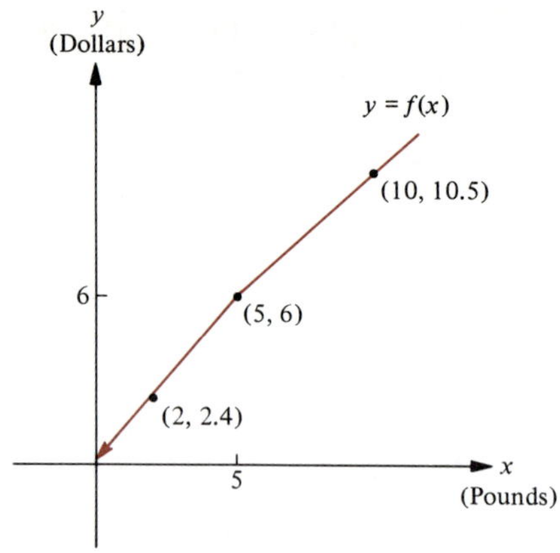

Figure 1.30

Definition 9

A function f is said to be **continuous** at a number c if

$$\lim_{x \to c} f(x) = f(c).$$

Otherwise f is said to be **discontinuous** at $x = c$.

In order for a function f to have a chance to be continuous at $x = c$, the domain of f must include c because of the presence of the symbol $f(c)$ in Definition 9. Similarly, $\lim_{x \to c} f(x)$ must exist in order for f to be continuous at $x = c$. Definition 9 says that $f(c)$ and $\lim_{x \to c} f(x)$ not only must each exist in order for f to be continuous at $x = c$, but that they must be *equal*. Recall from Section 1.8 that $\lim_{x \to c} f(x)$ can exist whether or not $f(c)$ is defined. See Figures 1.31 and 1.32 for some typical situations.

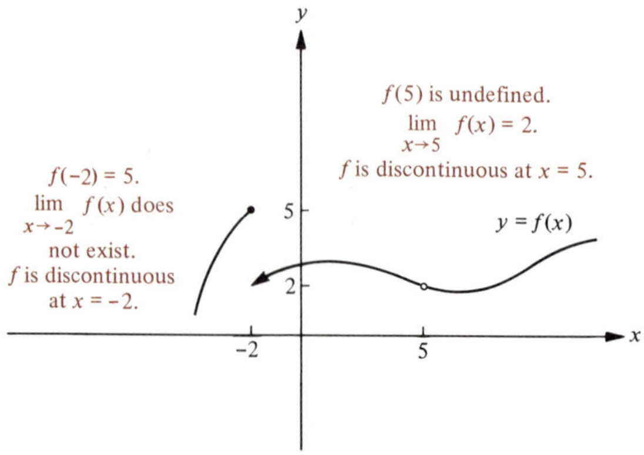

Figure 1.31

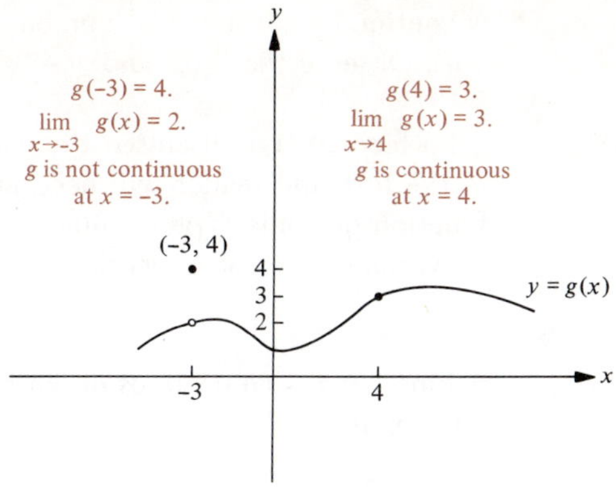

Figure 1.32

It is important that one be able to recognize whether certain familiar functions are continuous and also be able to detect any points of discontinuity without a great deal of effort. For example, is the quadratic function defined by $f(x) = x^2 + 1$ continuous at $x = 3$? We can answer the question affirmatively by showing that $\lim_{x\to 3} f(x) = f(3)$ as follows:

$$f(3) = (3)^2 + 1 = 10,$$

$$\lim_{x\to 3}(x^2 + 1) = \left(\lim_{x\to 3} x\right)\left(\lim_{x\to 3} x\right) + \lim_{x\to 3} 1$$

$$= 3(3) + 1 = 10.$$

This example indicates how Theorem 2 and Definition 9 can be used to prove Theorem 3.

► **Theorem 3**

(*i*) A polynomial function is continuous for all real numbers.
(*ii*) A rational function is continuous except at a number for which the denominator is zero.

As indicated above, the proof follows readily from Theorem 2 and Definition 9. ◄

Example 3

The polynomial function f defined by

$$f(x) = 7x^4 - 5x^3 + 2x^2 - 8x + 3$$

is continuous for all real numbers by Theorem 3(*i*). ▮

Example 4

The rational function g defined by

$$g(x) = \frac{x^2 + 5}{x^2 - 9}$$

is continuous for any real number except $x = 3$ or $x = -3$ by Theorem 3(*ii*). Observe that $g(3)$ and $g(-3)$ do not exist. ▮

So far we have discussed the continuity of a function at a point; later in the text, the reader will need an understanding of the continuity of a function on some type of interval. Therefore, we shall discuss continuity on various types of intervals for the remainder of this section.

▶ Definition 10

A function f is **continuous on an open interval** if it is continuous at each number in the interval.

Example 5

In Example 1 the function C is continuous at any positive number except $x = 8$ (see Figure 1.29). Therefore, by Definition 10, C is continuous on any open interval in its domain that does not include the number 8. Thus C is continuous on the intervals $(2, 5)$, $(5, 8)$, and $(8, 10)$. However, C is not continuous on the intervals $(5, 10)$ or $(7.9, 8.1)$. ▮

▶ Definition 11

A function f is **continuous on a closed interval** $[a, b]$, provided that the following three conditions are satisfied:

(*i*) f is continuous on the open interval (a, b).
(*ii*) If as $x \to a$ from within (a, b), then $f(x) \to f(a)$.
(*iii*) If as $x \to b$ from within (a, b), then $f(x) \to f(b)$.

Continuity on $[a, b)$ is defined in the same way, except that only (*i*) and (*ii*) are required, and continuity on $(a, b]$ requires only (*i*) and (*iii*).

Example 6

A function f with domain $[2, 7]$ is continuous on the open interval $(2, 7)$, and has a graph as shown in Figure 1.33. Definition 11(*i*) is satisfied. As $x \to 2$ from within $(2, 7)$, $f(x) \to 5$. However, $f(2) = 3$, and hence $f(x)$ does

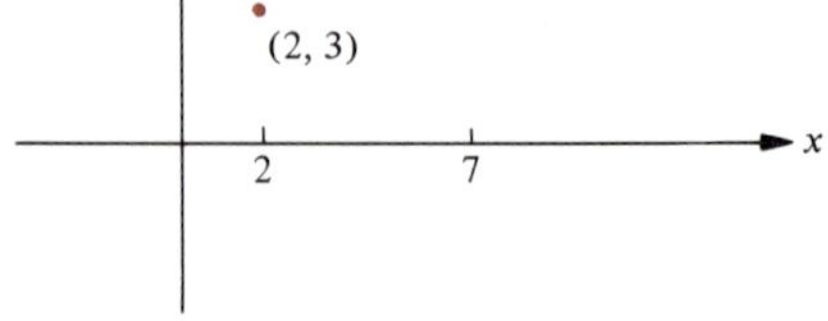

Figure 1.33

not approach $f(2)$ as x approaches 2 through values larger than 2. Thus Definition 11(*ii*) is not satisfied. As $x \to 7$ from within (2, 7), $f(x) \to f(7)$, where $f(7) = 6.5$. Hence Definition 11(*iii*) is satisfied. We can conclude that f is continuous on (2, 7], but that f is not continuous on [2, 7] or [2, 7). As one might expect intuitively, the jump in the function value at the endpoint $x = 2$ prevents f from being continuous on [2, 7] or [2, 7). ∎

Example 7

Suppose that an automobile manufacturer tests a new model car to determine the number y of miles per gallon the car can attain when driven at x miles per hour. Exhaustive tests are made for values of x in the closed interval [30, 80]; the results are shown in Figure 1.34. Certainly, y is a function of x, say $y = f(x) = -\frac{1}{550}x^2 + \frac{293}{11}$, and suppose that the nature of the practical situation has restricted the domain of f to the closed interval [30, 80]. The function f is continuous on the closed interval [30, 80] because it can be verified that all three parts of Definition 11 are satisfied. ∎

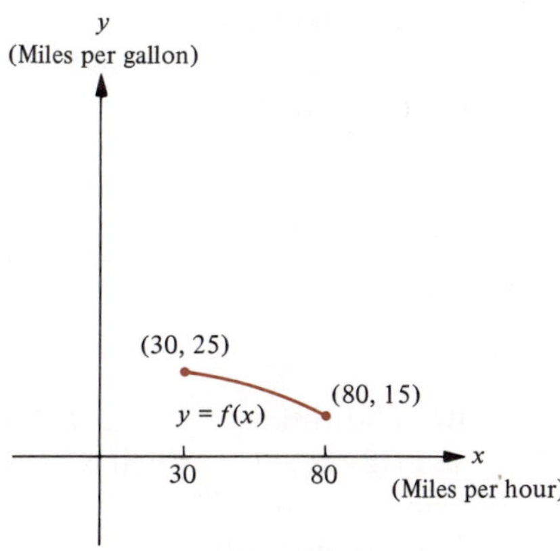

Figure 1.34

Because of practical restrictions, the domain of a function encountered in applications is often a closed interval $[a, b]$ as in Example 7. For future reference, important properties of a function that is continuous on a closed interval are stated in Exercises 12 and 13.

APPLICATIONS

Example 8

Suppose that a biologist is studying the effect of a new drug. He injects a known amount of the drug into a person and then measures the concentration y of the drug in the blood after x hours. Measurements are made for several people and for several values of x. The biologist seeks to find a mathematical model of the situation in the form of a function f such that $y = f(x)$. Since the concentration of the drug in the blood should be 0 at the beginning of the experiment, $f(0)$ should equal 0. The function

f should be continuous for $x \geq 0$ because the change in concentration is expected to be gradual rather than something that suddenly jumps from one value to another. After a period of time the effect of the drug should be negligible, and we should require that

$$\lim_{x \to \infty} f(x) = 0$$

be true. For the people tested, the concentration after 1 hour is about $\frac{2}{3}$ per cent, after 2 hours is between $\frac{3}{4}$ and $\frac{4}{5}$ per cent, and after 4 hours is down to about $\frac{2}{3}$ per cent again. Show that the function f given by

$$y = f(x) = \frac{x}{20x^2 + 50x + 80}, \qquad x \geq 0,$$

possesses the properties described above.

$$f(0) = \frac{0}{0 + 0 + 80} = 0.$$

The denominator $20x^2 + 50x + 80$ is always positive, as can be seen by completing the square and writing it in the form $20(x + \frac{5}{4})^2 + \frac{195}{4}$. Hence by Theorem 3(*ii*) of this section there are no values of x for which f is discontinuous. By use of the methods of Section 1.7 we find that

$$\lim_{x \to \infty} \frac{x}{20x^2 + 50x + 80} = \lim_{x \to \infty} \frac{1/x}{20 + (50/x) + (80/x^2)}$$
$$= \frac{0}{20 + 0 + 0} = 0.$$

By substitution, $f(1)$ and $f(4)$ can be shown to equal exactly $\frac{2}{3}$ per cent and $f(2) \approx .0077$, which is between $\frac{3}{4}$ and $\frac{4}{5}$ per cent. ($\approx$ means "is approximately equal to.")

There are still some other properties that f must possess in order for it to be satisfactory for the biologist to use as a means of predicting concentration of the drug in the blood. These properties will be discussed in Example 4 of Section 3.2. ▮

EXERCISES

Suggested minimum assignment: Exercises 2, 3, 6, 7, and 9(*b*).

1. Suppose that a cost function C is given by

$$C(x) = \begin{cases} 5x & \text{if } 0 < x < 5, \\ 6x & \text{if } x \geq 5. \end{cases}$$

Is C continuous at $x = 5$? Why?

2. Suppose that a revenue function R is given by

$$R(x) = \begin{cases} 3x & \text{if } 0 \leq x < 10, \\ 30 + 2x & \text{if } x \geq 10. \end{cases}$$

Is R continuous at $x = 10$? Why? Draw a graph.

3. Suppose that a profit function P is given by

$$P(x) = \tfrac{1}{5}x - 3, \qquad \text{provided that } 0 \leq x \leq 1000.$$

Is P continuous at $x = 25$? Why?

4. Why is it not appropriate to discuss continuity for the type of function known as a sequence?

5. Refer to the function f in Example 2 on p. 37.
(*a*) Is f continuous at $x = 10$?
(*b*) Is f continuous at $x = 40$?

6. Refer to the function f in Example 2 on p. 51.
(*a*) Is f continuous at $x = 2$? Why?
(*b*) Is f continuous on $(0, 2)$?
(*c*) Is f continuous on $[0, 2]$? Why?

7. Let f be defined by

$$\begin{cases} f(x) = \dfrac{x - 1}{x^2 - 1} & \text{if } x \neq 1, \\ f(1) = 3. \end{cases}$$

(*a*) Evaluate $\lim_{x \to 1} f(x)$.
(*b*) Explain why f is discontinuous at $x = 1$.
(*c*) Does f have any other points of discontinuity?

8. Sketch the graph of the function f defined by

$$f(x) = \begin{cases} x + 5 & \text{if } x < 2, \\ 6 & \text{if } x \geq 2. \end{cases}$$

(*a*) Is f continuous on $[0, 2)$?
(*b*) Is f continuous on $(1, 3)$? Why?

9. State all values of x for which each of the following rational functions is discontinuous.

(*a*) $f(x) = \dfrac{x}{x - 1}$. (*b*) $g(x) = \dfrac{x^2 - 16}{x^2 - 5x + 6}$.

(*c*) $h(x) = \dfrac{x^2 + 3x + 1}{x^2 + 3x}$.

10. Show that f is continuous at $x = 2$ by using the procedure illustrated just prior to Theorem 3, if $f(x) = x^2 + 7x + 2$.

11. If $y = f(x)$ and $y = g(x)$ are each continuous at $x = c$, use Theorem 2(*i*) to show that $y = f(x) + g(x)$ is continuous at $x = c$. In other words, show that the sum of two continuous functions is continuous. (Similar statements can be made about products, differences, and quotients, provided that in a quotient the denominator is not zero at $x = c$.)

12. (*a*) Draw a sketch to illustrate the following property of a continuous function on a closed interval: If f is continuous on $[a, b]$ and $f(a) \neq f(b)$, then for any number K between $f(a)$ and $f(b)$ there is at least one value of x, say $x = c$, such that $f(c) = K$.

(*b*) If $f(a)$ and $f(b)$ have opposite signs in part (*a*), what can be said about a root of the equation $f(x) = 0$?

13. (*a*) Suppose that f is not a constant function and that f is continuous on a closed interval $[a, b]$. Draw a sketch which illustrates the property that $f(x)$ must take on a least value m and a greatest value M on $[a, b]$.

    (*b*) Draw a sketch to illustrate that the statement of part (*a*) might not be true if $[a, b]$ were replaced by (a, b).

14. Suppose that f is continuous at $x = c$ and $f(c) > 0$. Show, by use of Definitions 8 and 9, that there is an open interval that contains c throughout which $f(x) > 0$. [It is also true that if f is continuous at $x = c$ and $f(c) < 0$, then there is an open interval that contains c, throughout which $f(x) < 0$. The facts stated in this exercise will be helpful in proving some theorems in Chapter 3.]

NEW VOCABULARY

real line 1.2
rational number 1.2
irrational number 1.2
inequality 1.2
open interval 1.2
closed interval 1.2
half-open interval 1.2
endpoints of an interval 1.2
interior points of an interval 1.2
absolute value 1.2
distance 1.2
rectangular (Cartesian) coordinate system 1.3
increment 1.3
distance formula 1.3
graph of an equation 1.3
curve 1.3
intercepts 1.3
function 1.4
image 1.4
domain 1.4
range 1.4
independent variable 1.4
dependent variable 1.4
graph of function 1.4
revenue function 1.4
cost function 1.4
fixed cost 1.4
profit function 1.4
break-even chart 1.4
break-even point 1.4
slope 1.5
point-slope form 1.5
intercept form 1.5
slope-intercept form 1.5
general form 1.5
sequence 1.6
terms of sequence 1.6
polynomial function of degree n 1.6
constant function 1.6
linear function 1.6
quadratic function 1.6
parabola 1.6
rational function 1.6
discrete variable 1.7
continuous variable 1.7
limit of function (as independent variable becomes infinite) 1.7
limit of function (as independent variable approaches finite number) 1.8
continuous function 1.9
discontinuous function 1.9
continuity on an interval 1.9

2 Differentiation

Scholars generally regard the development of calculus in the latter part of the 17th century, by Isaac Newton (1642–1727) and Gottfried Leibniz (1646–1716), as one of the most significant advances in the history of human thought. Two quotations attributed to Newton reveal the perspective in which he viewed his monumental achievement. In one quotation Newton stated that if he had seen farther than other men, it was only because he had stood on the shoulders of giants. He also said, "I do not know what I may appear to the world; but to myself I seem to have been only like a boy playing on the seashore, and diverting myself in now and then finding a smoother pebble or a prettier shell than ordinary, while the great ocean of truth lay all undiscovered before me." Time has shown that one of the more potent tools for the discovery of unknown truths has been Newton's own calculus.

Ordinarily, the study of calculus is organized into two parts—the differential calculus and the integral calculus. Fundamental to the study of the differential calculus is "the derivative of a function." The purpose of this chapter is to define, illustrate, and apply the idea of the derivative of a function.

Prerequisites: Chapter 1; 2.6 is not a prerequisite to optional Section 2.7.
Suggested Sections for Moderate Emphasis: 2.1–2.6
Suggested Sections for Minimum Emphasis: 2.1–2.5

2.1 Instantaneous Rates of Change and Tangent Lines

Let $y = f(x)$ express the relationship of the daily profit y of a theater to the price x of each ticket, where the units in both cases are in dollars. The theater manager wants to analyze the effect of price changes on the profit.

Assume that the domain of the function f consists of all real values of x in the open interval $(0, 8)$. The domain represents the set of ticket prices that the manager can charge.* Let x_1 be some value of x in the domain. Starting at x_1, we change the ticket price x by an amount $\Delta x \neq 0$ to get the number $x_1 + \Delta x$, which must also be in $(0, 8)$. The increment Δx can be either a positive or negative number. When x changes from x_1 to

* In reality, of course, the ticket price x will be some positive integral multiple of \$.01.

$x_1 + \Delta x$, y changes from $f(x_1)$ to $f(x_1 + \Delta x)$, as shown in Figure 2.1. The change in y that corresponds to a change of Δx in x is denoted by Δy and represents the change in the profit. Therefore,

$$\Delta y = f(x_1 + \Delta x) - f(x_1).$$

Of course, Δy can be positive, negative, or zero. In Figure 2.1 Δx is positive and Δy is negative.

If Δy is divided by Δx, the result represents the **average rate of change** in profit with respect to a change in the ticket price between x_1 and $x_1 + \Delta x$. We can write

$$\frac{\Delta y}{\Delta x} = \frac{f(x_1 + \Delta x) - f(x_1)}{\Delta x}.$$

Geometrically, this average rate of change can be represented by the slope of a line that passes through the two points $(x_1, f(x_1))$ and $(x_1 + \Delta x, f(x_1 + \Delta x))$, as shown in Figure 2.1; this line is called a **secant line.** Obviously, the location of this line, and hence its slope, will depend on the value x_1 and the change in x, namely, Δx.

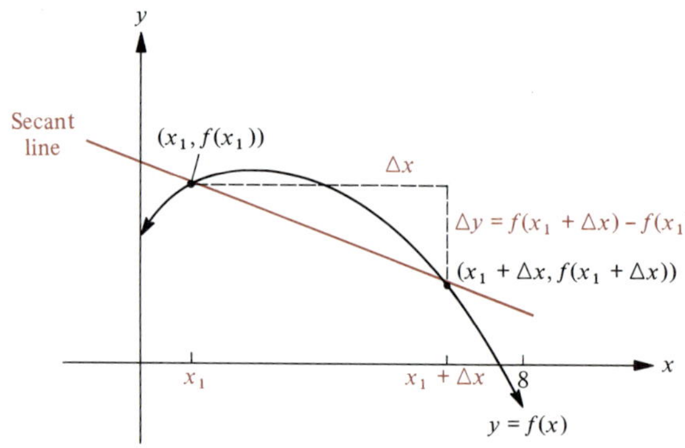

Figure 2.1

Example 1

Suppose that the profit function is given by $y = f(x) = -x^2 + 8x - 5$; the initial ticket price is \$1; and the price change is an increase of \$4. What is the corresponding change in profit, and what is the average rate of change of profit on the interval [1, 5]?

We are given $x_1 = 1$, $\Delta x = 4$, and asked to find the corresponding values of Δy and $\Delta y/\Delta x$.

$$\begin{aligned}\Delta y &= f(x_1 + \Delta x) - f(x_1)\\ &= f(1 + 4) - f(1)\\ &= f(5) - f(1)\\ &= [-(5)^2 + 8(5) - 5] - [-(1)^2 + 8(1) - 5]\\ &= 10 - 2\\ &= 8.\end{aligned}$$

Therefore,

$$\frac{\Delta y}{\Delta x} = \frac{8}{4} = 2.$$

Thus the change in profit is \$8, and the average rate of change in profit per unit change in ticket price on [1, 5] is \$2. The graph of the profit function, along with the secant line passing through (1, 2) and (5, 10) with slope of 2, is shown in Figure 2.2. The graph is a portion of a parabola. ∎

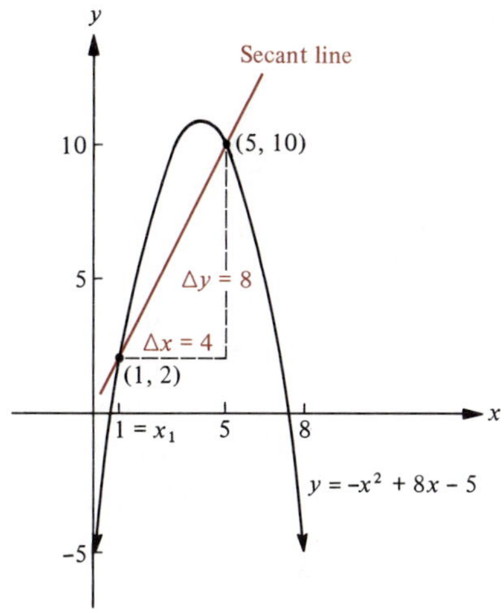

Figure 2.2

Example 2

As in Example 1 let $y = f(x) = -x^2 + 8x - 5$. This time, however, we shall let the initial ticket price be $x_1 = 2$ and let $\Delta x \neq 0$ be unspecified. Again we seek Δy and $\Delta y/\Delta x$.

$$\begin{aligned}\Delta y &= f(x_1 + \Delta x) - f(x_1)\\ &= f(2 + \Delta x) - f(2)\\ &= [-(2 + \Delta x)^2 + 8(2 + \Delta x) - 5] - [-(2)^2 + 8(2) - 5]\\ &= [-(4 + 4\,\Delta x + (\Delta x)^2) + 8(2 + \Delta x) - 5] - 7\\ &= -4 - 4\,\Delta x - (\Delta x)^2 + 16 + 8\,\Delta x - 5 - 7\\ &= 4\,\Delta x - (\Delta x)^2,\end{aligned}$$

and

$$\frac{\Delta y}{\Delta x} = 4 - \Delta x,$$

which completes the required calculation. ∎

For various values of Δx in Example 2 the corresponding values of $\Delta y/\Delta x$ are shown in Table 2.1. The interesting thing about the table is that $\Delta y/\Delta x$ is rather close to 4 when Δx is near 0 (notice particularly $\Delta x = -.01$ and

Table 2.1

Δx	-2	-1	$-.1$	$-.01$	$.01$	$.1$	1	2
$\frac{\Delta y}{\Delta x}$	6	5	4.1	4.01	3.99	3.9	3	2

$\Delta x = .01$). This suggests the idea of a limit and leads to the conjecture that, for $x_1 = 2$,

$$\frac{\Delta y}{\Delta x} \to 4 \qquad \text{as} \qquad \Delta x \to 0$$

or

$$\lim_{\Delta x \to 0} \frac{\Delta y}{\Delta x} = 4.$$

Using a previous limit theorem, we can verify this conjecture:

$$\lim_{\Delta x \to 0} \frac{\Delta y}{\Delta x} = \lim_{\Delta x \to 0} (4 - \Delta x) = \lim_{\Delta x \to 0} 4 - \lim_{\Delta x \to 0} \Delta x = 4 - 0 = 4.$$

Thus, for $x_1 = 2$, the average rate of change of profit with respect to the ticket price approaches 4 as Δx approaches 0. The limiting value $\left(\lim_{\Delta x \to 0} (\Delta y/\Delta x) = 4\right)$ represents the **instantaneous rate of change** of profit at $x_1 = 2$, as opposed to the average rate of change of profit between 2 and $2 + \Delta x$ which is represented by the ratio $\Delta y/\Delta x$.

The ideas just presented may also be interpreted geometrically. In Figure 2.3 two secant lines are drawn through (2, 7), corresponding to two of the

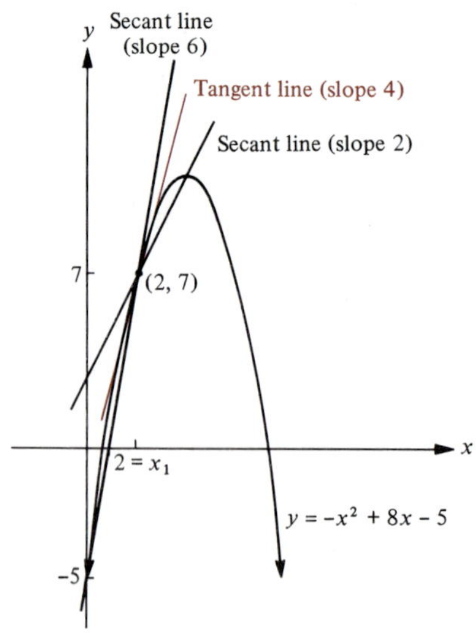

Figure 2.3

values of Δx in Table 2.1, namely, -2 and 2. The slopes of these lines as also given in Table 2.1 are 6 and 2, respectively. A third line, in color in Figure 2.3, is also drawn through (2, 7); it has a slope of 4, which is the limiting value of $\Delta y/\Delta x$ as $\Delta x \to 0$. The latter line is known as the **tangent line.**

We have just presented an intuitive introduction to the idea of the instantaneous rate of change of one variable with respect to another variable at a given point and to the geometric counterpart—the slope of the tangent line at the point. Let us now define precisely these important concepts.

▶ Definition 1

Suppose that a function f is defined by $y = f(x)$ throughout some open interval that contains the number x_1. The **instantaneous rate of change** of y with respect to x at $x = x_1$ is

$$\lim_{\Delta x \to 0} \frac{\Delta y}{\Delta x} = \lim_{\Delta x \to 0} \frac{f(x_1 + \Delta x) - f(x_1)}{\Delta x}$$

if this limit exists. The **tangent line** to the curve at $(x_1, f(x_1))$ is the line through $(x_1, f(x_1))$ with slope

$$\lim_{\Delta x \to 0} \frac{\Delta y}{\Delta x} = \lim_{\Delta x \to 0} \frac{f(x_1 + \Delta x) - f(x_1)}{\Delta x}$$

if this limit exists.

Figure 2.4 is presented to demonstrate that the slope of the tangent line is the limit of the slopes of the secant lines as Δx approaches zero, provided that the limit exists.

A notation is needed to show that the concepts just defined, the instantaneous rate of change and the slope of the tangent line, depend upon both

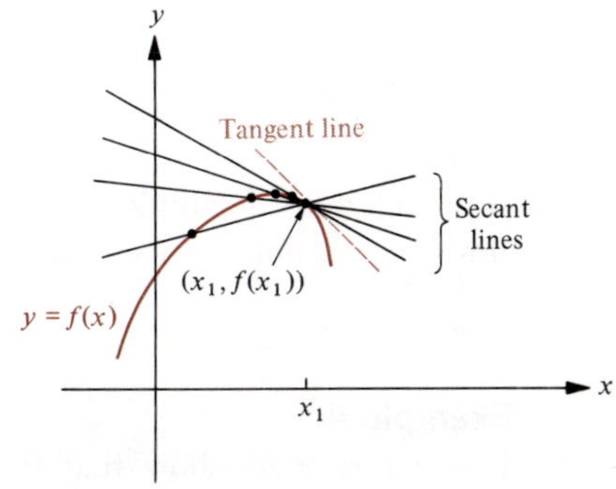

Figure 2.4. *Slopes of secant lines approach slope of tangent line as* $\Delta x \to 0$.

the function f and the value x_1; hence, if the limit exists, we choose to write

$$f'(x_1) = \lim_{\Delta x \to 0} \frac{\Delta y}{\Delta x} = \lim_{\Delta x \to 0} \frac{f(x_1 + \Delta x) - f(x_1)}{\Delta x}.$$

The way to read $f'(x_1)$ is "f prime of x_1."

Example 3 illustrates how the equation of the tangent line to a curve at a point can be determined.

Example 3

In the discussion following Example 2 we found that the tangent line through (2, 7) had a slope of $f'(2) = 4$, so its equation can be written using the point-slope form of the equation of a line,

$$y - 7 = 4(x - 2),$$

or in general form,

$$4x - y - 1 = 0. \blacksquare$$

In the majority of examples we shall consider, the instantaneous rate of change of y with respect to x at $x = x_1$ will exist, and hence it will be possible to draw a tangent line to $y = f(x)$ at $x = x_1$. There are exceptions, however. We shall learn in Section 2.4 that if f is *discontinuous* at $x = x_1$, then $f'(x_1)$ will *not* exist; although f is discontinuous at $x = x_1$, it is still possible to introduce increments $\Delta x \neq 0$ and Δy and to discuss the idea of secant line, but there will be no tangent line because the limit in Definition 1 will not exist. On the other hand, if f is *continuous* at $x = x_1$, then $f'(x_1)$ may or may not exist, as we illustrate next. First consider statement (2.1) which is of theoretical importance (an outline of its proof is given in Exercise 16).

If f is continuous at x_1 and Δx approaches zero, then Δy also approaches zero. (2.1)

A consequence of (2.1) is that the limit of Definition 1, $\lim_{\Delta x \to 0} (\Delta y/\Delta x)$, is a type of limit in which the numerator and denominator each approach zero as the limiting process takes place. Example 7 of Section 1.8 and the discussion prior to that example illustrate that it is quite possible for $\lim_{\Delta x \to 0} (\Delta y/\Delta x)$ to exist even though the numerator and denominator each approach zero; Example 4, however, illustrates that it is also possible for $\lim_{\Delta x \to 0} (\Delta y/\Delta x)$ *not* to exist when the numerator and denominator each approach zero.

Example 4

If $y = f(x) = |x|$, show that the instantaneous rate of change of y with respect to x at $x = 0$ does *not* exist. (See Figure 2.5.)

$$\frac{\Delta y}{\Delta x} = \frac{f(0 + \Delta x) - f(0)}{\Delta x} = \frac{|\Delta x| - 0}{\Delta x} = \frac{|\Delta x|}{\Delta x}.$$

If $\Delta x > 0$, $\Delta y/\Delta x$ is always equal to 1 but, if $\Delta x < 0$, $\Delta y/\Delta x$ is always equal to -1. In order for $f'(0) = \lim_{\Delta x \to 0} (\Delta y/\Delta x)$ to exist, $\Delta y/\Delta x$ would have to approach the same value whether Δx approaches zero through positive or negative values. However, $\Delta y/\Delta x \to 1$ if $\Delta x > 0$ and $\Delta x \to 0$, whereas $\Delta y/\Delta x \to -1$ if $\Delta x < 0$ and $\Delta x \to 0$. Therefore, the instantaneous rate of change of y with respect to x at $x = 0$ does not exist and, as one might guess from Figure 2.5, there is no tangent line to the curve at the origin. Note that f is continuous at $x = 0$ because $\lim_{x \to 0} f(x) = f(0) = 0$. ▮

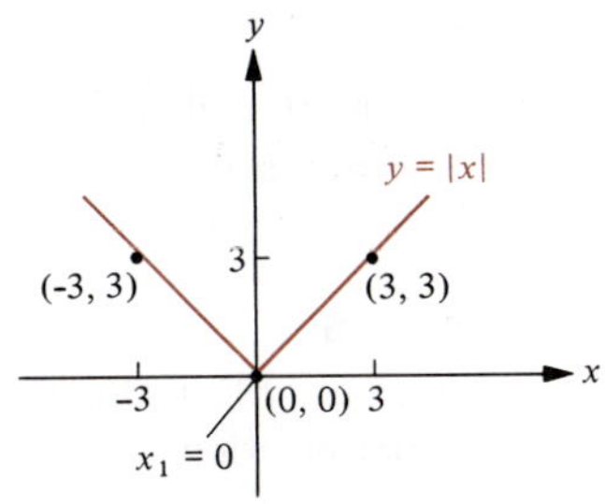

Figure 2.5

APPLICATIONS

Example 5

One reason that a theater manager might be interested in a mathematical analysis such as that begun in Examples 1 and 2 would be to determine the ticket price x_1 that would maximize (make largest) the profit y. In Chapter 3 we shall find that if a maximum exists for a polynomial function over an open interval, then the slope $f'(x_1)$ of the tangent line must be zero. For the profit function given by

$$y = f(x) = -x^2 + 8x - 5,$$

we can calculate

$$\begin{aligned}\Delta y &= f(x_1 + \Delta x) - f(x_1)\\ &= [-(x_1 + \Delta x)^2 + 8(x_1 + \Delta x) - 5] - [-x_1^2 + 8x_1 - 5]\\ &= [-x_1^2 - 2x_1\,\Delta x - (\Delta x)^2 + 8x_1 + 8\,\Delta x - 5] - [-x_1^2 + 8x_1 - 5]\\ &= -2x_1\,\Delta x - (\Delta x)^2 + 8\,\Delta x\end{aligned}$$

and

$$\frac{f(x_1 + \Delta x) - f(x_1)}{\Delta x} = -2x_1 - \Delta x + 8.$$

The slope $f'(x_1)$ of the tangent line at $x = x_1$ is computed as follows:

$$\begin{aligned} f'(x_1) &= \lim_{\Delta x \to 0} \frac{f(x_1 + \Delta x) - f(x_1)}{\Delta x} \\ &= \lim_{\Delta x \to 0} (-2x_1 - \Delta x + 8) \\ &= \lim_{\Delta x \to 0} (-2x_1) - \lim_{\Delta x \to 0} \Delta x + \lim_{\Delta x \to 0} 8 \\ &= -2x_1 + 8. \end{aligned}$$

We wish to find that value x_1 for which the slope of the tangent line is zero. Thus

$$\begin{aligned} 0 &= -2x_1 + 8, \\ 2x_1 &= 8, \\ x_1 &= 4. \end{aligned}$$

This means that if there is a maximum profit, then that maximum profit will occur at a ticket price of \$4. (In Exercise 6 of Section 3.4 it can be shown that maximum profit *will* occur at a ticket price of \$4.) ▮

Example 6

This example illustrates a use of an average rate of change for sequences. In a paper entitled "Innovation in the States: A Diffusion Study," Virginia Gray [11] examines the diffusion (spread) of ideas from state to state through the legal adoption of those ideas by the respective states. The policy areas selected for study are education, welfare, and civil rights.

Let A_t represent the cumulative proportion of adopters of a particular type of legislation (such as Fair Housing) t years after innovation of the law by some state. Since t is assumed to be a positive integer, $\{A_t\}$ is a sequence of numbers. The terms of $\{A_t\}$ are never greater than 1 because not over 100 per cent of the states can adopt legislation of a given type. Let L represent the maximum possible proportion of adopters of a certain type of legislation. If $L = 1$, all states have the potential for adoption; however, for some types of legislation there are reasons why $L < 1$.

Starting at a particular time t_1 and changing t by an amount Δt, we can denote the corresponding change in A_t by ΔA_t (Δt and ΔA_t play the roles of Δx and Δy, respectively, that were discussed earlier in this section). $\Delta A_t/\Delta t$ is the average rate of change of the cumulative proportion of adopters with respect to time from t_1 to $t_1 + \Delta t$ ($\Delta A_t/\Delta t$ is comparable to $\Delta y/\Delta x$).

One object of Gray's paper is to predict the rate of diffusion among the states—in other words, if A_1 is known for a particular type of legislation, we want to predict A_2, A_3, A_4, and so on. In order to predict A_2, A_3, A_4, and so on, which correspond to time changes of 1 year each, let us choose $\Delta t = 1$. The basic assumption made is that $\Delta A_t/\Delta t$ is directly proportional to A_t and $L - A_t$. Thus the mathematical model is

$$\frac{\Delta A_t}{\Delta t} = bA_t(L - A_t),$$

where b is the constant of proportionality. Because $\Delta t = 1$, the equation becomes

$$\Delta A_t = bA_t(L - A_t).$$

Such an equation involving a sequence and increments is known as a difference equation. Since $\Delta t = 1$, $\Delta A_t = A_{t+1} - A_t$, and the difference equation becomes

$$A_{t+1} = bA_t(L - A_t) + A_t. \tag{2.2}$$

Since A_{t+1} is expressed in terms of A_t, (2.2) makes it possible to find any term of $\{A_t\}$ if the previous term is known. For a particular type of legislation, b can be estimated using experience from other related types of legislation, L can be estimated, and after 1 year A_1 is known. Then (2.2) is used to find A_2. Once A_2 is known, (2.2) is used again to determine A_3, and this process can be continued. In fact, a computer program can be written to generate as many terms of the sequence $\{A_t\}$ as desired. ∎

EXERCISES

Suggested minimum assignment: Exercises 1(*a*), 3, 7, 9, 11, and 13.

1. Suppose that $y = x^2 - 4x + 7$ and $x_1 = 3$.
   (*a*) If $\Delta x = .5$, find Δy. (*b*) If $\Delta x = -.1$, find Δy.
2. Suppose that $y = \dfrac{1}{x + 5}$ and $x_1 = -2$.
   (*a*) If $\Delta x = \frac{1}{3}$, find Δy. (*b*) If $\Delta x = -1$, find Δy.
3. Let f be defined by $f(x) = 2 - x^2$. Suppose that $x_1 = 1$ and $\Delta x = .5$. Plot the secant line joining $(x_1, f(x_1))$ and $(x_1 + \Delta x, f(x_1 + \Delta x))$, and find the slope of this secant line.
4. Explain why the secant line between two points on the graph of a function is never a vertical line.
5. Suppose that $y = x^3$ and $x_1 = 1$.
   (*a*) If $\Delta x = .2$, find Δy. Draw the secant line and find its slope.
   (*b*) If $\Delta x = -1$, find Δy. Draw the secant line and find its slope.
   (*c*) Find the average rate of change of y with respect to x between $x = 1$ and $x = 1.1$.
6. Suppose that $y = x^2$ and $x_1 = 1$. Let Δx equal -1, $-.1$, $-.01$, $.01$, $.1$, and 1. Construct a table like Table 2.1.
7. Suppose that $y = f(x) = x^3$ and $x_1 = 2$.
   (*a*) If $\Delta x = .01$, find Δy. (*b*) If $\Delta x = -.01$, find Δy.
   (*c*) Make a conjecture as to the value of $\lim_{\Delta x \to 0} \dfrac{\Delta y}{\Delta x} = f'(2)$.
8. Suppose that the number y of bacteria in a culture after t hours is given by $y = f(t) = t^2 + 5t + 100$, provided that $0 \le t \le 24$. Let the initial time be $t_1 = 3$, and let $\Delta t \neq 0$ be unspecified. Find Δy. Find

$\frac{\Delta y}{\Delta t}$. Find the instantaneous rate of change of y with respect to t at $t = 3$.

9. If $y = x^2 - 5x$, find the average rate of change of y with respect to x between $x = -1$ and $x = 1$. Also find the instantaneous rate of change of y with respect to x at $x = -1$.
10. Let a profit function f be given by $y = f(x) = -x^2 + 10x - 3$, provided that $0 < x < 9$. Find the instantaneous rate of change of y with respect to x at $x = 3$.
11. If $y = f(x) = x^3$, show that $f'(2) = 12$. (See Exercise 7).
12. Show that the slope of the tangent line to $y = x^2$ at $(1, 1)$ is 2. (See Exercise 6.)
13. Find the equation of the tangent line to $y = x + x^2$ at $x = -2$.
14. Find the equation of the tangent line to $y = 6 - x^2$ at $x = 3$.
15. Sketch the graph of $y = f(x) = |x - 1|$. Explain why there is no tangent line to this curve at $(1, 0)$.
16. Prove that if f is continuous at x_1 and Δx approaches zero, then Δy approaches zero. [*Hint:* $\lim_{x \to x_1} (f(x) - f(x_1)) = 0$ follows from $\lim_{x \to x_1} f(x) = f(x_1)$. Let $x = x_1 + \Delta x$, and let Δx approach zero. Then demonstrate that $\lim_{\Delta x \to 0} (f(x_1 + \Delta x) - f(x_1)) = 0$.]
17. Use the method of Example 5 on the profit function given in Exercise 10 to find the value x_1 of x at which maximum profit may occur.

2.2 The Derivative and Its Interpretations

Most of the material presented so far has been given as a foundation so that the reader will be able to understand the concept that will be defined in this section. This concept is important in such diverse ways as the study of marginal analysis in economics, the study of rates of tissue change in biology, and the study of population problems in sociology.

In the last section we learned that the average rate of change of $y = f(x)$ with respect to x between $x = x_1$ and $x = x_1 + \Delta x$ is given by the ratio $\Delta y/\Delta x$ or $[f(x_1 + \Delta x) - f(x_1)]/\Delta x$. If Δx is allowed to approach zero in this ratio, we get the average rate of change over a smaller and smaller interval. In the limit we get what is called the instantaneous rate of change of y with respect to x at $x = x_1$. In other words, if $y = f(x)$, then the instantaneous rate of change of y with respect to x at $x = x_1$ is given by

$$f'(x_1) = \lim_{\Delta x \to 0} \frac{\Delta y}{\Delta x} = \lim_{\Delta x \to 0} \frac{f(x_1 + \Delta x) - f(x_1)}{\Delta x}, \qquad (2.3)$$

provided that this limit exists. The phrase **rate of change** will often be used instead of instantaneous rate of change and will be understood to have the same meaning.

We also learned in Section 2.1 that a geometric interpretation of (2.3), provided that the limit exists, is the slope of the tangent line to $y = f(x)$ at the point $(x_1, f(x_1))$.

Example 1

Find the rate of change of $y = f(x) = x^2 - x$ with respect to x at $x = 3$; also find the slope of the tangent line to $y = f(x) = x^2 - x$ at (3, 6).

By (2.3) with $x_1 = 3$ we can find both; we have

$$\begin{aligned} f'(3) &= \lim_{\Delta x \to 0} \frac{\Delta y}{\Delta x} = \lim_{\Delta x \to 0} \frac{f(3 + \Delta x) - f(3)}{\Delta x} \\ &= \lim_{\Delta x \to 0} \frac{[(3 + \Delta x)^2 - (3 + \Delta x)] - [(3)^2 - 3]}{\Delta x} \\ &= \lim_{\Delta x \to 0} \frac{9 + 6\,\Delta x + (\Delta x)^2 - 3 - \Delta x - 6}{\Delta x} \\ &= \lim_{\Delta x \to 0} \frac{5\,\Delta x + (\Delta x)^2}{\Delta x} \\ &= \lim_{\Delta x \to 0} (5 + \Delta x) \\ &= 5. \end{aligned}$$

See Figure 2.6 for a graph. ∎

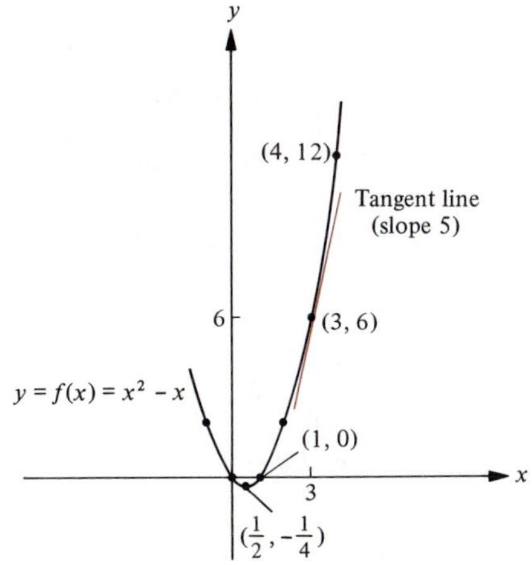

Figure 2.6

The limit given in (2.3) and illustrated in Example 1 leads to the definition of the **derivative** of a function, one of the most important concepts in calculus.

► **Definition 2**

The **derivative** (or **derived function**) of the function f is the function f' defined by

$$f'(x) = \lim_{\Delta x \to 0} \frac{f(x + \Delta x) - f(x)}{\Delta x},$$

where x is any real number for which this limit exists.

The limit given in Definition 2 and the limit given in (2.3) differ in that Definition 2 defines a function f', whereas (2.3) yields a value of f' at the specific value $x = x_1$. It is customary to yield to the convention of also using the term derivative to represent the value $f'(x)$.

Notice in Definition 2 that $f(x)$ appears in the definition of $f'(x)$, and therefore $f'(x)$ cannot exist unless $f(x)$ exists—in other words, the domain of f' is a subset of the domain of f. The function f is said to be **differentiable** at $x = x_1$ if $f'(x_1)$ exists.

Example 2

Let the function f be defined by $y = f(x) = x^2 - x$. Find $f'(x)$. Find $f'(3)$.

$$\begin{aligned}
f'(x) &= \lim_{\Delta x \to 0} \frac{f(x + \Delta x) - f(x)}{\Delta x} \\
&= \lim_{\Delta x \to 0} \frac{((x + \Delta x)^2 - (x + \Delta x)) - (x^2 - x)}{\Delta x} \\
&= \lim_{\Delta x \to 0} \frac{x^2 + 2x\,\Delta x + (\Delta x)^2 - x - \Delta x - x^2 + x}{\Delta x} \\
&= \lim_{\Delta x \to 0} \frac{2x\,\Delta x + (\Delta x)^2 - \Delta x}{\Delta x} \\
&= \lim_{\Delta x \to 0} (2x + \Delta x - 1) \\
&= 2x - 1.
\end{aligned}$$

The calculation above is valid for any real number x—that is, the domain of f' is the set of all real numbers. Hence $f'(3)$ can be found by substituting $x = 3$ as follows:

$$f'(3) = 2(3) - 1 = 5. \blacksquare$$

One could calculate $f'(3)$ in Example 2 without first finding $f'(x)$; this was done in Example 1 of this section. However, by making the general calculation of $f'(x)$ first, it is possible to find $f'(x_1)$ quickly for *any* number x_1 in the domain of f'. For example, $f'(1) = 2(1) - 1 = 1$. The graph of f is given in Figure 2.7, and the graph of f' is given in Figure 2.8; notice the relationship between the two graphs as illustrated by $f'(3) = 5$ and $f'(1) = 1$.

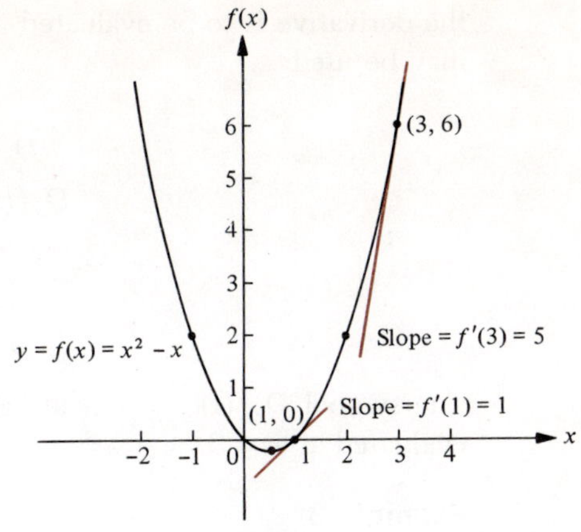

Figure 2.7

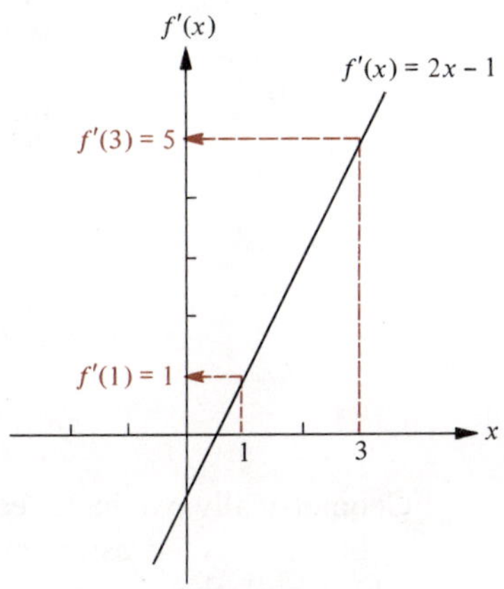

Figure 2.8

Let $y = f(x) = x^2 - x$ be used again to demonstrate some other notation which is often used in working with the derivative. Each of the following has the same meaning:

$$f'(x) = 2x - 1,$$
$$D_x y = 2x - 1,$$
$$D_x f(x) = 2x - 1,$$
$$y' = 2x - 1,$$
$$\frac{dy}{dx} = 2x - 1.$$

The symbols $D_x y$ and dy/dx are read "the derivative of y with respect to x"; the dy/dx notation will be explained more fully in Section 2.6. If the derivative is to be evaluated at $x = 3$, any of the following notations may be used:

$$f'(3) = 5,$$
$$D_x y|_{x=3} = 5,$$
$$D_x f(x)|_{x=3} = 5,$$
$$y'|_{x=3} = 5,$$
$$\left.\frac{dy}{dx}\right|_{x=3} = 5.$$

The symbol $D_x f(x)|_{x=3}$ is read "the derivative of $f(x)$ with respect to x evaluated at $x = 3$."

Example 3

Find y' if $y = 1/2x$.

$$\begin{aligned} y' &= \lim_{\Delta x \to 0} \frac{1/[2(x + \Delta x)] - (1/2x)}{\Delta x} \\ &= \lim_{\Delta x \to 0} \frac{x - (x + \Delta x)}{2x(x + \Delta x)} \frac{1}{\Delta x} \\ &= \lim_{\Delta x \to 0} \frac{-\Delta x}{2x(x + \Delta x)(\Delta x)} \\ &= \lim_{\Delta x \to 0} \frac{-1}{2x(x + \Delta x)} \\ &= -\frac{1}{2x^2}. \ \blacksquare \end{aligned}$$

Geometrically, we have seen that the derivative of a function at a point may be interpreted as a slope. In general, the derivative gives the rate of change of the dependent variable with respect to the independent variable. In the next two examples we indicate some specialized interpretations.

Example 4

In this example we point out an economic interpretation of the derivative of a function. Suppose that $R(x)$ and $C(x)$ are given, where R is a revenue function and C is a cost function. **Marginal revenue** is defined as the derivative of R with respect to x. **Marginal cost** is defined as the derivative of C with respect to x. **Marginal profit** [recall that $P(x) = R(x) - C(x)$] is defined as the derivative of P with respect to x.

Suppose, for example, that

$$R(x) = x^2 - x.$$

Then, from Example 2 of this section, the marginal revenue $R'(x)$ is given by

$$R'(x) = 2x - 1.$$

The marginal revenue function R' can be evaluated at a number, say, $x = 3$, in its domain to obtain

$$R'(3) = 5.$$

In Example 2 of Section 2.6 we shall attempt to explain the economic meaning of marginal revenue, marginal cost, and marginal profit. In this example we are just trying to indicate one important use of the derivative. ∎

Example 5

Another interpretation of the derivative occurs in the study of the motion of a particle along a line. The particle might be a spacecraft going to the moon, an object on a production line, or an automobile in a traffic pattern. The line along which the particle moves is called the s-axis and can be visualized as a horizontal line with positive direction to the right (see Figure 2.9). The location of the particle at various times is given by an **equation of motion**

$$s = f(t).$$

$s = t^2 - t$	t
0	1
2	2
6	3
12	4

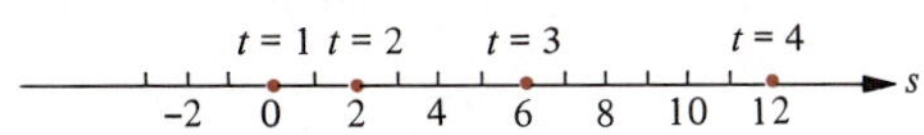

Figure 2.9

Assume in this example that the independent variable t is measured in seconds and that the dependent variable s is measured in feet. Starting at a particular time $t = t_1$, we suppose that t is changed by the amount $\Delta t \neq 0$ and that the corresponding change in s is denoted by Δs. It is natural to call

$$\frac{\Delta s}{\Delta t} = \frac{f(t_1 + \Delta t) - f(t_1)}{\Delta t}$$

the **average velocity** of the particle between $t = t_1$ and $t = t_1 + \Delta t$ and to call

$$f'(t_1) = \lim_{\Delta t \to 0} \frac{f(t_1 + \Delta t) - f(t_1)}{\Delta t},$$

if it exists, the **instantaneous velocity** of the particle at $t = t_1$. If $s = t^2 - t$, then $D_t s = 2t - 1$ (see Example 2). The instantaneous velocity at $t = 3$ is given by

$$D_t s|_{t=3} = 2(3) - 1 = 5 \text{ feet per second.}$$

Instantaneous velocity will often be referred to as just **velocity.** A positive velocity indicates that the particle is moving to the right along the s-axis, whereas a negative velocity signifies movement to the left. ▮

APPLICATIONS

Example 6

In this section we have learned that the derivative of a function may be interpreted in a general sense as a rate of change of one variable with respect to another variable. We have also discussed well-known geometric, economic, and physical interpretations of the derivative. Other specialized interpretations arise in various disciplines; in this example we give an illustration adapted from pp. 18–19 of Thrall et al. [23].

Suppose that a substance is kept at constant pressure but that its temperature is allowed to vary. The specific heat of the substance is a function of the temperature (the specific heat of a substance is defined as the amount of heat needed to raise the temperature of a unit mass of that substance 1 degree). The derivative of the specific heat of a substance with respect to the temperature gives the rate of change of specific heat with respect to the temperature, and might be useful in problems concerning heats of reaction.

Let T be the temperature of ethyl alcohol in degrees Celsius, and let y be the specific heat of ethyl alcohol. For $0 \leq T \leq 60$, it has been found experimentally that y can be approximated by

$$y = (5.068)(10)^{-1} + (2.68)(10)^{-3}T + (5.4)(10)^{-6}T^2.$$

Definition 2 can be used to show (after some calculation) that

$$D_T y = (2.68)(10)^{-3} + (1.08)(10)^{-5}T.$$

This result can be evaluated for various values of T if desired. For example,

$$\begin{aligned} D_T y|_{T=20} &= (2.68)(10)^{-3} + (1.08)(10)^{-5}(20) \\ &= (2.896)(10)^{-3}. \end{aligned}$$

This means that if ethyl alcohol were being heated, at the *instant* the temperature reached 20 degrees Celsius, the specific heat would be increasing by $(2.896)(10)^{-3}$ unit per degree of increase in the temperature. ▮

★Example 7 Probability Density Function

Consider a certain type of machine that has a life span of no more than 20 years. Thus a sample space of the life span of this type of machine is $U = \{t : 0 \leq t \leq 20\}$, where t is a real number. A measurement is defined on U in such a way that to each value of t there is associated a real number s representing the proportion of maximum age at the time of breakdown. The range of the measurement is $S = \{s : 0 \leq s \leq 1\}$ and provides a new sample space. Here we have a *continuous* measurement because the range S is an interval of real numbers. Suppose that a fairly accurate cumulative probability distribution function F has been found to be given by

$$F(x) = \begin{cases} 0 & \text{if } x < 0, \\ x^2 & \text{if } 0 \leq x \leq 1, \\ 1 & \text{if } x > 1. \end{cases}$$

The graph of F is given in Figure 2.10. Notice from Figure 2.10 that F has the necessary properties to be a cumulative probability distribution function. That is, $\lim_{x \to -\infty} F(x) = 0$, $\lim_{x \to \infty} F(x) = 1$, the graph of F never decreases as x increases, and the fourth requirement that $\lim_{\Delta x \to 0} F(x + \Delta x) = F(x)$ for any real number x, where $\Delta x > 0$, is satisfied. Suppose that we seek the probability of event $E = \{s : .3 < s \leq .5\}$. Let $E_1 = \{s : s \leq .5\}$ and $E_2 = \{s : s \leq .3\}$ so that $P(E) = P(E_1) - P(E_2)$. Recalling Example 8 of Section 1.8, we can say $P(E_1) = F(.5)$ and $P(E_2) = F(.3)$. Therefore,

$$\begin{aligned} P(E) = P(E_1) - P(E_2) &= F(.5) - F(.3) \\ &= (.5)^2 - (.3)^2 \\ &= .25 - .09 \\ &= .16. \end{aligned}$$

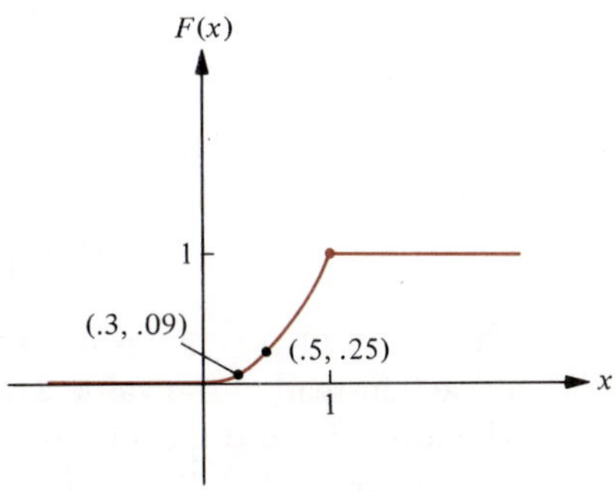

Figure 2.10. *Graph of cumulative probability distribution function F.*

Thus we have found that the probability is .16 that a machine will break down between 30 and 50 per cent of its maximum life span.

Now suppose for our machine illustration that we think of probability as one unit of some sort of quantity that is distributed in some manner over the sample space $S = [0, 1]$. The probability $P(E)$, where $E = \{s : c < s \leq d\}$, can then be thought of as the fraction of that one unit of quantity that is distributed over the subinterval $(c, d]$ of $[0, 1]$. If we divide this subquantity $P(E) = F(d) - F(c)$ by $d - c$ (the length of the interval), we have a measure of the concentration or density of the quantity that is distributed to the interval $(c, d]$. In other words,

$$\begin{pmatrix} \text{the probability} \\ \text{density in } (c, d] \end{pmatrix} = \frac{F(d) - F(c)}{d - c}.$$

Suppose that we let the interval $(c, d]$ get smaller and smaller in length by considering intervals of the form $(c, x]$, where $c < x < d$ and $x \to c$.

We say that

$$\lim_{x \to c} \frac{F(x) - F(c)}{x - c},$$

if it exists, represents the concentration or density of probability at $x = c$. Let $c + \Delta x = x$; then $\Delta x \to 0$ as $x \to c$ and

$$\lim_{x \to c} \frac{F(x) - F(c)}{x - c} = \lim_{\Delta x \to 0} \frac{F(c + \Delta x) - F(c)}{\Delta x}.$$

From the material of this section we recognize the latter limit as the derivative of F evaluated at $x = c$.

There is no assurance that F' exists for all possible values $x = c$, but if it does exist, we call $F'(c) = p(c)$ the value of the **probability density function** p at $x = c$. For any values of x for which F' does not exist, the function p is assigned values so that the graph of p does not have any isolated points. The domain of p is the set of all real numbers.

In our illustration it can be shown that

$$\lim_{x \to c} \frac{F(x) - F(c)}{x - c} \quad \text{is} \quad \begin{cases} 2c & \text{if } 0 \le c < 1, \\ \text{nonexistent} & \text{if } c = 1, \\ 0 & \text{otherwise.} \end{cases}$$

We choose to define function p according to

$$p(x) = \begin{cases} 2x & \text{if } 0 \le x \le 1, \\ 0 & \text{otherwise.} \end{cases}$$

The graph of p is given in Figure 2.11. We could have assigned $p(1)$ the value 0 instead of the value 2, but our choice makes p continuous on $[0, 1]$. The area of the trapezoid in Figure 2.11 is .16, representing the fact that .16 unit of probability is distributed to $(.3, .5]$.

The probability density function p plays a role for continuous measurements that is analogous to the role of the weighting function w for discrete measurements; this is true in the sense that p and w each display how the total probability of 1 is distributed over the sample space S. ▮

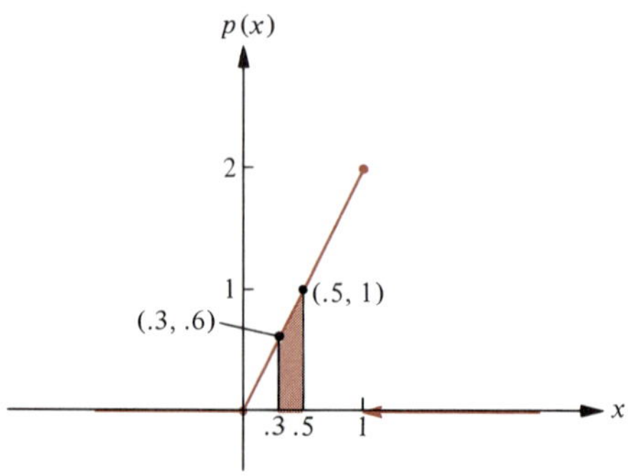

Figure 2.11. *Graph of probability density function p.*

EXERCISES

Suggested minimum assignment: Exercises 1, 3, 5, 7, 13, and 15(*b*).

1. Suppose that $y = f(x) = 2x^2$.
   (*a*) Find the rate of change of y with respect to x at $x = -1$.
   (*b*) What is the slope of the tangent line at $(-1, 2)$?
2. Suppose that $y = f(x) = x^3 - x$.
   (*a*) Find the average rate of change of y with respect to x between $x = 2$ and $x = 2.5$.
   (*b*) Find the instantaneous rate of change of y with respect to x at $x = 2$.
3. If $f(x) = 3x^2 - 5x$, find $f'(x)$ using Definition 2. What is the domain of the function f'?
4. If $f(x) = \dfrac{1}{x-1}$, find $f'(x)$. What is the domain of the function f'?
5. Let f be defined by $f(x) = \dfrac{x}{x-1}$. Is f differentiable at $x = 4$? If so, find $f'(4)$. Is f differentiable at $x = 1$? If so, find $f'(1)$.
6. If $y = \dfrac{1}{x} - 3x$, find y'.
7. If $y = 1 - x - x^2$, find $D_x y$. Also find $D_x y|_{x=0}$.
8. Suppose that $y = f(x) = 3 - x^2$. Sketch the graph of f and also sketch the graph of f'. Explain the relationship between the graphs, focusing your attention at the point where $x = 2$.
9. If $f(x) = \dfrac{x-3}{2x+1}$, find $D_x f(x)$.
10. Explain why $f'(x_1)$, if it exists, could be calculated using the formula
$$f'(x_1) = \lim_{x \to x_1} \frac{f(x) - f(x_1)}{x - x_1}.$$
11. If $f(x) = 3x^2 - 8$, calculate $f'(-2)$ using the formula in Exercise 10.
12. A revenue function is given by $R(x) = 5x - \frac{1}{6}x^2$. Find the marginal revenue $R'(x)$.
13. A cost function is given by $C(x) = 2x + 1$. Find the marginal cost $C'(x)$.
14. A profit function is given by $P(x) = 10x - x^2$. Find the marginal profit $P'(x)$. Also find $P'(3)$.
15. A particle moves along a line according to the equation of motion $s = t^3$, where s is measured in feet and t in seconds.
   (*a*) Find the velocity of the particle at $t = 1$.
   (*b*) Find the velocity of the particle at $t = 2$.
   (*c*) Find the average velocity of the particle between $t = 1$ and $t = 2$.
16. A particle moves along a line according to the equation of motion $s = 8 - t^2$, where s is measured in meters and t in seconds. Find ds/dt and $ds/dt|_{t=2}$. Is the particle moving to the right or to the left along the s-axis when $t = 2$?

17. A ball is thrown directly upward at time $t = 0$ seconds. Its height s in feet above the point where it is released is given by

$$s = 64t - 16t^2,$$

provided that $0 \leq t \leq 4$. How long does it take for the ball to reach its highest point? (*Hint:* The velocity is 0 at its highest point.)

18. A cumulative probability distribution function F is given by

$$F(x) = \begin{cases} 0 & \text{if } x < 0, \\ \frac{1}{4}x^2 & \text{if } 0 \leq x \leq 2, \\ 1 & \text{if } x > 2. \end{cases}$$

Find the corresponding probability density function p.

19. In Example 7 what is the probability that a machine will break down between 60 and 70 per cent of its maximum life span?

2.3 Basic Differentiation Formulas

When $f(x)$ is given, the process of finding $f'(x)$, if it exists, is called **differentiation.** In Section 2.2 the differentiation of some carefully chosen functions was carried out by directly calculating the limit—a rather tedious method. Rather than apply the definition of the derivative of a function each time a function is to be differentiated, we can use formulas that make it possible to write the answer quickly. In this section we shall state some of these formulas, give some examples to illustrate their use, and then discuss briefly the method of obtaining the formulas. The process will be repeated in the next section for some additional formulas. In Theorem 1 five basic differentiation formulas are stated—the formulas are valid if it is assumed that all indicated derivatives exist.

▶ Theorem 1

(*i*) If $f(x) = k$, where k is a real number, then $D_x f(x) = 0$.
(*ii*) If n is a real number, $D_x x^n = nx^{n-1}$.
(*iii*) If k is a real number, then $D_x(kf(x)) = kD_x f(x)$.
(*iv*) $D_x(f(x) + g(x)) = D_x f(x) + D_x g(x)$.
(*v*) Let $u = f(x)$. If $y = u^n$, where n is a real number, then $D_x y = nu^{n-1} D_x u$. (**Power rule for differentiation**)

A discussion of the proof of this theorem is given at the end of this section. ◀

Example 1

Consider the constant function defined by $f(x) = 5$. By Theorem 1(*i*), $D_x f(x) = 0$. Alternatively, we can write $f'(x) = 0$, or, if $y = 5$ is given, we can write $y' = 0$ or $dy/dx = 0$. Part (*i*) says that the derivative of a constant function is zero—this agrees with the fact that the graph is a horizontal line that has slope zero at every point. ▮

Example 2
Use Theorem 1(*ii*) to differentiate (*a*) $y = x^8$, (*b*) $f(x) = x^{-\pi}$, (*c*) $f(x) = \sqrt{x^3}$, (*d*) $y = x^0$.

Solutions:

(*a*) The exponent 8 is multiplied by x to an exponent that is 1 less than the original exponent of x. Hence $y' = 8x^7$.

(*b*) $D_x f(x) = -\pi x^{-\pi-1}$.

(*c*) Rewrite the stated problem as $f(x) = x^{3/2}$. Then $f'(x) = \frac{3}{2}x^{1/2} = \frac{3}{2}\sqrt{x}$.

(*d*) $y' = 0x^{-1} = 0$. Since $x^0 = 1$, part (*i*) instead of (*ii*) could also be used to get the same answer. ▮

Example 3
Let a revenue function be given by $R(x) = 8x^3$. Find the marginal revenue $R'(x)$.

By Theorem 1(*iii*), $R'(x) = 8D_x x^3$. Then, by part (*ii*), $R'(x) = 8(3x^2) = 24x^2$. Part (*iii*) says that the derivative of a real number times a function is the real number times the derivative of the function. ▮

Part (*iv*) of Theorem 1 will be illustrated in Example 4. Part (*iv*) says that the derivative of the sum of two functions is the sum of their derivatives, but part (*iv*) can easily be extended to the case where there are any finite number of functions rather than just two functions.

Example 4
Find (*a*) $D_x(x^{5/2} + x)$, (*b*) $D_x(x^3 - 8x^2)$, and (*c*) $D_x(3x^4 + x^6 - x^2)$.

Solutions:

(*a*) $D_x(x^{5/2} + x^1) = \frac{5}{2}x^{3/2} + 1x^0 = \frac{5}{2}x^{3/2} + 1$.

(*b*) $D_x(x^3 - 8x^2) = 3x^2 - 16x$.

(*c*) $D_x(3x^4 + x^6 - x^2) = 12x^3 + 6x^5 - 2x$.

Notice in (*b*) we used the fact that

$$x^3 - 8x^2 = x^3 + (-8x^2),$$

and hence that Theorem 1(*iv*) applies to differences of functions as well as sums of functions. ▮

Example 5
If $y = (3x + 1)^8$, find $D_x y$.

If we let $u = 3x + 1$, then $y = u^8$ and Theorem 1(*v*) applies with $n = 8$.

$$D_x y = 8u^7 D_x u = 8u^7(3) = 24u^7 = 24(3x + 1)^7.$$

With a little practice the problem can be done without even writing $u = 3x + 1$, but the reader must note that $8(3x + 1)^7$ is *not* the answer—remember to multiply by $D_x u$. ▮

If $u = x$ in Theorem 1(*v*), then $D_x u = 1$ and part (*v*) reduces to part (*ii*). Hence part (*ii*) could be omitted from Theorem 1, and the power rule for differentiation, part (v), could be used instead in Examples 2, 3, and 4.

Example 6

Find the slope of the tangent line to $y = (x^3 - x^2 + x)^3$ at $x = 1$.

By Theorem 1(v),

$$y' = 3(x^3 - x^2 + x)^2 D_x(x^3 - x^2 + x).$$

Then by parts (ii), (iii), and (iv),

$$y' = 3(x^3 - x^2 + x)^2(3x^2 - 2x + 1).$$

The slope of the tangent line is

$$y'|_{x=1} = 3(1 - 1 + 1)^2(3 - 2 + 1) = 6. \blacksquare$$

Example 7

Find the rate of change of y with respect to x at $x = 2$ if $y = (x^2 + 2)^2/x^2$.

First write

$$y = \frac{x^4 + 4x^2 + 4}{x^2} = x^2 + 4 + 4x^{-2}.$$

Then

$$D_x y = 2x + 0 - 8x^{-3} = 2x - \frac{8}{x^3},$$

and the rate of change of y with respect to x at $x = 2$ is

$$D_x y|_{x=2} = 2(2) - \frac{8}{2^3} = 3. \blacksquare$$

The remainder of this section is devoted to discussion of the proof of Theorem 1. Although some parts are harder to prove than others, the main idea is to use the definition of derivative (Definition 2 of Section 2.2). Part (i) is proved as follows:

$$D_x f(x) = \lim_{\Delta x \to 0} \frac{f(x + \Delta x) - f(x)}{\Delta x} = \lim_{\Delta x \to 0} \frac{k - k}{\Delta x}$$

$$= \lim_{\Delta x \to 0} \frac{0}{\Delta x} = \lim_{\Delta x \to 0} (0) = 0.$$

Proofs of parts (iii) and (iv) are similar and also require the use of Definition 2; they are left as exercises. If n is a positive integer, part (ii) can be proved in a similar manner by using the binomial theorem to expand $(x + \Delta x)^n$—see Exercise 22; the case in which n is a negative integer can be handled after Theorem 2 of the next section has been studied (Exercise 19 of Section 2.4). If $n = 0$, part (ii) is a special case of part (i). Thus we have indicated how to prove part (ii) if n is any integer. The proof of part (ii) for any real number n will be given in Section 6.3, provided that $x > 0$. Part (v) will be proved in Example 5 of the next section.

APPLICATIONS

Example 8

In many applied problems it is necessary to solve various types of equations. In this example we discuss a general method, called **Newton's method,** for approximating a real root of an equation of the form

$$f(x) = 0.$$

We shall see that one of the prerequisites for using Newton's method is the ability to find $f'(x)$ by using rules studied in this section.

Suppose that a portion of the graph of $y = f(x)$ is as shown in Figure 2.12. A root r of $f(x) = 0$ is a number r for which $f(r) = 0$; such a number is the x-coordinate of a point of intersection of $y = f(x)$ with the x-axis. We first make an estimate x_1 of the desired root—this number should be fairly close to the root r. As a second, and usually better, estimate of r we choose the x-coordinate x_2 of the point where the tangent line to $y = f(x)$ at $(x_1, f(x_1))$ crosses the x-axis. The slope of this tangent line is $f'(x_1)$, and the tangent line passes through $(x_1, f(x_1))$; thus its equation in point-slope form is

$$y - f(x_1) = f'(x_1)(x - x_1).$$

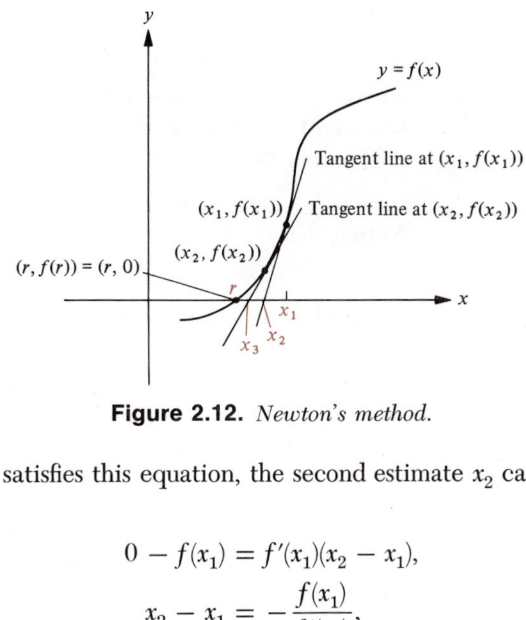

Figure 2.12. *Newton's method.*

Since $(x_2, 0)$ satisfies this equation, the second estimate x_2 can be found as follows:

$$0 - f(x_1) = f'(x_1)(x_2 - x_1),$$

$$x_2 - x_1 = -\frac{f(x_1)}{f'(x_1)},$$

$$x_2 = x_1 - \frac{f(x_1)}{f'(x_1)}.$$

The process is repeated by letting the third estimate x_3 of the root r be the x-coordinate of the point at which the tangent line at $(x_2, f(x_2))$ intersects the x-axis. This yields

$$x_3 = x_2 - \frac{f(x_2)}{f'(x_2)}.$$

In general, successive estimates of r are given by

$$x_n = x_{n-1} - \frac{f(x_{n-1})}{f'(x_{n-1})},$$

for $n = 2, 3, 4, \ldots$.

If we assume that the successive estimates of r get very near to r, it follows that r can be approximated to any number of decimal places of accuracy by repeating the estimation process enough times. The repetitive nature of the procedure makes it possible to program a digital computer to do Newton's method. We have not stated the precise conditions under which Newton's method is valid, but clearly f must be differentiable at points near r and $f'(x_1)$, $f'(x_2)$, etc., cannot be zero.

Let the procedure be illustrated by estimating a root of the equation

$$f(x) = x^3 - x - 4 = 0.$$

Since

$$f(1) = 1 - 1 - 4 = -4 < 0$$

and

$$f(2) = 8 - 2 - 4 = 2 > 0,$$

and the polynomial function f is continuous, there is a root of the equation between 1 and 2 (see Exercise 12 of Section 1.9). Let the first approximation of this root be

$$x_1 = 2.$$

Noting that

$$f'(x) = 3x^2 - 1,$$

we substitute to find x_2 as follows:

$$x_2 = 2 - \frac{f(2)}{f'(2)} = 2 - \frac{2}{3(2)^2 - 1} = \frac{20}{11}.$$

Suppose that one more estimate x_3 is computed:

$$\begin{aligned} x_3 &= \frac{20}{11} - \frac{f(\frac{20}{11})}{f'(\frac{20}{11})} \\ &= \frac{20}{11} - \frac{(\frac{20}{11})^3 - \frac{20}{11} - 4}{3(\frac{20}{11})^2 - 1}. \end{aligned}$$

After some arithmetic we find that

$$x_3 = \tfrac{21,324}{11,869},$$

which is very near the exact root r. In fact, $x_3 \approx 1.7966$, whereas the correct value of r to four decimal places can be shown to be 1.7963 by repeating the method more times.

Newton's method can be used to estimate roots of equations other than polynomial equations. For example, after we learn how to differentiate 2^x, we can approximate the root between 1 and 2 of the equation

$$2^x + x - 4 = 0.$$

(See Exercise 9 of Section 6.3.)

Newton's method can be used as a technique for calculating square roots of positive real numbers to any desired accuracy (see Exercise 26). ▮

EXERCISES

Suggested minimum assignment: Exercises 1, 3, 7, 9, 13, 15, 16, and 19.

In Exercises 1–6 find $D_x f(x)$.

1. $f(x) = 8.$
2. $f(x) = \pi.$
3. $f(x) = x^4.$
4. $f(x) = x^{-3} + x^{\sqrt{3}}.$
5. $f(x) = \sqrt[3]{x} + x.$
6. $f(x) = x^2(2x^3 + 3).$

In Exercises 7–10 find y'.

7. $y = 5x^3 - 8x^2 + 3x - 4.$
8. $y = \dfrac{1}{x^2} + \dfrac{2x}{3}.$
9. $y = (x^2 + x)^5.$
10. $y = (x^4 + 3)^4.$

In Exercises 11–13 find the slope of the tangent line to the given curve at the given value of x.

11. $y = 3x^3 - 8x + 2,\ x = 2.$
12. $y = (1 - 3x)^4,\ x = -1.$
13. $y = \dfrac{1}{x},\ x = 3.$

In Exercises 14–15 find the rate of change of y with respect to x at the given value of x.

14. $y = \dfrac{x^2 + x^4}{x},\ x = 1.$
15. $y = \sqrt{3x + 1},\ x = 5.$
16. For the cost function given by $C(x) = \dfrac{x^2 + 10}{5}$, find the marginal cost $C'(x)$.
17. For the revenue function given by $R(x) = 20x - \dfrac{x^2}{10}$, find the marginal revenue $R'(x)$.
18. For the profit function given by $P(x) = x + 2x^2 - x^3$, find the marginal profit $P'(x)$.

In Exercises 19–21 a particle is moving along a line according to the given equation of motion which is assumed to be valid if $t \geq 1$. Find the velocity of the particle when $t = 4$.

19. $s = \dfrac{(t^3 + 1)^2}{t^3}$.

20. $s = \sqrt{t^2 + 9}$.

21. $s = (t^2 - t)^2$.

22. Prove Theorem 1(*ii*) for the case that n is a positive integer.
23. Prove Theorem 1(*iii*).
24. Prove Theorem 1(*iv*).
25. Use Newton's method (Example 8) to approximate the root between 1 and 2 of the equation

$$x^3 + x^2 - x - 2 = 0.$$

Use $x_1 = 1$ and find x_2 and x_3.

26. Use Newton's method (Example 8) to approximate $\sqrt{5}$ by approximating the root between 2 and 3 of the equation

$$x^2 - 5 = 0.$$

Use $x_1 = 2$ and find x_2 and x_3. (As a special project, if digital computer facilities are available, one could attempt to write a program for finding successive approximations to $\sqrt{5}$ by Newton's method.)

2.4 Basic Differentiation Formulas (Continued)

In this chapter we have learned that the ability to find the derivative of a function is useful in problems concerning slope of a tangent line to a curve, rate of change, marginal cost, marginal revenue, marginal profit, or velocity. The derivative can be found more quickly using the formulas of Theorem 1 than by applying Definition 2. In Theorem 2 we state three more very important differentiation formulas, which are valid under the assumption that all indicated derivatives exist and all denominators are not zero.

► **Theorem 2**

(*i*) $D_x(f(x)g(x)) = f(x)\,D_x g(x) + g(x)\,D_x f(x)$. **(Product rule)**

(*ii*) $D_x\left(\dfrac{f(x)}{g(x)}\right) = \dfrac{g(x)\,D_x f(x) - f(x)\,D_x g(x)}{(g(x))^2}$. **(Quotient rule)**

(*iii*) If $y = g(u)$ and $u = f(x)$, then $\dfrac{dy}{dx} = \dfrac{dy}{du}\dfrac{du}{dx}$. **(Chain rule)**

A discussion of the proof of this theorem is given at the end of this section. ◄

Example 1

Apply Theorem 2(*i*) to differentiate the product $(x^3 + x^2)(x^2 + 8)$.

$$\begin{aligned} D_x[(x^3 + x^2)(x^2 + 8)] &= (x^3 + x^2)\, D_x(x^2 + 8) + (x^2 + 8)\, D_x(x^3 + x^2) \\ &= (x^3 + x^2)(2x) + (x^2 + 8)(3x^2 + 2x) \\ &= 5x^4 + 4x^3 + 24x^2 + 16x. \ \blacksquare \end{aligned}$$

Theorem 2(*i*) says that *the derivative of the product of two functions is the first times the derivative of the second plus the second times the derivative of the first.* It is possible to avoid using the product rule in Example 1 by multiplying the two functions together at the beginning.

Theorem 2(*ii*) can be used whenever the function to be differentiated can be expressed as a quotient of two functions. In words, it says that *the derivative of the quotient of two functions is the denominator times the derivative of the numerator minus the numerator times the derivative of the denominator, all divided by the square of the denominator.* Observe the use of this quotient rule in the next example.

Example 2

Find the slope of the tangent line to $y = x^2/(3x + 1)$ at $x = 1$.

By Theorem 2(*ii*),

$$\begin{aligned} y' &= \frac{(3x + 1)\, D_x(x^2) - x^2 D_x(3x + 1)}{(3x + 1)^2} \\ &= \frac{(3x + 1)(2x) - x^2(3)}{(3x + 1)^2} \\ &= \frac{3x^2 + 2x}{(3x + 1)^2}. \end{aligned}$$

Another procedure that will lead to the same result for y' is to first write $y = x^2(3x + 1)^{-1}$ and then to use the product rule.

The slope of the tangent line at $x = 1$ is

$$\begin{aligned} y'|_{x=1} &= \frac{3(1)^2 + 2(1)}{[3(1) + 1]^2} \\ &= \frac{5}{16}. \ \blacksquare \end{aligned}$$

Example 3

If $y = g(u) = 3u + 1$ and $u = f(x) = x^2 + 4$, use Theorem 2(*iii*) to find dy/dx.

$$\begin{aligned} \frac{dy}{dx} &= \frac{dy}{du}\frac{du}{dx} \\ &= [D_u(3u + 1)][D_x(x^2 + 4)] \\ &= 3(2x) \\ &= 6x. \ \blacksquare \end{aligned}$$

Some additional discussion and explanation is needed in connection with Example 3. If y is a function of u, say $y = g(u)$, and u is a function of

x, say $u = f(x)$, then a combination of these functional relationships makes y a function of x. In general we have

$$y = g(u) = g(f(x)).$$

A function given by $y = g(f(x))$ is called a **composite function.** The domain of the composite function given by $y = g(f(x))$ consists of the elements x in the domain of f such that $f(x)$ is in the domain of g. In Example 3 $g(u) = 3u + 1$ and $f(x) = x^2 + 4$; therefore,

$$y = g(f(x)) = g(x^2 + 4) = 3(x^2 + 4) + 1 = 3x^2 + 13,$$

and the domain of this composite function is the set of all real numbers. The answer to Example 3 can be checked as follows:

$$\frac{dy}{dx} = D_x\, g(f(x)) = D_x\,(3x^2 + 13) = 6x.$$

However, we want to emphasize the method used in Example 3, since the chain rule will be useful in several sections later in this text.

Example 4

If $z = (w^2 - 1)^3$ and $w = 3/(t - 1)$, find the rate of change of z with respect to t when $t = 2$.

By the chain rule,

$$\frac{dz}{dt} = \frac{dz}{dw}\frac{dw}{dt}.$$

dz/dw can be found by the power rule for differentiation:

$$\frac{dz}{dw} = 3(w^2 - 1)^2(2w).$$

dw/dt can be found by the quotient rule:

$$\frac{dw}{dt} = \frac{(t - 1)(0) - 3(1)}{(t - 1)^2} = \frac{-3}{(t - 1)^2}.$$

Hence

$$\frac{dz}{dt} = [6w(w^2 - 1)^2]\frac{-3}{(t - 1)^2}. \tag{2.4}$$

When $t = 2$, $w = 3/(t - 1) = 3/(2 - 1) = 3$. Substituting $t = 2$ and $w = 3$ in (2.4), we find that

$$\left.\frac{dz}{dt}\right|_{t=2} = 6(3)(9 - 1)^2\frac{-3}{(2 - 1)^2} = -3456. \blacksquare$$

Example 5

Suppose that $y = u^n$, where n is a real number, and $u = f(x)$. Find $D_x y$.

By the chain rule,

$$\frac{dy}{dx} = \frac{dy}{du}\frac{du}{dx}$$

or

$$D_x y = D_u y \, D_x u.$$

By Theorem 1(*ii*),

$$D_u y = nu^{n-1}.$$

Therefore,

$$D_x y = nu^{n-1} \, D_x u.$$

This is the power rule for differentiation, which is just a special case of the chain rule for the case when $y = u^n$. ▮

Before we can discuss the proof of Theorem 2, we need to make the statement of a very well-known theorem.

▶ **Theorem 3**

If a function f is differentiable at $x = x_1$, then f is continuous at $x = x_1$.

A hint for the proof of this important relationship between differentiability and continuity is given in Exercise 24. ◀

One consequence of Theorem 3 is that if a function f is discontinuous at $x = x_1$, then f is not differentiable at $x = x_1$. The converse of Theorem 3 is not always true; in Example 4 of Section 2.1 an example is given of a function that is continuous at a point but not differentiable at that point.

The remainder of this section is devoted to some discussion of the proof of Theorem 2. Theorem 2(*i*) can be proved as follows:

$$D_x(f(x)g(x)) = \lim_{\Delta x \to 0} \frac{f(x + \Delta x)g(x + \Delta x) - f(x)g(x)}{\Delta x}$$

$$= \lim_{\Delta x \to 0} \frac{f(x + \Delta x)g(x + \Delta x) - f(x + \Delta x)g(x) + f(x + \Delta x)g(x) - f(x)g(x)}{\Delta x}$$

$$= \lim_{\Delta x \to 0} f(x + \Delta x)\left[\frac{g(x + \Delta x) - g(x)}{\Delta x}\right] + \lim_{\Delta x \to 0} g(x)\left[\frac{f(x + \Delta x) - f(x)}{\Delta x}\right]$$

$$= f(x) \, D_x g(x) + g(x) \, D_x f(x).$$

An explanation of why $\lim\limits_{\Delta x \to 0} f(x + \Delta x) = f(x)$ needs to be made. We are assuming that f is differentiable at x; therefore, by Theorem 3, f is continuous at x. By (2.1) of Section 2.1, $\lim\limits_{\Delta x \to 0} [f(x + \Delta x) - f(x)] = 0$, and therefore $\lim\limits_{\Delta x \to 0} f(x + \Delta x) = f(x)$.

The proof of Theorem 2(*ii*) is similar to the proof of part (*i*) and is left as Exercise 18.

For Theorem 2(*iii*) suppose that x is given an increment $\Delta x \neq 0$. This causes a change Δu in u, which in turn causes a change Δy in y. If $\Delta u \neq 0$,

$$\frac{\Delta y}{\Delta x} = \frac{\Delta y}{\Delta u}\frac{\Delta u}{\Delta x}$$

and

$$\frac{dy}{dx} = \lim_{\Delta x \to 0} \frac{\Delta y}{\Delta x} = \lim_{\Delta x \to 0} \frac{\Delta y}{\Delta u}\frac{\Delta u}{\Delta x}$$
$$= \left(\lim_{\Delta x \to 0} \frac{\Delta y}{\Delta u}\right)\left(\lim_{\Delta x \to 0} \frac{\Delta u}{\Delta x}\right).$$

Since we are assuming that f is differentiable, f is also continuous and hence $\Delta u \to 0$ as $\Delta x \to 0$. Therefore,

$$\frac{dy}{dx} = \left(\lim_{\Delta u \to 0} \frac{\Delta y}{\Delta u}\right)\left(\lim_{\Delta x \to 0} \frac{\Delta u}{\Delta x}\right) = \frac{dy}{du}\frac{du}{dx}.$$

Unfortunately, this is not a complete proof of the chain rule because we assumed that $\Delta u \neq 0$. If $\Delta u = 0$, the proof is more complicated and the reader is referred to a more advanced text.

APPLICATIONS

Example 6

Let $C(x)$ give the cost of producing x items. The average cost per item is $C(x)/x$. The function f defined by

$$f(x) = \frac{C(x)}{x}$$

is called the **average cost function.** It is of some importance in economics because a manufacturer wants the average cost to be as small as possible. We shall learn in the next chapter that in order to find when the average cost function is smallest, it is necessary to find its derivative (when the derivative exists). By the quotient rule of this section,

$$f'(x) = \frac{xC'(x) - C(x)}{x^2}.$$

Suppose that x_1 items have been produced at a cost of $C(x_1)$. The fraction

$$\frac{[C(x_1 + \Delta x) - C(x_1)]/C(x_1)}{\Delta x/x_1}$$

is the ratio of the percentage change in total cost to a given percentage change in quantity produced. The ratio is called the elasticity of cost on the interval between x_1 and $x_1 + \Delta x$ and may also be written

$$\frac{\Delta C/\Delta x}{C(x_1)/x_1}.$$

If $\Delta x \to 0$, the ratio becomes

$$\frac{C'(x_1)}{C(x_1)/x_1},$$

which is the marginal cost at x_1 divided by the average cost of producing

x_1 items and is called the **elasticity of cost at $x = x_1$**. In general, the function E which represents the elasticity of cost is determined by

$$E(x) = \frac{C'(x)}{C(x)/x}.$$

Further interpretation of the meaning of elasticity of cost will be given in Example 5 of Section 2.6. ▮

EXERCISES

Suggested minimum assignment: Exercises 1, 5, 7, 11, 12, 15, and 17.

Find y' in Exercises 1–6 using Theorem 2(*i*) or (*ii*).

1. $y = x^2(2x^3 + 3)$.
2. $y = x(2x + 3)^4$.
3. $y = x\sqrt{2x + 1}$.
4. $y = \dfrac{x}{2x + 7}$.
5. $y = \dfrac{x^2 + 1}{x^2 + 2}$.
6. $y = \dfrac{1}{x^2 - 1}$.

Find dy/dx in Exercises 7 and 8 using Theorem 2(*iii*).

7. $y = 1 - 2u,\ u = x^3 - 2x^2 + 6$.
8. $y = \dfrac{1}{u},\ u = x^2 - 8x + 2$.
9. Find the slope of the tangent line to $y = (x^2 + 2x - 3)(x^3 - x^2 + 6)$ at $x = -1$.
10. By repeated use of the product rule show that if $y = f(x)g(x)h(x)$, then $y' = f'(x)g(x)h(x) + f(x)g'(x)h(x) + f(x)g(x)h'(x)$.
11. If $y = \dfrac{2}{x - 3}$, find the rate of change of y with respect to x at $x = 5$.
12. Suppose that a cost function is given by $C(x) = \dfrac{x^2 + 5}{2x + 1}$. Find the marginal cost $C'(x)$.
13. Suppose that a revenue function is given by $R(x) = \dfrac{1}{x + 1}(5x^2 + 2)$. Find the marginal revenue $R'(x)$.
14. Suppose that a profit function is given by $P(x) = \sqrt{x}(x - 1)$. Find the marginal profit $P'(x)$.
15. A particle moves along a line according to the given equation of motion. Find the velocity when $t = 4$.

    (*a*) $s = t(t^2 - 15)^4$. (*b*) $s = \dfrac{t}{\sqrt{2t + 1}}$.
16. Find the rate of change of y with respect to x when $x = 4$, if $y = \sqrt{2u - 1}$ and $u = \sqrt{2x + 1}$.

17. If $y = g(u) = u^2 + 1$ and $u = f(x) = 3x + 7$, find $g(f(x))$ and then find $D_x g(f(x))$. Check by using the chain rule.
18. Prove Theorem 2(*ii*).
19. Prove Theorem 1(*ii*) of Section 2.3 for the case that n is a negative integer. (*Hint:* Write x^n as $1/x^{-n}$, where $-n$ is a positive integer, and use the quotient rule.)
20. Find $D_x z$ if $z = \dfrac{y}{y+1}$ and $y = x + \dfrac{1}{x}$.
21. Find $D_t s$ if $s = \dfrac{(t^3+1)^2}{t}$.
22. Find $D_v w$ if $w = \dfrac{v}{(v^2+2)^3}$.
23. Sketch $y = f(x) = |x| + 1$. Is f continuous at $x = 0$? Is f differentiable at $x = 0$?
24. Prove Theorem 3. [*Hint:*

$$\lim_{x \to x_1} (f(x) - f(x_1)) = \lim_{x \to x_1} (x - x_1)\frac{f(x) - f(x_1)}{x - x_1} = (0)(f'(x_1)) = 0.$$

See Exercise 10 of Section 2.2.]
25. Suppose that a cost function is given by $C(x) = x^2 + 1$. Find the derivative of the average cost function. Also find the elasticity of cost $E(x)$. (See Example 6.)

2.5 Higher Derivatives

In Definition 2 of Section 2.2 the derivative f' of a function f was defined. The function f' is also called the **derived function** or the **first derivative** of f. One can begin again with the function f' and define its derivative, which is denoted by f'' and which is called the **second derivative** of f. In other words, the second derivative is the derivative of the first derivative—the limit defining $f''(x)$ is

$$f''(x) = \lim_{\Delta x \to 0} \frac{f'(x + \Delta x) - f'(x)}{\Delta x},$$

provided that this limit exists. The process can be continued—the **third derivative** is the derivative of the second derivative, the **fourth derivative** is the derivative of the third derivative, and in general for any positive integer n the **nth derivative** is the derivative of the $(n-1)$st derivative. Derivatives beyond the first derivative are called **higher derivatives.** Example 1 illustrates the correct notation to use when writing higher derivatives (except that the higher derivative notation corresponding to the dy/dx notation will not be presented until the next section).

Example 1
Suppose that $y = f(x) = x^6$. Then

$$\begin{aligned}
y' &= f'(x) = D_x y = D_x f(x) = 6x^5,\\
y'' &= f''(x) = D_x^2 y = D_x^2 f(x) = 30x^4,\\
y''' &= f'''(x) = D_x^3 y = D_x^3 f(x) = 120x^3,\\
y^{(4)} &= f^{(4)}(x) = D_x^4 y = D_x^4 f(x) = 360x^2,\\
y^{(5)} &= f^{(5)}(x) = D_x^5 y = D_x^5 f(x) = 720x,\\
y^{(6)} &= f^{(6)}(x) = D_x^6 y = D_x^6 f(x) = 720,\\
y^{(7)} &= f^{(7)}(x) = D_x^7 y = D_x^7 f(x) = 0.
\end{aligned}$$

All derivatives of x^6 higher than the sixth derivative of x^6 are equal to zero. ▮

As an illustration of how to read the notation in Example 1, $D_x^4 y$ is read "the fourth derivative of y with respect to x." One should think of $D_x^4 y$ as a single symbol rather than trying to assign any meaning to the component parts—notice where the dependent variable, y, is written in this notation and observe also the location of the independent variable, x. Since the fourth derivative is the derivative of the third derivative, the following notations have the same meaning as $D_x^4 y$:

$$D_x y''', D_x(f'''(x)), D_x(D_x^3 y), D_x(D_x^3 f(x)).$$

If the fourth derivative is evaluated at $x = 1$ in Example 1, the correct notations are

$$y^{(4)}|_{x=1} = f^{(4)}(1) = D_x^4 y|_{x=1} = D_x^4 f(x)|_{x=1} = 360.$$

Next we mention two interpretations that can be associated with the second derivative; the reader can be assured that several other uses of higher derivatives will be encountered later in this text. The second derivative is the rate of change of the first derivative; since the first derivative gives the slope of a tangent line to a curve, we may say that the second derivative gives the rate of change of slope. Also, for a particle moving along a line, the second derivative gives the rate of change of the velocity of the particle, which is called the **acceleration** of the particle.

Example 2

Find the rate of change of slope at the point (3, 4) on the semicircle $y = \sqrt{25 - x^2}$.

By the power rule for differentiation,

$$y' = \left(\frac{1}{2}\right)(25 - x^2)^{-1/2}(-2x) = \frac{-x}{(25 - x^2)^{1/2}}.$$

By the quotient rule,

$$\begin{aligned}
y'' &= \frac{(25 - x^2)^{1/2}(-1) - (-x)(\frac{1}{2})(25 - x^2)^{-1/2}(-2x)}{25 - x^2}\\
&= \frac{[-(25 - x^2) - x^2]/(25 - x^2)^{1/2}}{25 - x^2}\\
&= \frac{-25}{(25 - x^2)^{3/2}}.
\end{aligned}$$

Therefore,

$$y''|_{x=3} = \frac{-25}{(16)^{3/2}} = -\frac{25}{64}.$$

This means that if one could move from left to right along the semicircle, then at the instant the point (3, 4) is reached, the slope is becoming smaller by $\frac{25}{64}$ unit per each unit of change in x. See Figure 2.13. ▮

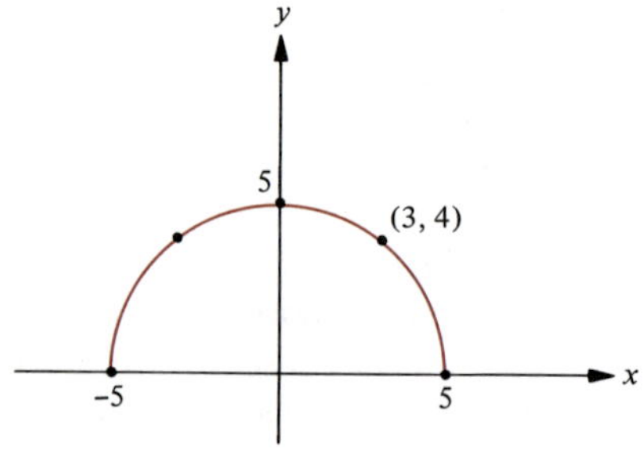

Figure 2.13. $y = \sqrt{25 - x^2}$.

Example 3

A particle moves along a line according to the equation of motion $s = t^3 - 15t^2$. Find the velocity and acceleration at $t = 6$ seconds. (Assume that s is measured in feet and t in seconds.)

$$D_t s = 3t^2 - 30t,$$
$$D_t^2 s = 6t - 30.$$

Since $D_t s|_{t=6} = 3(36) - 180 = -72$, the velocity is -72 feet per second. The units for acceleration are feet per second per second, and therefore the acceleration is $D_t^2 s|_{t=6} = 6$ feet per second per second. If the particle is visualized as moving back and forth along a horizontal line with positive direction to the right, the positive acceleration means that a force is being exerted on the particle in the positive direction when $t = 6$; however, because the velocity is negative, the particle is actually moving to the left, but it is slowing down, since the acceleration is positive. In a situation in which velocity and acceleration have opposite signs, the particle is slowing down, whereas the particle is speeding up when these signs are the same. ▮

Example 4

If $D_x^4 y = (x^2 + 3)^5$, find $D_x^6 y$.

By the power rule for differentiation

$$D_x^5 y = D_x (x^2 + 3)^5 = 5(x^2 + 3)^4(2x) = (10x)(x^2 + 3)^4.$$

$D_x^6 y$ is the derivative of $D_x^5 y$ and is found by applying the product rule, the power rule for differentiation, and simplifying the result.

$$
\begin{aligned}
D_x^6 y &= D_x[(10x)(x^2+3)^4] \\
&= (10x)\, D_x(x^2+3)^4 + (x^2+3)^4\, D_x(10x) \\
&= (10x)(4)(x^2+3)^3(2x) + 10(x^2+3)^4 \\
&= 10(x^2+3)^3[8x^2 + (x^2+3)] \\
&= 10(x^2+3)^3(9x^2+3) \\
&= 30(x^2+3)^3(3x^2+1). \ \blacksquare
\end{aligned}
$$

APPLICATIONS

Example 5

Suppose that a manufacturer sells an item by the ton (or partial ton). If he sells x tons per week, he has found that the price $p(x)$ that he must charge per ton depends upon the number x of tons sold. His weekly revenue is given by

$$R(x) = xp(x).$$

Using the product rule of Section 2.4, the marginal revenue $R'(x)$ is given by

$$
\begin{aligned}
R'(x) &= xp'(x) + (p(x))(1) \\
&= xp'(x) + p(x).
\end{aligned}
$$

The rate of change of the marginal revenue $R'(x)$ with respect to x is the derivative of $R'(x)$, which is designated $R''(x)$. If we apply the product rule again, we find that

$$
\begin{aligned}
R''(x) &= xp''(x) + (p'(x))(1) + p'(x) \\
&= xp''(x) + 2p'(x).
\end{aligned}
$$

We could call $R'(x)$ the velocity in revenue and $R''(x)$ the acceleration in revenue. Under normal conditions the velocity in revenue is positive, because the more tons that are sold, the greater the gross revenue from sales. Then if the acceleration in revenue is also positive, it means that the rate of increase of revenue is speeding up as more tons are sold; a negative acceleration in revenue indicates a slowing down in the rate of increase of revenue. ▮

EXERCISES

Suggested minimum assignment: Exercises 1, 2, 8, 11, 13, and 15(*b*).

1. If $y = 7x^4$, find y', y'', and y'''.
2. If $f(x) = 7x^3 - x^5$, find $f'(2)$, $f''(2)$, and $f'''(2)$.
3. If $y = \sqrt{x}$, find $D_x y$, $D_x^2 y$, $D_x^3 y$, and $D_x^4 y$.
4. If $y = u^3$ and $u = x + \sqrt{x}$, find $D_x y$ and $D_x^2 y$.
5. If $f(x) = \dfrac{3}{x-4}$, find $f'(x)$ and $f''(x)$.

6. If $f(x) = x(2x + 3)^6$, find $f'(x)$ and $f''(x)$.
7. If $y = \frac{1}{8}x^8 + 4000x^3 + 89x^2$, find $y^{(4)}$.
8. If $f(x) = (2x + 1)^{10}$, find $D_x^4 f(x)$.
9. If $y = \sqrt{7x + 8}$, find $y''|_{x=4}$.
10. If $f'(x) = 3x^2$, find $f''(x)$ by using $f''(x) = \lim_{\Delta x \to 0} \dfrac{f'(x + \Delta x) - f'(x)}{\Delta x}$.
11. If $D_x^7 y = \dfrac{1}{2x}$, find $D_x^8 y$.
12. If $y = x^5 - x^4$, find $D_x(D_x^3 y)$.
13. If $y = 2x^4$, find the rate of change of slope at $x = 1$.
14. If $y = \sqrt{x^2 + 3}$, find the rate of change of slope at $(1, 2)$.
15. In each of the following, a particle moves along a line according to the given equation of motion. Find the velocity and acceleration at $t = 3$ seconds. Assume that s is measured in meters.
    (*a*) $s = t^3 - 3t^2 - 8t + 31$. (*b*) $s = (t^2 - 7)^4$.
16. In each of the following, a particle moves along a line according to the given equation of motion. At $t = 2$ seconds, tell whether the particle is moving in the positive or negative direction along the s-axis and tell whether it is speeding up or slowing down.
    (*a*) $s = t^2$. (*b*) $s = t^3 - 10t^2$.
    (*c*) $s = 30t - t^3$.
17. A revenue function is given by $R(x) = x^3 + x^2$. Find the rate of change of marginal revenue at $x = 2$. At $x = 2$ is the rate of increase of revenue speeding up or slowing down?

2.6 Differentials

In Example 2 of this section we shall try to gain a better understanding of marginal revenue, marginal cost, and marginal profit. Before this can be done, we need to introduce a concept of considerable importance.

▶ Definition 3

Suppose that $y = f(x)$ and that f is differentiable at x. If x is changed by an increment $\Delta x \neq 0$, then the **differential** $\boldsymbol{dx}$ of the independent variable x is defined by

$$dx = \Delta x,$$

and the **differential** $\boldsymbol{dy}$ of the dependent variable y is defined by

$$dy = f'(x)\,\Delta x \qquad \text{or} \qquad dy = f'(x)\,dx.$$

Definition 3 involves two separate definitions; in order to show that these two definitions are consistent with each other, we examine the function defined by $y = f(x) = x$. Then $dy = f'(x)\,\Delta x = (1)\,\Delta x = \Delta x$. However, $dy = dx$, since $y = x$. Hence $dx = \Delta x$.

The geometric interpretation of dy is important (see Figure 2.14). Since

$$dy = f'(x)\,dx,$$

we obtain, by dividing by $dx \neq 0$, the equation

$$f'(x) = \frac{dy}{dx}.$$

Since dy/dx is the slope of the tangent line to $y = f(x)$ at (x, y) and since $dx = \Delta x$, it follows that in Figure 2.14 dy must be the distance from the horizontal up to the tangent line.

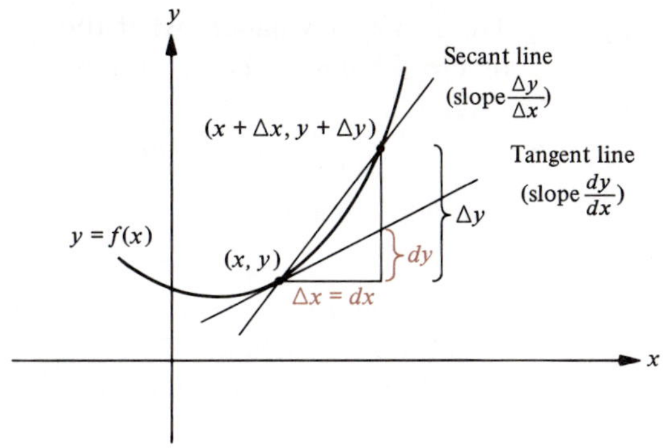

Figure 2.14

As we learned in Section 2.1, Δy is the exact change in y corresponding to the change of Δx in x. In general, both Δy and dy can be either positive, negative, or zero, depending upon the figure. Notice that, in general, Δy does *not* equal dy, but by definition Δx *does* equal dx. If $|\Delta x|$ is relatively small, then dy and Δy usually are almost the same and dy can be calculated as a close approximation to Δy, as we illustrate in Example 1.

Example 1

If $y = f(x) = x^2$, calculate Δy and dy if $x = 2$ and $\Delta x = dx = .01$.

$$\begin{aligned}
\Delta y &= f(x + \Delta x) - f(x) \\
&= f(2.01) - f(2) \\
&= (2.01)^2 - (2)^2 \\
&= 4.0401 - 4 \\
&= .0401. \\
dy &= f'(x)\,dx = 2x\,dx = 2(2)(.01) = .04.
\end{aligned}$$

Notice that dy is easier to calculate than Δy and that dy is a close approximation to Δy when $|\Delta x|$ is a relatively small number; several useful applications of dy result from the latter fact. ▮

Example 2

Suppose that a motorcycle manufacturer produces x motorcycles on a single work day. The number x depends upon several factors, such as the number of workers available and the frequency of breakdown of machinery, but x is always an integer from 0 to 100 inclusive. The cost in dollars of producing x motorcycles in one day is known to be given by the cost function C, where

$$C(x) = 2000 + 300x - \frac{x^2}{2}.$$

If 50 motorcycles come off the assembly line in 1 day, the cost (dollar signs omitted) is

$$C(50) = 2000 + (300)(50) - \frac{(50)^2}{2} = 15{,}750.00.$$

The manager wonders what the cost would have been if $x = 51$, instead of $x = 50$, and so he computes

$$C(51) = 2000 + (300)(51) - \frac{(51)^2}{2} = 15{,}999.50.$$

The difference is

$$C(51) - C(50) = 249.50,$$

which would have been the exact cost of one additional motorcycle.

Let us also calculate the marginal cost at $x = 50$ as follows (we assume now that x is any real number so that C can be differentiated):

$$C(x) = 2000 + 300x - \frac{x^2}{2},$$

$$C'(x) = 300 - x,$$

$$C'(50) = 250.$$

The answer for $C'(50)$ is almost the same as for $C(51) - C(50)$ and is much easier to calculate—let us see why this is no coincidence. If

$$y = C(x),$$

with $x = 50$ and $\Delta x = 1$, then

$$\Delta y = C(x + \Delta x) - C(x) = C(51) - C(50).$$

Also, using Definition 3 of this section,

$$dy = C'(x)\,\Delta x = [C'(50)](1) = C'(50).$$

Since dy is an approximation to Δy, it follows that $C'(50)$ is an approximation to $C(51) - C(50)$. Hence the marginal cost $C'(50)$ gives approximately the cost of the next, or 51st, motorcycle.

In general, the marginal cost $C'(k)$ gives the approximate cost of the $(k + 1)$st item. In a similar way it can be shown that the marginal revenue $R'(k)$ is approximately the revenue that will be received for the $(k + 1)$st item, assuming that the first k items have already been sold. The marginal profit $P'(k)$ gives the approximate profit derived from the sale of the $(k + 1)$st item and

$$P'(k) = R'(k) - C'(k).$$

Suppose in our motorcycle problem that the revenue function R is given by

$$R(x) = 400x + \frac{x^2}{10}.$$

Then the marginal revenue is

$$R'(x) = 400 + \frac{x}{5},$$

and at $x = 50$ the marginal revenue is

$$R'(50) = 400 + \frac{50}{5} = 410.$$

The approximate profit for the sale of the 51st motorcycle is the marginal profit $P'(50)$, where

$$P'(50) = R'(50) - C'(50) = 410 - 250 = 160. \blacksquare$$

If $y = f(x)$, we learned at the beginning of this section that

$$f'(x) = \frac{dy}{dx},$$

which justifies our previous use of dy/dx as a notation for "the derivative of y with respect to x." Since meanings have been assigned to dy and dx, the symbol dy/dx can also be thought of as a quotient. If dy/dx is used for the first derivative of y with respect to x, then the notation for higher derivatives is

$$\frac{d^2y}{dx^2}, \frac{d^3y}{dx^3}, \ldots.$$

No meaning is assigned to d^2y, dx^2, d^3y, dx^3, ..., so that the symbols for higher derivatives *cannot* be thought of as quotients but must be regarded as entire symbols; for example, the symbol d^2y/dx^2 is "the second derivative of y with respect to x," even though d^2y and dx^2 by themselves have no meaning. Observe carefully the notation used in the next example.

Example 3

If $y = f(x) = x^{-1}$, find dy, dy/dx, d^2y/dx^2, and d^3y/dx^3.

$$dy = f'(x)\,dx = -x^{-2}\,dx = -\frac{1}{x^2}dx,$$

$$\frac{dy}{dx} = f'(x) = -x^{-2} = -\frac{1}{x^2},$$

$$\frac{d^2y}{dx^2} = \frac{d}{dx}\left(\frac{dy}{dx}\right) = \frac{d}{dx}(y') = \frac{dy'}{dx} = \frac{d(-1/x^2)}{dx} = -(-2)x^{-3} = \frac{2}{x^3},$$

$$\frac{d^3y}{dx^3} = \frac{d}{dx}\left(\frac{d^2y}{dx^2}\right) = \frac{d}{dx}(y'') = \frac{dy''}{dx} = \frac{d(2/x^3)}{dx} = 2(-3)x^{-4} = -\frac{6}{x^4}. \blacksquare$$

Example 4

Show that the differential of the product of two differentiable functions is the first times the differential of the second plus the second times the differential of the first.

Let

$$h(x) = f(x)g(x).$$

By the product rule,

$$h'(x) = f(x)g'(x) + g(x)f'(x).$$

Therefore,

$$\begin{aligned} d(h(x)) = h'(x)\,dx &= (f(x))(g'(x)\,dx) + (g(x))(f'(x)\,dx) \\ &= (f(x))(d(g(x))) + (g(x))(d(f(x))). \end{aligned}$$

Other rules for derivatives carry over in a similar way to corresponding rules for differentials. ▮

APPLICATIONS

Example 5

In Example 6 of Section 2.4 the elasticity of cost at a point x_1 was defined as the marginal cost at x_1 divided by the average cost of producing x_1 items. Now that we have a better understanding of the meaning of marginal cost, let us use the cost function of Example 2 to try to gain a better understanding of elasticity of cost. The average cost of producing a motorcycle after 50 have been produced is

$$\frac{C(50)}{50} = \frac{15{,}750}{50} = 315.$$

The approximate cost of making one additional motorcycle is the marginal cost

$$C'(50) = 250,$$

which is somewhat less than the average cost of the first 50 motorcycles. The ratio

$$E(50) = \frac{C'(50)}{C(50)/50} = \frac{250}{315} \approx .8$$

is the elasticity of cost at $x = 50$ and indicates that the cost of the next motorcycle is only about $\frac{8}{10}$ of the average cost of the first 50 motorcycles. In general, an elasticity of cost that is less than 1 indicates that the next item will cost less than the average of the cost of the items already produced, and therefore that the average cost will become smaller if another item is produced. An elasticity of cost greater than 1 indicates that production of another item will raise the average cost per item. A more comprehensive discussion of the role of elasticity in economics may be found in Chapter 6 of Kim [14]. ▮

EXERCISES

Suggested minimum assignment: Exercises 1(*f*), 3, 9, 10, and 11.

1. Find dy in each of the following.

   (*a*) $y = x^2$. (*b*) $y = \dfrac{x+3}{x}$. (*c*) $y = x$.

   (*d*) $y = (1 + x^2)^3$. (*e*) $y = 7$. (*f*) $y = x^3 - 4x^2$.

2. Draw several figures similar to Figure 2.14 and label Δx, Δy, and dy on each. Be sure to include cases in which Δy is negative and cases in which Δx is negative.
3. If $y = f(x) = x^3$, calculate Δy and dy if $x = 2$ and $\Delta x = dx = .1$.
4. If $y = f(x) = 3 - 4x + 2x^2$, calculate Δy and dy if $x = -2$ and $\Delta x = dx = -.1$.
5. If $V = \frac{4}{3}\pi r^3$, calculate ΔV and dV if $r = 1$ and $\Delta r = .2$.
6. The volume of a cube with side x is given by $V = x^3$. If the side, originally thought to equal 10 inches, is found to be 10.3 inches, use differentials to find approximately how much is added to the volume by the discovery of the change in x. Also find the exact change ΔV.
7. If $y = 4x^5 - 8x^3 + 7x^2 - 8x + 3$, find $\dfrac{dy}{dx}$, $\dfrac{d^2y}{dx^2}$, and $\dfrac{d^3y}{dx^3}$.
8. A cost function is given by $C(x) = \dfrac{5x^3}{x^2 + 2}$. Find $C'(8)$ and give an interpretation of your answer.
9. A manufacturer of storm doors produces x storm doors on a single work day, where x is an integer from 0 to 50 inclusive. Suppose that the cost function for this job is given by

   $$C(x) = 200 + 40x - \frac{x^2}{4}.$$

   Find $C'(10)$. What is the exact cost of the 11th storm door? Suppose also that the revenue function is given by

   $$R(x) = 50x + \frac{x^2}{20}.$$

   Find $P'(10)$, the approximate profit on the sale of the 11th storm door.
10. The revenue from the sale of x items is given by $R(x) = 5x + \dfrac{10}{x+1}$. Use differentials to find the approximate revenue from the sale of the seventh item. Also find the exact revenue from the sale of the seventh item.
11. If $y = (3x + 4)^9$, find $\dfrac{dy}{dx}$ and $\dfrac{d^2y}{dx^2}$.
12. Show that the differential of the sum of two differentiable functions is the sum of their differentials.

13. Approximate $\sqrt{82}$ without tables by calculating $y + dy$ if $y = \sqrt{x}$, $x = 81$, and $\Delta x = 1$.
14. Suppose that $y = g(u)$ and $u = f(x)$, where g and f are differentiable. Show that $dy = g'(u)\,du$ is valid even though u is dependent upon x. (*Hint:* Use the chain rule.)
15. If $y = k$, where k is a real number, does dy always equal zero? Why?

2.7 Implicit Differentiation (Optional)

In the last few sections we have begun with an equation of the form

$$y = f(x),$$

in which y has been expressed as a function of x, and we have learned rules for calculating dy/dx and higher derivatives of y with respect to x. When $y = f(x)$ is given, we say that y is expressed **directly** or **explicitly** as a function of x. Many equations, such as

$$x^2 + 4y^2 = 64,$$
$$y^3 + x^2y = 5,$$

and

$$y + \sqrt{x^2 + y^2} = x + 6,$$

are not solved for y in terms of x, but certainly a relationship between x and y is implied. Even though we may not attempt to solve for y, we shall assume that equations of this type do define one or more differentiable functions f such that $y = f(x)$; we say that y is defined **implicitly** as a function of x. If

$$x^2 + 4y^2 = 64,$$

it is easy to solve for y in terms of x in order to see that there are two possible ways in which the equation defines y as a differentiable function of x; we obtain

$$y = +\tfrac{1}{2}\sqrt{64 - x^2}$$

and

$$y = -\tfrac{1}{2}\sqrt{64 - x^2}.$$

Geometrically, the two equations above represent the upper and lower halves of an oval-shaped curve called an **ellipse** (see Figure 2.15). We are interested in finding dy/dx when y is defined implicitly as a function of x, but we cannot depend upon being able to solve a given equation for y, as this may be difficult or impossible. In an equation in which there are a mixture of x's and y's, we may treat x as the independent variable and y as the dependent variable and calculate dy/dx by the following technique, known as **implicit differentiation:**

> Differentiate each term in the equation with respect to x, treating y as a function of x. Then solve for $\dfrac{dy}{dx}$.

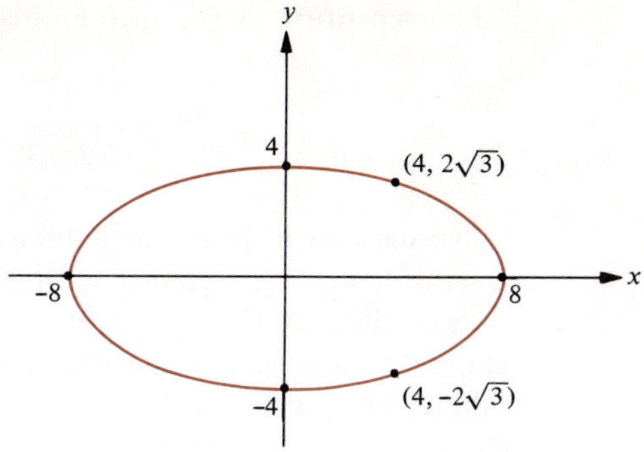

Figure 2.15. $x^2 + 4y^2 = 64$.

The reader will find that the hardest thing to understand in Examples 1 through 3, which follow, is the technique to be used when the dependent variable y occurs in a term to be differentiated. For example, suppose that the term y^2 is encountered; let $z = y^2$, where y in turn is a function of x, and apply the chain rule to obtain

$$\frac{d}{dx}(y^2) = \frac{dz}{dx} = \frac{dz}{dy}\frac{dy}{dx} = 2y\frac{dy}{dx}.$$

As another example, suppose that the term $2y/x$ is encountered. To differentiate this term with respect to x, the quotient rule and chain rule are applied as follows:

$$\begin{aligned} D_x\left(\frac{2y}{x}\right) &= \frac{x\,D_x(2y) - 2y\,D_x x}{x^2} \\ &= \frac{x[D_y(2y)]\,(dy/dx) - 2y\,D_x x}{x^2} \\ &= \frac{x(2)\,(dy/dx) - 2y(1)}{x^2} \\ &= \frac{2x\,(dy/dx) - 2y}{x^2}. \end{aligned}$$

Example 1
Find the slope of the tangent line to the ellipse $x^2 + 4y^2 = 64$ at the point $(4, 2\sqrt{3})$.

First find dy/dx by implicit differentiation:

$$2x + 8y\frac{dy}{dx} = 0,$$

$$\frac{dy}{dx} = -\frac{x}{4y}.$$

Then substitute $(4, 2\sqrt{3})$ to find that the desired slope is

$$\left.\frac{dy}{dx}\right|_{(4,2\sqrt{3})} = -\frac{4}{4(2\sqrt{3})} = -\frac{1}{2\sqrt{3}}\frac{\sqrt{3}}{\sqrt{3}} = -\frac{\sqrt{3}}{6}. \blacksquare$$

Notice that in Example 1 the expression for dy/dx involves *both* x and y—this happens frequently with implicit differentiation. If only the value of x is given in Example 1, the slope cannot be determined because the slope depends on whether the point is on the upper or lower part of the ellipse. However,

$$\frac{dy}{dx} = -\frac{x}{4y}$$

can also be used to find the correct slope at $(4, -2\sqrt{3})$, and implicit differentiation in general will give the correct derivative of *each* differentiable function defined by a given equation. The next example illustrates how higher derivatives can be determined after dy/dx has been found by implicit differentiation (some uses for higher derivatives were given in Section 2.5 and others will arise in Chapter 3).

Example 2
Find d^2y/dx^2 and d^3y/dx^3 if $x^2 + 4y^2 = 64$.
From Example 1,

$$\frac{dy}{dx} = -\frac{x}{4y}.$$

Differentiate again, substitute for dy/dx, and simplify:

$$\begin{aligned}
\frac{d^2y}{dx^2} &= -\frac{(4y)(1) - x(4)\,(dy/dx)}{(4y)^2} \\
&= -\frac{4y - 4x(-x/4y)}{16y^2} \\
&= -\frac{16y^2 + 4x^2}{64y^3} \\
&= -\frac{4(4y^2 + x^2)}{64y^3} \\
&= -\frac{4(64)}{64y^3} \qquad \text{(from the given equation)} \\
&= -\frac{4}{y^3}.
\end{aligned}$$

In a similar manner d^3y/dx^3 is computed:

$$\begin{aligned}\frac{d^3y}{dx^3} = \frac{d}{dx}\left(\frac{d^2y}{dx^2}\right) &= \frac{d}{dx}(-4y^{-3})\\ &= \left[\frac{d}{dy}(-4y^{-3})\right]\frac{dy}{dx}\\ &= 12y^{-4}\frac{dy}{dx}\\ &= 12y^{-4}\left(-\frac{x}{4y}\right)\\ &= -\frac{3x}{y^5}. \quad \blacksquare\end{aligned}$$

Example 3

If $y^3 + x^2y = 5$, calculate the rate of change of y with respect to x at (2, 1).

Notice that (2, 1) does satisfy the given equation. First calculate dy/dx by implicit differentiation and then evaluate at (2, 1); we choose to use the notation y' instead of dy/dx.

$$\begin{aligned} 3y^2y' + x^2y' + y(2x) &= 0,\\ y'(3y^2 + x^2) &= -2xy,\\ y' &= -\frac{2xy}{3y^2 + x^2}.\end{aligned}$$

Therefore,

$$y'|_{(2,1)} = -\frac{2(2)(1)}{3 + 4} = -\frac{4}{7}. \quad \blacksquare$$

When an equation contains a mixture of x's and y's, there is no reason why we cannot choose x instead of y as the dependent variable. In Example 4 each term is differentiated with respect to y, and x is treated as a function of y.

Example 4

Find $dx/dy|_{(2,1)}$ if $y^3 + x^2y = 5$.

$$\begin{aligned} 3y^2 + x^2(1) + y\left(2x\frac{dx}{dy}\right) &= 0,\\ 2xy\frac{dx}{dy} &= -(3y^2 + x^2),\\ \frac{dx}{dy} &= -\frac{3y^2 + x^2}{2xy},\\ \left.\frac{dx}{dy}\right|_{(2,1)} &= -\frac{3+4}{4} = -\frac{7}{4}. \quad \blacksquare\end{aligned}$$

The reader probably observed that the answers to Examples 3 and 4 are reciprocals of each other. From geometrical considerations it should not be surprising that

$$\frac{dx}{dy} = \frac{1}{dy/dx}, \qquad \text{provided that } \frac{dy}{dx} \neq 0,$$

since it would seem that the rate of change of x with respect to y should be the reciprocal of the rate of change of y with respect to x.

APPLICATIONS

Example 5

The marketing department of a large corporation has determined an equation that expresses the annual demand z for its product as a function of its price x per item and the price y per item of a rival synthetic product. Suppose that the equation is

$$z = 3000 - 30x^2 + 20y^2,$$

provided that $1 \leq x \leq 10$ and $1 \leq y \leq 8$. Both x and y are measured in dollars per item, and z is measured in units of production; x, y, and z are each functions of time t. Notice from the equation above that the demand for the product will increase if its price is decreased and will also increase if the price of the competing synthetic product is increased. Suppose that the rates of change of the prices x and y are the maximum allowed by federal guidelines, say 4 and 3 per cent, respectively, per year. Find the expected rate of change (per year) of the demand when $x = \$6$ and $y = \$4$. In other words, find dz/dt when $x = 6$, $y = 4$, $dx/dt = 6(.04) = .24$ and $dy/dt = 4(.03) = .12$.

If the given equation is differentiated implicitly with respect to t, we get

$$\frac{dz}{dt} = 0 - 60x\frac{dx}{dt} + 40y\frac{dy}{dt},$$

which can be evaluated as

$$\frac{dz}{dt} = -60(6)(.24) + 40(4)(.12)$$

$$= -67.2 \text{ units per year.}$$

The significance of the negative answer is that the demand is decreasing under the given pricing practices. ■

EXERCISES

Suggested minimum assignment: Exercises 3, 7, 11, 17, and 23.

In Exercises 1–10 find $\dfrac{dy}{dx}$ by implicit differentiation.

1. $x^2 + y^2 = 9$.
2. $x^2 - y^2 = 7$.
3. $xy = 5$.
4. $xy^2 + yx^2 = 2$.
5. $y^2 - 8x = xy$.
6. $y^2 - 8x = \frac{x}{y}$.
7. $y^5 + xy^3 + x^4 - 3 = 0$.
8. $2x^2 + 3y^2 = 12$.
9. $\sqrt{x} + \sqrt{y} = 4$.
10. $(x^2 + y^2)^5 - 32 = 0$.

In Exercises 11–14 find $\frac{dy}{dx}$ and $\frac{d^2y}{dx^2}$. Use implicit differentiation.

11. $x^2 + y^2 = 4$.
12. $x^2 + xy + y^2 - 3 = 0$.
13. $xy + y^2 = 12$.
14. $x^2 + 4y^2 = 5$.
15. If $y^2 - x^2 = 16$, find $D_x y$, $D_x^2 y$, and $D_x^3 y$.

In Exercises 16–18 find the slope of the tangent line at the given point.

16. $x^2y^2 = 12 - x$; $(3, 1)$.
17. $y + \sqrt{x^2 + y^2} = x + 6$; $(3, 4)$.
18. $x^{2/3} + y^{2/3} = 2$; $(1, 1)$.
19. Find $D_t s$ if $s^2 + t^2 + st = 10$.
20. Explain why $x^2 + y^2 = 16$ defines two differentiable functions f_1 and f_2 such that $y = f_1(x)$ and $y = f_2(x)$.
21. The location of a particle moving along a line is determined by the equation $st^2 - st^3 + 8s - 1 = 0$. Find s when $t = 1$. Then use implicit differentiation to find the velocity of the particle when $t = 1$.
22. If $x^3 + y^3 - 3xy = 13$, verify that the rate of change of y with respect to x at $(-1, 2)$ is the reciprocal of the rate of change of x with respect to y.
23. If $2\pi r^2 + 2\pi rh = 15$, find $\frac{dh}{dr}$ by implicit differentiation. Also find $\frac{dr}{dh}$ by implicit differentiation. Verify that $\frac{dr}{dh}$ is the reciprocal of $\frac{dh}{dr}$.

NEW VOCABULARY

average rate of change 2.1
secant line 2.1
instantaneous rate of change 2.1
tangent line 2.1
rate of change 2.2
derivative 2.2
derived function 2.2
differentiable 2.2
marginal revenue 2.2
marginal cost 2.2
marginal profit 2.2
equation of motion 2.2
average velocity 2.2
instantaneous velocity 2.2
velocity 2.2
differentiation 2.3
power rule for differentiation 2.3
product rule 2.4
quotient rule 2.4
chain rule 2.4

composite function 2.4
first derivative 2.5
second derivative 2.5
higher derivatives 2.5
acceleration 2.5
differential dx 2.6
differential dy 2.6
explicit functional relationship 2.7
implicit functional relationship 2.7
ellipse 2.7
implicit differentiation 2.7

3 Maximum and Minimum Problems

In many applied problems there are functions that must be maximized or minimized. Such problems are called optimization problems and are approached in a variety of ways, depending upon the nature of the problem. Some optimization problems, such as those illustrated in this chapter, are particularly vulnerable to the techniques of differentiation. Other optimization problems, such as linear programming problems, are usually solved by an algebraic algorithm. Still others, which will be studied in Chapter 7, will require further knowledge of the calculus.

Historically, optimization problems have provided considerable incentive for the discovery of new mathematics. There is evidence that questions pertaining to (1) the maximum height of a projectile in free flight near the earth, and (2) the maximum distance from one planet to another or to the sun provided some motivation for the discovery of calculus. Today optimization problems are pertinent to many disciplines other than the physical sciences, and the need for solving these problems contributes significantly to the interest of these disciplines in the techniques of calculus.

Prerequisites: Chapter 1; 2.1–2.5. Section 3.4 is not a prerequisite to 3.5 and 3.6. Sections 3.4–3.6 are not needed in optional section 3.7. The only prerequisite to the first part of 3.6 (symmetry and asymptotes) is Chapter 1.
Suggested Sections for Moderate Emphasis: 3.1–3.4 and the first part of 3.6
Suggested Sections for Minimum Emphasis: 3.1–3.3 and the first part of 3.6

3.1 Introduction

Problems in which one tries to maximize profits or to minimize expenses are illustrations of optimization problems. According to Webster's definition, an optimum is a most favorable point, degree, or amount. In this chapter we shall discuss various uses of the derivative in solving some optimization problems.

Example 1
As an illustration we introduce a problem that will be solved in Section 3.4. A store manager sells an item, say a certain style of small street lamp,

for x dollars, where $5 \leq x \leq 15$. Each lamp costs him \$5, and hence his profit per lamp is $(x - 5)$ dollars. The number of lamps sold depends upon the selling price x. By keeping records, the manager has observed that $(150 - 10x)$ is a good estimate of the number of lamps per month which are sold, and hence the monthly profit is given by

$$P(x) = \text{(profit per lamp)(number sold per month)}$$

or

$$P(x) = (x - 5)(150 - 10x).$$

If the selling price is \$8, then the profit is \$3 per lamp on sales of $150 - 10(8) = 70$ lamps, for a profit of \$210. If the selling price is \$11.50, the profit per lamp is \$6.50, but the number sold is only $150 - (10)(11.5) = 35$ lamps. However, the profit is

$$P(11.50) = (6.50)(35) = \$227.50,$$

which is better than when $x = 8$. The problem is to find, if possible, the value of x in the closed interval [5, 15] that will maximize the profit. ▮

In this section we shall present some background material that is needed in order to solve maximum and minimum problems similar to the one just discussed. In Section 1.2 we studied closed intervals such as [4, 7], open intervals such as (4, 7), and half-open intervals such as (4, 7] and [4, 7); these intervals are called **finite intervals** because their length (7 − 4) is a finite number. On the other hand, the set of numbers

$$\{x : x \geq 5\}$$

is an example of an **infinite interval,** and will be denoted by

$$[5, \infty).$$

The plus infinity symbol ∞ does not represent a specific real number, but the $[5, \infty)$ notation is used to indicate that the set includes *all* real numbers that are 5 or greater. See Figure 3.1. Examples of infinite open intervals are

$$(-\infty, 2),$$

which designates the set

$$\{x : x < 2\},$$

and

$$(-3, \infty),$$

which designates the set

$$\{x : x > -3\}.$$

The set of all real numbers can be denoted by $(-\infty, \infty)$.

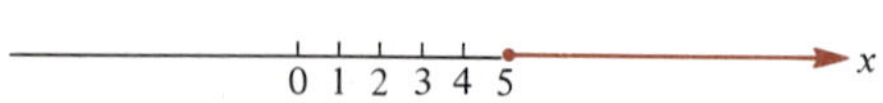

Figure 3.1. *Infinite interval* $[5, \infty)$.

In Example 2 we introduce the idea of calculating the sign of an expression on various intervals. This process will be repeated frequently in this chapter.

Example 2

Suppose that $y = f(x) = x^2 - 2x - 8$. For what values of x is $f(x)$ positive, and for what values of x is $f(x)$ negative?

First write $f(x)$ in the factored form

$$f(x) = (x + 2)(x - 4).$$

Since $f(x)$ is the product of the factors $(x + 2)$ and $(x - 4)$, we can determine the sign of $f(x)$ for a given value of x by finding the sign (plus or minus) of each factor. The factor $(x + 2)$ is negative $(-)$ if $x < -2$ and positive $(+)$ if $x > -2$, and we say the factor $(x + 2)$ **changes sign** at $x = -2$. The factor $(x - 4)$ changes sign at $x = 4$ because it is negative if $x < 4$, but it has the opposite sign (positive) if $x > 4$. We use the points -2 and 4 to divide the real line into the intervals $(-\infty, -2)$, $(-2, 4)$, and $(4, \infty)$. (We shall always list the intervals from left to right along the real line.) Instead of considering a large number of particular values of x, we need to deal with only the three cases represented by the three intervals above—a factor can change sign at an endpoint (-2 or 4) but not in the interior of any of these intervals. Our method of solving the problem is to construct Table 3.1, from which it can be seen how $f(x)$ changes sign as the individual factors change sign. ▮

Table 3.1

Interval	Calculation of sign of $f(x) = (x + 2)(x - 4)$	Result	Conclusion
$(-\infty, -2)$	$(-)(-)$	$+$	$f(x)$ positive
$(-2, 4)$	$(+)(-)$	$-$	$f(x)$ negative
$(4, \infty)$	$(+)(+)$	$+$	$f(x)$ positive

In Example 2 the polynomial function f is continuous for all real numbers, and the points at which a factor changes sign are the roots of the equation $f(x) = 0$. In general, a continuous function cannot change sign (either from $+$ to $-$ or from $-$ to $+$) without taking on the value zero; however, a continuous function does *not* always change sign at a point where its value is zero. Figures 3.2 and 3.3 are presented to clarify these facts (see also Exercise 12 of Section 1.9). For a function f that is continuous for all real numbers, we can conclude that the only possible points at which a sign change *may* occur are obtained by solving the equation $f(x) = 0$.

Next we learn how to find the sign of $f(x)$ on various intervals, where f is a rational function (remember that a polynomial function is a special type of rational function). *Unless otherwise stated, we shall consider only the rational functions in this chapter.* Consider the function f defined by $f(x) = (x - 1)/(x - 2)$. The only solution of $f(x) = 0$ is $x = 1$, and it can

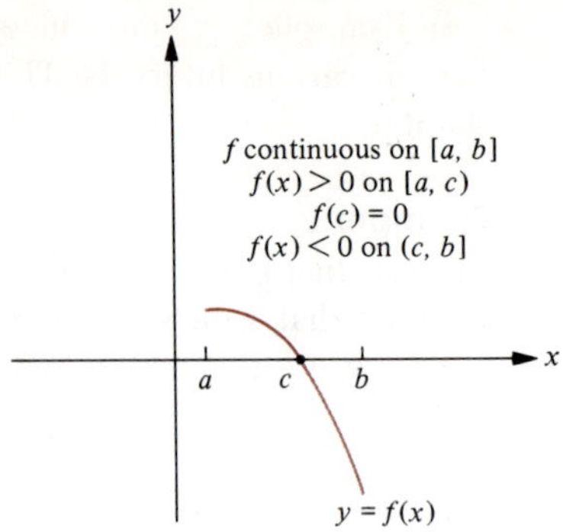

Figure 3.2. *$f(x)$ changes sign from + to − at $x = c$.*

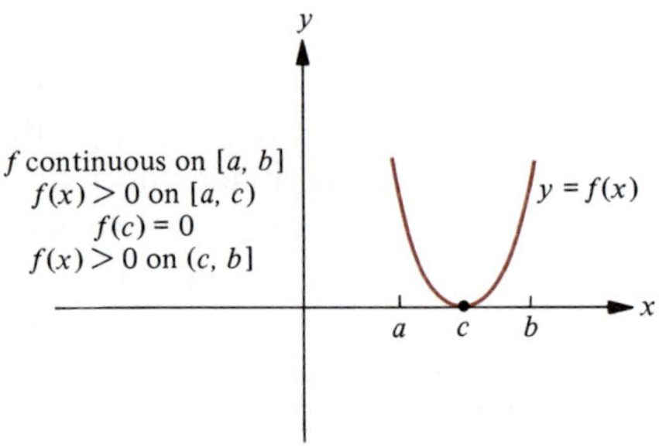

Figure 3.3. *$f(x)$ does not change sign at $x = c$.*

be shown that $f(x)$ changes sign at $x = 1$ [note that $f(.9) > 0$ and that $f(1.1) < 0$]. However, $f(x)$ also changes sign at $x = 2$ [note that $f(1.9) < 0$ and that $f(2.1) > 0$], despite the fact that 2 is not in the domain of f. In general, a rational function f *may* change sign at any point for which the numerator is zero [which is a solution of $f(x) = 0$], and also it *may* change sign at any point for which the denominator is zero (which is a point of discontinuity). Points at which either the numerator or denominator is zero are called **dividing points** of f because in making a table like Table 3.1 the intervals are determined by "cutting" the real line at the dividing points.

Given a rational function f in this chapter, we often shall need to make tables like Table 3.1 in order to determine the sign of either $f'(x)$ or $f''(x)$ on various intervals. The functions f' and f'' will also be rational functions [$f'(x)$ and $f''(x)$ can be obtained by the quotient rule], and the dividing points for f' and f'' are found again by finding points at which either the numerator or denominator is zero. When we are given a rational function f, the points of discontinuity will be the same for f, f', and f''; therefore, we have a function that is continuous and whose derivatives f' and f'' exist and are continuous except at points for which the denominator of $f(x)$ is zero.

Many functions met in applications are of the type discussed above, but often the domain is restricted for practical reasons to some finite or infinite interval (such as $[0, 5]$ or $(0, \infty)$). This must be remembered in making a table like Table 3.1. For example, suppose that $f(x) = (x - 1)/(x - 2)$ for $0 \leq x \leq 5$, provided that $x \neq 2$, and we want to compute the sign of $f(x)$ on various intervals. The dividing points of f are 1 and 2. The intervals that are considered are the open intervals (0, 1), (1, 2), and (2, 5)—we always elect not to include the endpoints or dividing points in the intervals.

Example 3

Suppose that

$$f(x) = \frac{(x + 1)^2(3 - x)}{(2x - 9)(x^2 + 4)}.$$

Find the open intervals on which $f(x)$ is positive and the open intervals on which $f(x)$ is negative.

Any number that makes any factor equal to zero (in either the numerator or denominator) is a dividing point of f. The dividing points are -1, 3, and $\frac{9}{2}$; there is no dividing point for $(x^2 + 4)$, since this quantity is never zero. The problem is solved in Table 3.2. One way to get the signs in Table 3.2 is to work with the individual factors. For example, $2x - 9 < 0$ when $x < \frac{9}{2}$, and hence $(-)$ appears for this factor in all intervals for which $x < \frac{9}{2}$; if, however, $x > \frac{9}{2}$, then $2x - 9 > 0$, and therefore $(+)$ appears for $(2x - 9)$ in the interval $(\frac{9}{2}, \infty)$. Another method of getting the signs is to substitute for x any chosen number in a particular interval. If $x = 0$, $2x - 9 = -9 < 0$, and therefore $(-)$ appears for $(2x - 9)$ in the interval $(-1, 3)$ that includes $x = 0$. ∎

Table 3.2

Interval	Calculation of sign of $f(x)$	Result	Conclusion
$(-\infty, -1)$	$\frac{(-)^2(+)}{(-)(+)}$	$-$	$f(x)$ negative
$(-1, 3)$	$\frac{(+)^2(+)}{(-)(+)}$	$-$	$f(x)$ negative
$(3, \frac{9}{2})$	$\frac{(+)^2(-)}{(-)(+)}$	$+$	$f(x)$ positive
$(\frac{9}{2}, \infty)$	$\frac{(+)^2(-)}{(+)(+)}$	$-$	$f(x)$ negative

APPLICATIONS

Example 4

A company is taken over by a new president, who immediately embarks on a cost-reduction program. He finds that several items are produced at one plant in a complicated manufacturing operation. The plant has been making very little profit, so he orders that a study be made to determine which items are profitable and which ones are not. Item A is analyzed by a cost accountant and an engineer. Item A has been sold during the past 2 years for prices that have fluctuated frequently between \$5.95 and \$8.95. The cost accountant determines that the daily sales S depend primarily upon the price p, and that S can be estimated quite well by the formula

$$S = 46 + 9p - p^2, \qquad 5.95 \leq p \leq 8.95.$$

The cost accountant also determines that the company has been losing money on item A whenever it is sold for less than \$7.40. The engineer studies the efficiency of the production operation, and he decides that more than 60 of item A should be manufactured daily—otherwise item A should be discontinued so that the labor and machinery can be used for another item. Should the new president order that item A be discontinued?

In order that sales keep pace with production, it is necessary that

$$\begin{aligned} 46 + 9p - p^2 &> 60, \\ -14 + 9p - p^2 &> 0, \\ p^2 - 9p + 14 &< 0, \\ (p - 2)(p - 7) &< 0. \end{aligned}$$

In other words, the sign of $(p - 2)(p - 7)$ must be negative. By making a table similar to Tables 3.1 and 3.2 (see Exercise 9), it can be shown that the restriction $(p - 2)(p - 7) < 0$ requires the price p to be in the open interval $(2, 7)$. In order to produce and sell more than 60 of item A daily (to satisfy the engineer), the price p must be less than \$7; in order to make a profit (to satisfy the cost accountant), p must be at least \$7.40. Hence the new president discontinues production of item A. ▮

EXERCISES

Suggested minimum assignment: Exercises 1, 2(*a*) (*c*), 3(*b*), 4, 5(*a*) (*c*), and 7.

1. In Example 1 find the monthly profit if the selling price of each lamp is \$9.30.
2. Change each of the following intervals to set notation. Show the set on a drawing of the real line.
   (*a*) $(-\infty, 3)$. (*b*) $(0, \infty)$.
   (*c*) $[1, \infty)$. (*d*) $(-\infty, -8]$.
3. Change from set notation to interval notation.
   (*a*) $\{x : x \geq 6\}$. (*b*) $\{x : x > -1\}$.
   (*c*) $\{x : x < 20\}$. (*d*) $\{x : x \leq 15\}$.
4. Sketch the graph of $y = f(x) = x^2 - 2x - 8$ (see Example 2). Explain how your graph agrees with the conclusion column of Table 3.1.

In Exercises 5 and 6 find the open intervals on which $f(x)$ is positive and the open intervals on which $f(x)$ is negative. Make a table similar to Tables 3.1 and 3.2.

5. (*a*) $f(x) = (x - 3)(x - 5)$, provided that $x \geq 0$.

(*b*) $f(x) = (x - 1)(x - 2)(x - 3)$, provided that $0 \leq x \leq 5$.

(*c*) $f(x) = \dfrac{(x - 4)(1 - x)}{x + 8}$.

6. (*a*) $f(x) = x^2 - 7x - 8$. (*b*) $f(x) = \dfrac{(x - 5)(6 - x)}{x - 7}$.

(*c*) $f(x) = \dfrac{(x - 1)^2(x - 2)}{2x + 1}$.

7. Suppose that $f(x) = \dfrac{2x}{x - 1}$.

(*a*) Find the dividing points of f.
(*b*) Find $f'(x)$ and then find the dividing points of f'.
(*c*) Find $f''(x)$ and then find the dividing points of f''.
(*d*) The functions f, f', and f'' all exist and are continuous, except for one value of x. What is that value?
(*e*) Does $f'(x)$ change sign at $x = 1$?

8. If $f(x) = \dfrac{(x - 1)(x - 2)^2}{(x - 3)^3}$, find the dividing points of f. Then find the points at which $f(x)$ changes sign.
9. Find the values of p for which $f(p) = (p - 2)(p - 7)$ is negative.

3.2 Increasing and Decreasing Functions

Let us assume that consumer demand for a certain item depends only upon the price of the item. If a store reduces the price of an item, the quantity of the item demanded by customers over a certain time period is ordinarily expected to increase. On the other hand, if the price is raised, the demand is ordinarily expected to decrease. In general, if two variables are related by $y = f(x)$ and if y decreases when x increases, we say that f is a **decreasing function.** If, however, y increases when x increases, we say that f is an **increasing function.** A study of increasing and decreasing functions in this section will lead directly to the solution of maximum and minimum problems.

Example 1

If the relationship between a store's selling price of an item and the number of the item demanded by consumers over a certain period of time is plotted, the result is known as a **demand curve.** It is customary in economics to plot price on the y-axis and demand on the x-axis. A typical demand curve $y = f(x)$ is shown in Figure 3.4, and f is a *decreasing* function throughout its domain. The point (1000, 35) represents the fact that 1000 of the item can be sold at \$35 each. The point (0, 50) represents the fact that no one

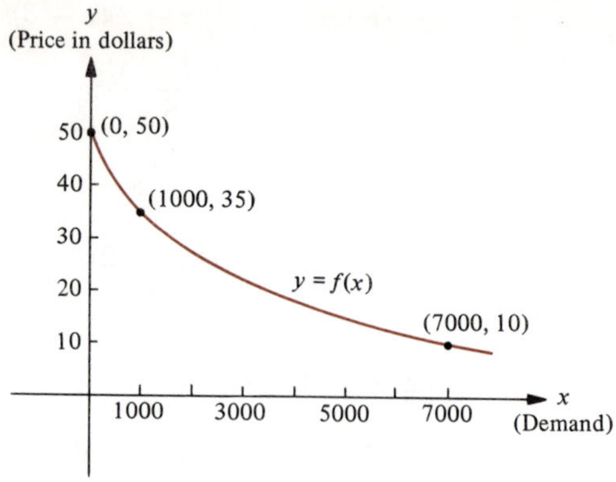

Figure 3.4. *Demand curve.*

will pay \$50 for the item. Since price and demand are nonnegative quantities, a demand curve is restricted to the first quadrant.

The quantity of an item that various producers will supply to a store over a certain period of time also depends upon the selling price of the item. A higher selling price is expected to spur greater production of the item because the producers can make more profit. In this situation it is also customary to plot price on the y-axis, with supply on the x-axis. The resulting graph is called a **supply curve** and is restricted to the first quadrant; ordinarily, the graph reveals an *increasing* function, such as the one shown in Figure 3.5. (Figures 3.4 and 3.5 are for *different* items.) For the item studied in Figure 3.5, a quantity of 400 will be supplied if the selling price is \$300 per item, but none will be supplied by any manufacturer if the selling price is \$100 or less. ∎

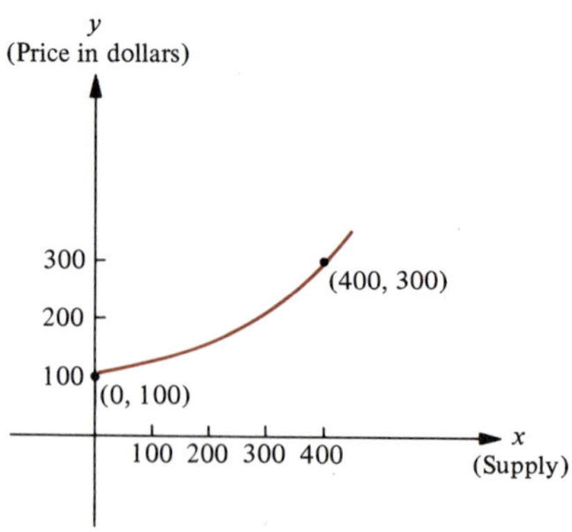

Figure 3.5. *Supply curve.*

We now define precisely what is meant by an increasing, or decreasing, function on an interval.

▶ **Definition 1**

Suppose that f is defined throughout an interval. Then

(*i*) f is said to be **increasing on the interval** if $f(x_1) < f(x_2)$ whenever $x_1 < x_2$, where x_1 and x_2 are any numbers in the interval.

(*ii*) f is said to be **decreasing on the interval** if $f(x_1) > f(x_2)$ whenever $x_1 < x_2$, where x_1 and x_2 are any numbers in the interval.

Although the interval referred to in Definition 1 may be open, closed, or half-open, and may or may not be an infinite interval, we shall be especially interested in the case when it is an open interval. Geometrically, Definition 1 says that a function is increasing on an interval if its graph is rising as x increases, whereas a function is decreasing on an interval if its graph is falling as x increases. In Figure 3.6 a function f is assumed to be defined on $[-2, \infty)$; f is increasing on the intervals $(-2, -1)$ and $(2, \infty)$, whereas f is decreasing on the interval $(-1, 2)$.

Theorem 1 is of practical importance in determining on what open intervals a function is increasing and on what open intervals it is decreasing.

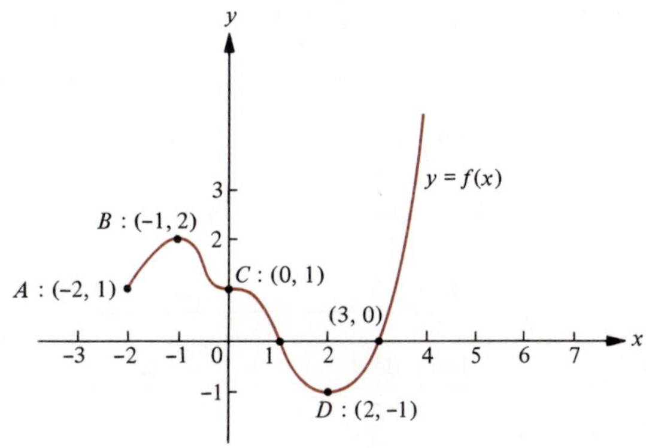

Figure 3.6

▶ **Theorem 1**

(*i*) If $f'(x) > 0$ at each point of (a, b), then f is increasing on (a, b).

(*ii*) If $f'(x) < 0$ at each point of (a, b), then f is decreasing on (a, b).

Theorem 1 seems reasonable from geometric considerations. If $f'(x) > 0$ at each point of (a, b), then the curve has a positive slope at each point of (a, b). See Figure 3.7. Hence it seems that the values of f should increase as x increases in (a, b). The argument above, especially when accompanied by a picture, may seem convincing; however, it does not constitute a proof. A proof will be given in optional Section 3.7. ◀

In Examples 2 and 3 we determine the sign of $f'(x)$ on various open intervals, using methods learned in Section 3.1; then we apply Theorem 1 to determine intervals on which f is increasing and intervals on which

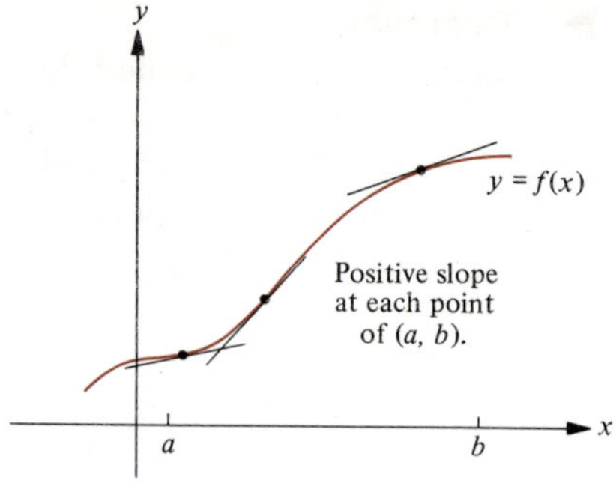

Figure 3.7

f is decreasing. (It is valid to apply Theorem 1 to infinite intervals as well as finite intervals.)

Example 2

Determine on which open intervals f is increasing and on which open intervals f is decreasing if $y = f(x) = \frac{2}{3}x^3 - \frac{1}{2}x^2 - 6x + 3$.

First find $f'(x)$ in factored form as follows:

$$f'(x) = 2x^2 - x - 6 = (2x + 3)(x - 2).$$

Hence $-\frac{3}{2}$ and 2 are the dividing points of f' at which a change in sign of $f'(x)$ may occur. The problem is solved in Table 3.3; notice that to use Theorem 1 we calculate the *sign of* $f'(x)$ rather than the sign of $f(x)$. ▮

Table 3.3

Interval	Calculation of sign of $f'(x)$	Result	Conclusion
$(-\infty, -\frac{3}{2})$	$(-)(-)$	$+$	f increasing
$(-\frac{3}{2}, 2)$	$(+)(-)$	$-$	f decreasing
$(2, \infty)$	$(+)(+)$	$+$	f increasing

Example 3

Repeat Example 2 for $y = f(x) = x^3$.

$$f'(x) = 3x^2 = (3x)(x).$$

The only dividing point of f' at which a change in sign of $f'(x)$ may occur is zero. The problem is solved in Table 3.4, and a graph of $y = x^3$ is shown in Figure 3.8. ▮

Table 3.4

Interval	Calculation of sign of $f'(x)$	Result	Conclusion
$(-\infty, 0)$	$(-)(-)$	$+$	f increasing
$(0, \infty)$	$(+)(+)$	$+$	f increasing

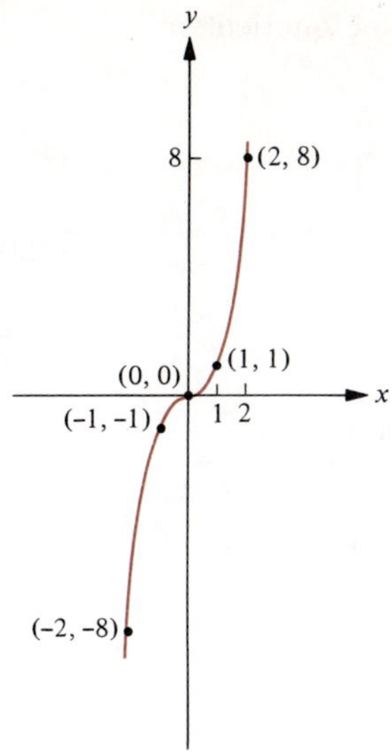

Figure 3.8. $y = f(x) = x^3$.

In Example 3 observe that the slope of the curve at the origin is not positive because

$$f'(0) = 3(0)^2 = 0.$$

Notice, however, that Definition 1(*i*) is satisfied for any open interval in the domain of f, including those that contain $x = 0$. These observations illustrate that the converse of the parts of Theorem 1 is not always true.

APPLICATIONS

Example 4

In Example 8 of Section 1.9 we showed that the function f, defined by

$$y = f(x) = \frac{x}{20x^2 + 50x + 80}, \qquad x \geq 0,$$

possessed several desirable properties of a function that would model the concentration y of a drug in the blood x hours after an injection of a known amount of the drug. In this example we shall first show, by use of the methods described in Section 3.1, that $f(x)$ is positive when x is positive (the function would be a poor model if it ever gave a negative concentration). Because the denominator is always positive (explained in Example 8 of Section 1.9), it is not difficult to construct Table 3.5.

It is quite important that the function f be an increasing function from the time of the injection ($x = 0$) until a later time when the effect of the drug begins to wear off, after which f should be a decreasing function. Let us apply the methods of this section to determine on which open intervals f is increasing and on which open intervals f is decreasing. By the quotient rule,

$$f'(x) = \frac{(20x^2 + 50x + 80)(1) - x(40x + 50)}{(20x^2 + 50x + 80)^2}$$

$$= \frac{-20x^2 + 80}{(20x^2 + 50x + 80)^2}$$

$$= \frac{(-20)(x + 2)(x - 2)}{(20x^2 + 50x + 80)^2}.$$

Table 3.5

Interval	Calculation of sign of $f(x)$	Result	Conclusion
$(0, \infty)$	$\frac{+}{+}$	$+$	$f(x)$ positive

As we stated before, $20x^2 + 50x + 80$ is always positive, so the only values of x at which a change in sign of $f'(x)$ may occur are $x = -2$ and $x = 2$. Since the domain of f is $\{x : x \geq 0\}$, we choose 2 as the only dividing point of f' (rather than -2 and 2). See Table 3.6. According to the table the given model indicates that the concentration of the drug in the blood increases for the first 2 hours and decreases after that. This conclusion agrees, of course, with our previous calculations of the concentration after 0, 1, 2, and 4 hours. ∎

Table 3.6

Interval	Calculation of sign of $f'(x)$	Result	Conclusion
$(0, 2)$	$\frac{(-)(+)(-)}{(+)^2}$	$+$	f increasing
$(2, \infty)$	$\frac{(-)(+)(+)}{(+)^2}$	$-$	f decreasing

EXERCISES

Suggested minimum assignment: Exercises 1, 3, 13, 16, and 20.

In each of Exercises 1–17 determine on which open intervals f is increasing and on which open intervals f is decreasing; in each case make a table similar to Tables 3.3 and 3.4.

1. $y = f(x) = x^2 - 8x,\ x > 0.$
2. $y = f(x) = \frac{1}{2}x^2 + 6x - 10.$
3. $y = f(x) = \frac{1}{3}x^3 - \frac{9}{2}x^2 + 20x - 4.$
4. $y = f(x) = 2x^3 + 1.$
5. $y = f(x) = 1 - 2x^3.$
6. $y = f(x) = x^4 - 2x^2 - 3.$

7. $y = f(x) = \dfrac{x^4}{2} - \dfrac{2x^3}{3} - 6x^2.$
8. $y = f(x) = x^4,\ x \geq -2.$
9. $y = f(x) = x^5 + 2.$
10. $y = f(x) = 7x + 3,\ 0 \leq x \leq 10.$
11. $y = f(x) = \frac{1}{3}x^3 + 2x^2 + x - 6.$ [*Hint:* $f'(x) = (x - r_1)(x - r_2)$, where r_1 and r_2 can be found by quadratic formula.]
12. $y = f(x) = x^3 - 9x + 5.$
13. $y = f(x) = \frac{1}{4}x^4 - \frac{8}{3}x^3 + 8x^2 - 10.$
14. $y = f(x) = \dfrac{3}{x}.$
15. $y = f(x) = \dfrac{1}{x^2}.$
16. $y = f(x) = \dfrac{x}{x^2 + 1}.$
17. $y = f(x) = (x - 1)^3(x + 1)^2.$
18. Find an example of a function f such that f is decreasing on any open interval and such that $f'(0) = 0$.
19. The equation of a demand curve is $y = 100/(x + 1), 0 \leq x \leq 20$. Sketch this curve and show that y' is negative on (0, 20).
20. The equation of a supply curve is

$$y = 50 + \frac{x^2}{100}, \qquad 0 \leq x \leq 100.$$

Sketch this curve, and show that y' is positive on (0, 100).

3.3 First Derivative Test for Maximum and Minimum Points

We now have the background to study maximum and minimum problems. Ordinarily, the graph of a function defined by $y = f(x)$ rises or falls as x increases. *We seek to find those points whose y-coordinates are larger (or smaller) than the y-coordinates of any other nearby points.* In this section we shall learn techniques of finding these **maximum points** and **minimum points.** In other words, we shall find all **maxima** and **minima** (these terms are plural). Otherwise stated, we shall find all **extreme points** (either a maximum point or a minimum point is called an extreme point). Then, in Section 3.4, we shall use the techniques of this section to solve some applied problems.

As a preview of what is to come, consider the function f with domain $[a, b]$ that is shown in Figure 3.9. The points $(x_1, f(x_1))$ and $(b, f(b))$ are maximum points, since their y-coordinates are greater than the y-coordinates of all other points on the curve in the immediate vicinity. The maximum point $(b, f(b))$ is called an **endpoint maximum** because it occurs

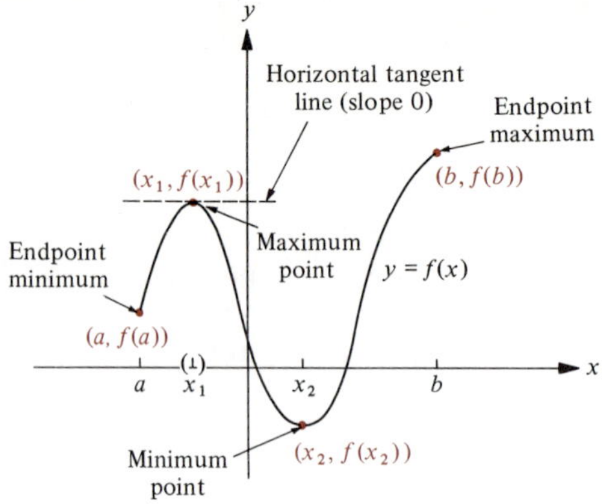

Figure 3.9

at an endpoint of the domain of f. Similarly, $(x_2, f(x_2))$ is a minimum point and $(a, f(a))$ is an **endpoint minimum.**

Because of the methods used to determine maximum and minimum points, these points will be studied in two groups; such points will occur at either

1. Interior points of the domain (such as x_1 and x_2 above), or
2. Endpoints of the domain (such as a and b above).

The domain of the rational functions that we consider will be a specified interval (with the omission of points at which the denominator is zero). If the interval does not include any endpoints, then obviously the function cannot have an endpoint maximum or an endpoint minimum. We shall focus our attention first at finding all maxima and minima that occur at interior points of the domain.

Definition 2

Suppose that x_1 is an interior point of the domain of a function f. The point $(x_1, f(x_1))$ is a **relative maximum point** if there exists an open interval containing x_1 such that $f(x_1) > f(x)$ for all $x \neq x_1$ in this open interval. The point $(x_1, f(x_1))$ is a **relative minimum point** if there exists an open interval containing x_1 such that $f(x_1) < f(x)$ for all $x \neq x_1$ in this open interval.

The word *relative* is included above to emphasize that the y-coordinate of a relative maximum point does *not* necessarily equal the greatest value of the function throughout its domain, but only that it equals the greatest function value in the immediate vicinity of the point. With this understanding *we shall omit the word relative in the future;* for instance, when we say maximum point, we shall mean relative maximum point.

As stated previously, the point $(x_1, f(x_1))$ in Figure 3.9 is a maximum point. On the x-axis in Figure 3.9 we have shown one possible open interval (containing x_1) of the type referred to in Definition 2. Any such open interval in the domain of f will suffice as long as $f(x_1)$ is the largest function value taken on throughout the interval. At the maximum point $(x_1, f(x_1))$

in Figure 3.9 the slope of the curve is zero because the tangent line at this point is horizontal—this fact is more than a mere coincidence, according to Theorem 2.

Theorem 2

Suppose that x_1 is an interior point of the domain of a rational function f, and suppose that $(x_1, f(x_1))$ is either a maximum or minimum point. Then $f'(x_1) = 0$.

Outline of Proof: Since x_1 is an interior point of the domain of the *rational* function f, $f'(x_1)$ exists. Suppose that $(x_1, f(x_1))$ is a maximum point. Because

$$f'(x_1) = \lim_{\Delta x \to 0} \frac{f(x_1 + \Delta x) - f(x_1)}{\Delta x},$$

it can be shown that $f'(x_1) \leq 0$ and $f'(x_1) \geq 0$ by consideration of the cases that $\Delta x > 0$ and $\Delta x < 0$, respectively. Therefore, $f'(x_1) = 0$. If $(x_1, f(x_1))$ is a minimum point, the proof is similar. ◀

From Theorem 2 we know that points at which the slope is zero are important; therefore, they are given a special name.

Definition 3

If $f'(x_1) = 0$, where x_1 is an interior point of the domain of f, then x_1 is called a **critical number** of the function f. The corresponding point $(x_1, f(x_1))$ is called a **critical point** of the function f.

Example 1

Refer to Figure 3.6. The critical numbers are -1, 0, and 2, because the slopes $f'(-1)$, $f'(0)$, and $f'(2)$ are zero. The corresponding points B, C, and D are critical points. Notice that B is a maximum point, D is a minimum point, and C is neither a maximum nor a minimum point. (Point A is an endpoint minimum.) ▮

Theorem 2 tells us that the critical numbers are the only numbers in the interior of the domain of a rational function that are *candidates* for consideration in looking for maximum or minimum points. At a critical number there may be a maximum point, there may be a minimum point, or there may be neither a maximum nor a minimum point (as we saw in Example 1). If the function increases just to the left of a critical point and decreases just to the right, then the critical point is a maximum point; if the function decreases just to the left of a critical point and increases just to the right, then the critical point is a minimum point. If the function increases (or if it decreases) on both sides of a critical point, then the critical point is neither a maximum nor a minimum point. These facts are stated formally in Theorem 3, which is called the **first derivative test** for maximum and minimum points.

Theorem 3

Let x_1 be a critical number of a rational function f. Let $x \neq x_1$ belong to an open interval (in the domain of f) that contains x_1 as its only critical number.

(*i*) $(x_1, f(x_1))$ is a maximum point if $f'(x) > 0$ for $x < x_1$ and $f'(x) < 0$ for $x > x_1$.

(*ii*) $(x_1, f(x_1))$ is a minimum point if $f'(x) < 0$ for $x < x_1$ and $f'(x) > 0$ for $x > x_1$.

(*iii*) $(x_1, f(x_1))$ is neither a maximum nor minimum point if $f'(x) > 0$ for both $x < x_1$ and $x > x_1$, or if $f'(x) < 0$ for both $x < x_1$ and $x > x_1$.

The proof of Theorem 3 is left as Exercise 18.

Figures 3.10, 3.11, and 3.12 illustrate Theorem 3 and should be studied to give the reader an intuitive understanding of the theorem.

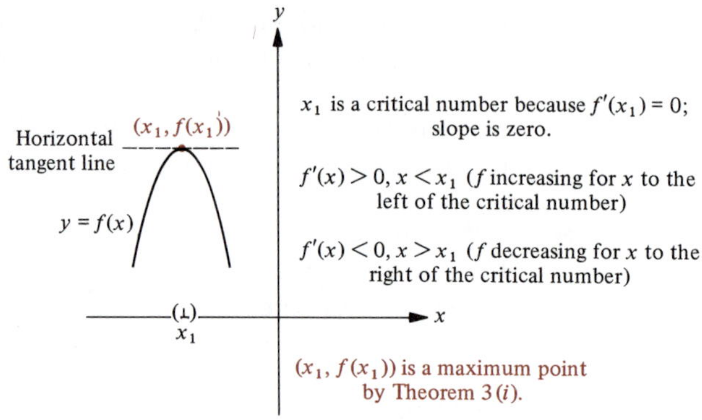

Figure 3.10. *Illustration of Theorem 3(i).*

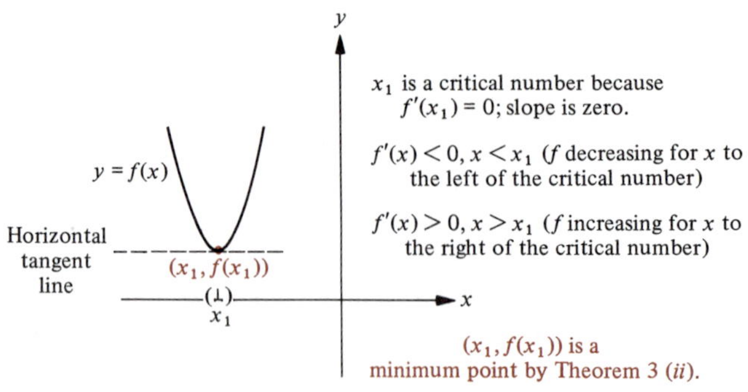

Figure 3.11. *Illustration of Theorem 3(ii).*

Example 2

Find all maximum and minimum points of f if

$$y = f(x) = \frac{x^4}{4} - \frac{4x^3}{3} + 2x^2 + 3.$$

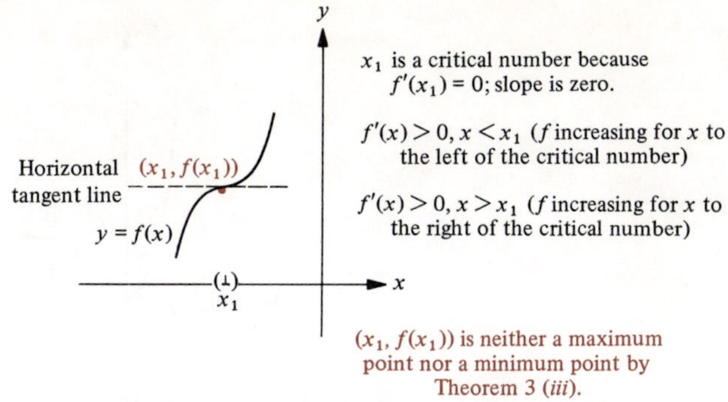

Figure 3.12. *Illustration of Theorem 3(iii).*

The domain of f is the set of all real numbers. We calculate $f'(x)$ as follows:

$$f'(x) = x^3 - 4x^2 + 4x = x(x^2 - 4x + 4) = x(x - 2)(x - 2).$$

The equation $f'(x) = x(x - 2)(x - 2) = 0$ yields the critical numbers $x = 0$ and $x = 2$, which are also the only dividing points of f'. The corresponding values of y are

$$f(0) = 3 \qquad \text{and} \qquad f(2) = \frac{16}{4} - \frac{4(8)}{3} + 2(4) + 3 = \frac{13}{3},$$

and therefore $(0, 3)$ and $(2, \frac{13}{3})$ are the only points that are candidates for maximum or minimum points. See Table 3.7. ∎

Table 3.7

Interval	Calculation of sign of $f'(x)$	Result	Conclusion
$(-\infty, 0)$	$(-)(-)(-)$	$-$; f decreasing	
			$(0, 3)$ is minimum point
$(0, 2)$	$(+)(-)(-)$	$+$; f increasing	
			$(2, \frac{13}{3})$ is neither maximum nor minimum point
$(2, \infty)$	$(+)(+)(+)$	$+$; f increasing	

Since we have the advantage of knowing that *all* maximum and minimum points have been located in Example 2, a reasonably accurate graph of $y = f(x)$ can be obtained by plotting only a very few points (see Figure 3.13).

We now turn our attention to finding endpoint extreme points. Suppose that the domain of a rational function f is an interval that includes the left endpoint—$[a, b]$, $[a, b)$, or $[a, \infty)$. Then $f(x) \to f(a)$ as $x \to a$ from within the interval, and f must be continuous on an interval of which a

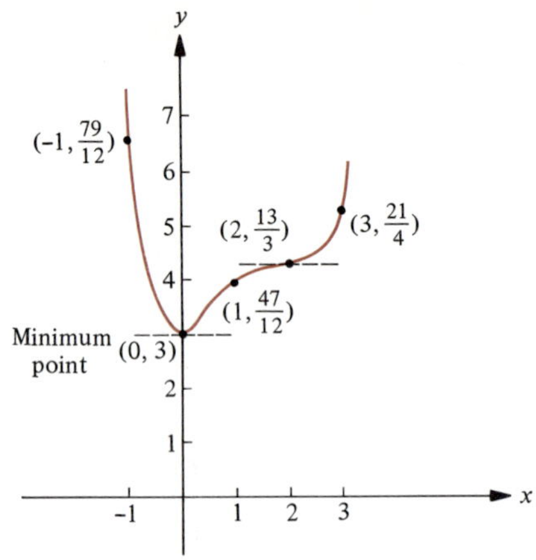

Figure 3.13. $y = \frac{x^4}{4} - \frac{4x^3}{3} + 2x^2 + 3.$

is the left endpoint (see Definition 11 of Section 1.9). If f is decreasing just to the right of a, then $(a, f(a))$ is an endpoint (relative) maximum because it is "higher" than any nearby point [(1) of Figure 3.14]. If f is increasing just to the right of a, then $(a, f(a))$ is an endpoint (relative) minimum [(2) of Figure 3.14]. Endpoint maximum or minimum points at the right endpoint of the domain of a rational function are detected in an analogous manner [(3) and (4) of Figure 3.14]. Any of these situations can be spotted in the process of making our standard table (like Table 3.7), and all endpoint maxima and minima can be included, as we demonstrate in Example 3.

Example 3

Find all maximum and minimum points of f if

$$y = f(x) = \frac{x^2 - 5}{x - 3}, \qquad 0 \le x \le 4.$$

The domain of the rational function f has been restricted to [0, 4], and $x = 3$ must be excluded from the domain so that the denominator is not zero. Notice that the left endpoint 0 and the right endpoint 4 are included in the domain. By the quotient rule,

$$\begin{aligned} f'(x) &= \frac{(x-3)(2x) - (x^2-5)(1)}{(x-3)^2} \\ &= \frac{x^2 - 6x + 5}{(x-3)^2} \\ &= \frac{(x-1)(x-5)}{(x-3)^2}. \end{aligned}$$

The number 1 is the only critical number because the only root of $f'(x) = 0$ in the domain is $x = 1$ (recall that 5 is not in the domain). The only interior

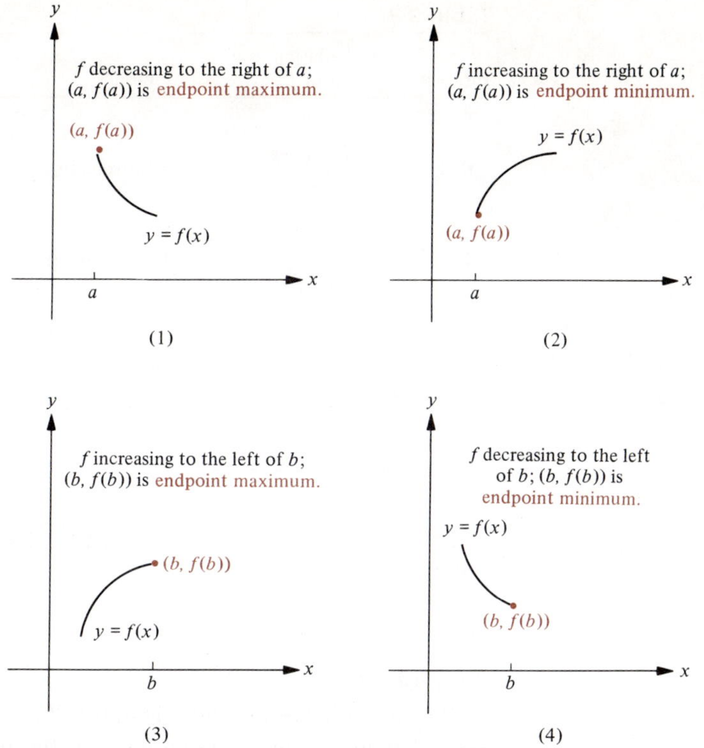

Figure 3.14. *Endpoint maxima and minima.*

point that is a candidate for being a maximum or minimum point is $(1, f(1)) = (1, 2)$. Also it will have to be determined whether $(0, \frac{5}{3})$ is an endpoint maximum or endpoint minimum, and the same thing will have to be determined for $(4, 11)$. The only dividing points of f' in $[0, 4]$ are $x = 1$ [where the numerator of $f'(x)$ is zero] and $x = 3$ [where the denominator of $f'(x)$ is zero]. The problem is solved in Table 3.8 (sketch postponed until Exercise 5 of Section 3.6). ∎

The following steps summarize the method that can always be used to find maximum and minimum points of a rational function f:

1. After noting the domain of f, calculate $f'(x)$.
2. Solve the equation $f'(x) = 0$ to find all critical numbers in the domain of f. Calculate $f(x_1)$ for each critical number x_1, and list all points $(x_1, f(x_1))$ that are candidates for being maximum or minimum points in the interior of the domain of f. Also list points, if any, that can be endpoint maxima or minima.
3. Note all dividing points of f'[critical numbers together with numbers for which the denominator of $f'(x)$ is zero] together with any finite endpoints, in order to determine the intervals to use in the first column of tables such as Tables 3.7 and 3.8. Complete such a table; use Theorem 3

Table 3.8

Interval	Calculation of sign of $f'(x)$	Result	Conclusion
			$(0, \frac{5}{3})$ is endpoint minimum
$(0, 1)$	$\frac{(-)(-)}{(-)^2}$	$+$; f increasing	
			$(1, 2)$ is maximum point
$(1, 3)$	$\frac{(+)(-)}{(-)^2}$	$-$; f decreasing	
			No maximum or minimum point at $x = 3$, since 3 not in domain
$(3, 4)$	$\frac{(+)(-)}{(+)^2}$	$-$; f decreasing	
			$(4, 11)$ is endpoint minimum

and the explanation in Figure 3.14 to detect all maximum and minimum points.

APPLICATIONS

Example 4

In this example we present one application of the material in this section; several more examples will appear in the next section.

The density of water at its freezing point is 1 gram per cubic centimeter, but the density varies slightly as the temperature increases. Assume that the temperature is measured on the Celsius scale. According to Thrall et al. [23], pp. 20–21, the specific weight (density) s of water at temperature t is given by

$$s = f(t) = 1 + (5.3)(10)^{-5}t - (6.53)(10)^{-6}t^2 + (1.4)(10)^{-8}t^3,$$
$$\text{provided that } 0 \leq t < 100.$$

Let us find all maximum and minimum points of f on $[0, 100)$.

$$\begin{aligned} f'(t) &= (5.3)(10)^{-5} - (2)(6.53)(10)^{-6}t + (3)(1.4)(10)^{-8}t^2 \\ &= (4.2)(10)^{-8}\left[t^2 - \frac{1306}{4.2}t + \frac{5300}{4.2}\right]. \end{aligned}$$

By the quadratic formula the roots of $f'(t) = 0$ are approximately $t = 4.1$ and $t = 306.8$, and therefore $f'(t)$ may be written

$$f'(t) \approx (.000000042)(t - 4.1)(t - 306.8).$$

The number 4.1 is the only critical number, because 306.8 is outside the domain. If $t = 4.1$, s can be calculated to be approximately 1.0001085, and hence $(4.1, 1.0001085)$ is a critical point. If $t = 0$, then $s = 1$ and the endpoint $(0, 1)$ must be considered. The only dividing point of f' in $[0, 100)$ is 4.1. See Table 3.9 for the answers to the problem.

Table 3.9

Interval	Calculation of sign of $f'(t)$	Result	Conclusion
			(0, 1) is endpoint minimum
(0, 4.1)	$(+)(-)(-)$	$+$; f increasing	
			(4.1, 1.0001085) is maximum point
(4.1, 100)	$(+)(+)(-)$	$-$; f decreasing	

The maximum point at the temperature of about 4.1°C is of interest. As water is warmed starting at $t = 0$, the breaking down of some hydrogen bonds enables water molecules to crowd more closely together, which makes the density tend to increase. However, the increased energy of the water molecules causes them to spread apart, which makes the density tend to decrease. These two phenomena have opposite effects. Up to $t = 4.1$ the first phenomenon is dominant and the density increases to a maximum; above $t = 4.1$ the second phenomenon is dominant and the density of water decreases. ▮

EXERCISES

Suggested minimum assignment: Exercises 1, 3, 5, 7, 11, and 13.

In Exercises 1–4 find all critical numbers of the given function.

1. $y = f(x) = x^3 - 12x$.
2. $y = f(x) = x^4 - 4x^3 + 1$.
3. $y = f(x) = \dfrac{x^2}{x - 1}$.
4. $y = f(x) = \dfrac{x^2}{x^2 - 9}$, $-2 < x < 5$.
5. Find all maximum and minimum points in Exercise 1. Sketch.
6. Find all maximum and minimum points in Exercise 2. Sketch.
7. Find all maximum and minimum points in Exercise 3.
8. Find all maximum and minimum points in Exercise 4.

In Exercises 9–14 find all maximum and minimum points of the given function. Make a table similar to Tables 3.7 and 3.8.

9. $y = f(x) = x^4$, $-1 \leq x \leq 1$.
10. $y = f(x) = 3 - x^4$, $x \geq -1$.
11. $y = f(x) = (x - 2)^5$, $0 \leq x \leq 3$.
12. $y = f(x) = (2 - x)^6$.
13. $y = P(x) = \dfrac{40x}{x^2 + 4}$, $0 \leq x \leq 4$, where P is a profit function.
14. $y = f(x) = (x + 2)^2(x - 3)^3$.

In Exercises 15–17 find all maximum and minimum points of the given function. Sketch. Make a table similar to Tables 3.7 and 3.8.

15. $y = f(x) = x^3 - \frac{3}{2}x^2 - 18x + 8$.

16. $y = f(x) = \dfrac{x^4}{4} - 3x^3 + 9x^2 - 1$.

17. $y = f(x) = x^4 - \frac{20}{3}x^3$.

18. Prove Theorem 3.

19. The graph of the constant function defined by $y = f(x) = 3$ is a horizontal line. Some authors consider every point on the line to be both a maximum point and a minimum point, whereas others say that the function has no maximum or minimum points. According to Definition 2, which viewpoint do we take in this text? What minor changes could be made in Definition 2 to force us to adopt the other viewpoint?

3.4 Applied Maximum and Minimum Problems

Many of the important applications of mathematics involve maximizing or minimizing some quantity. If a formula for such a quantity can be written as a rational function of just one variable, then the procedures of Section 3.3 can be used.

Example 1

Return to Example 1 of Section 3.1. A store manager wants to maximize his monthly profit that is given by

$$P(x) = (x - 5)(150 - 10x),$$

where the domain of P is the closed interval $[5, 15]$.

$$P(x) = -10x^2 + 200x - 750,$$
$$P'(x) = -20x + 200 = (-20)(x - 10).$$

Equating $P'(x)$ to zero yields the critical number $x = 10$. Then

$$P(10) = (10 - 5)(150 - (10)(10)) = 250,$$

and hence $(10, 250)$ is a candidate for being a maximum or minimum point. The points $(5, 0)$ and $(15, 0)$ may be endpoint maximum or minimum points. We see from Table 3.10 that maximum profit will be attained if the selling price of each lamp is \$10. This makes the monthly profit \$250, which is a larger profit than can be attained with any other selling price. Notice that there is zero profit when the selling price is either \$5 or \$15. ▮

The behavior of the profit function in Example 1 is typical, in the sense that a profit function usually has a maximum point. Up to a certain point, the greater the quantity produced of an item (during a certain period of time), the greater the profit. Beyond that point the extra quantity that is produced may not be sold or may be sold at a loss. As we have learned

Table 3.10

Interval	Calculation of sign of $P'(x)$	Result	Conclusion
			(5, 0) is endpoint minimum
(5, 10)	$(-)(-)$	$+$; P increasing	
			(10, 250) is maximum point
(10, 15)	$(-)(+)$	$-$; P decreasing	
			(15, 0) is endpoint minimum

previously, a profit function P for an item is often given by

$$P(x) = R(x) - C(x),$$

where R is the revenue function and C is the cost function. The maximum profit, if it exists, occurs at a critical number obtained by solving the equation $P'(x) = 0$. If $P'(x) = 0$, then

$$R'(x) - C'(x) = 0$$

and

$$R'(x) = C'(x).$$

Therefore, *maximum profit, if it exists, occurs only if marginal revenue equals marginal cost.* This result is an important theorem in economics.

Example 2

The equation of the demand curve for an item is

$$y = f(x) = 14 - \tfrac{1}{3}x^2.$$

Assume throughout this example that $0 \leq x \leq 5$. The revenue is the number of the item sold times the price received per item; that is,

$$\begin{aligned} R(x) &= xf(x) \\ &= x(14 - \tfrac{1}{3}x^2) \\ &= 14x - \tfrac{1}{3}x^3. \end{aligned}$$

Suppose that the cost function is given by

$$C(x) = 5x + 2.$$

We seek the value of x (critical number) at which maximum profit may occur, and therefore we solve the equation $R'(x) = C'(x)$ as follows:

$$\begin{aligned} 14 - x^2 &= 5, \\ x^2 &= 9, \\ x &= 3. \end{aligned}$$

(The solution $x = -3$ of $x^2 = 9$ is omitted because -3 is not a permissible value of x.) In Exercise 4 the reader is asked to verify that the profit function P really has a *maximum* point at $x = 3$. ∎

In Examples 3 and 4 the quantity to be maximized or minimized will be formulated first in terms of two variables, but an auxiliary equation will make possible the elimination of one variable.

Example 3

A land corporation is willing to spend \$1200 to fence in a rectangular portion of its property (see Figure 3.15). Its property already has a fence along one side; this existing fence will be used for one side of the fenced rectangle. On the sides of length x (as shown in Figure 3.15) the corporation wants wire fence that costs \$2 per foot to buy and install, and on the side of length y, a wood fence that costs \$3 per foot to buy and install is desired. Find x and y so that the enclosed area will be as large as possible.

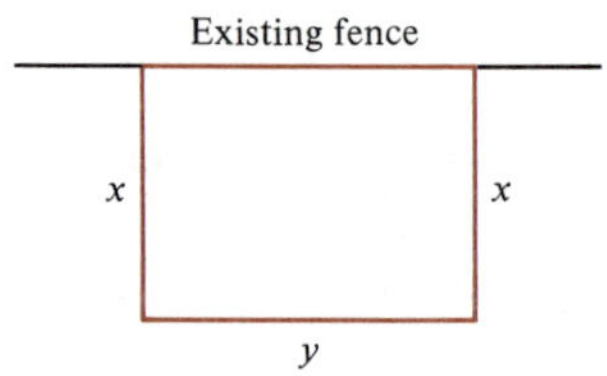

Figure 3.15

The quantity to be maximized is the area

$$A = xy,$$

where $x > 0$ and $y > 0$. An auxiliary equation can be written expressing the total cost of the fence, namely,

$$2x + 2x + 3y = 1200,$$

from which we obtain

$$y = \frac{1200 - 4x}{3}.$$

Therefore,

$$A = x\left(\frac{1200 - 4x}{3}\right) = 400x - \frac{4}{3}x^2.$$

Now A is expressed as a function of a single variable x, and the domain of the function is $(0, 300)$. (If $x \geq 300$, there is not enough money to purchase both types of fencing material.)

$$\frac{dA}{dx} = 400 - \frac{8}{3}x = \left(-\frac{8}{3}\right)(x - 150).$$

Setting dA/dx equal to zero to obtain the critical number, we find that

$$x = 150 \text{ feet.}$$

If $x = 150$, then

$$y = \frac{1200 - 4(150)}{3} = \frac{1200 - 600}{3} = 200 \text{ feet}$$

and

$$A = 150(200) = 30{,}000 \text{ square feet.}$$

The solution (see Table 3.11) is $x = 150$ feet and $y = 200$ feet, which yields a maximum possible enclosed area of 30,000 square feet under the restriction that only $1200 can be spent. ∎

Table 3.11

Interval	Calculation of sign of $\frac{dA}{dx}$	Result	Conclusion
(0, 150)	$(-)(-)$	$+$; A increasing	(150, 30,000) is maximum point
(150, 300)	$(-)(+)$	$-$; A decreasing	

Example 4

A certain company decides to manufacture aluminum cans (in the shape of a right circular cylinder). Each can is to have a volume of 16π cubic inches. Find the radius r and height h of each can so as to make the surface area (and thus the cost) as small as possible. See Figure 3.16.

The total surface area S that is to be minimized is the sum of the areas of the top, the bottom, and the side. Therefore,

$$S = \pi r^2 + \pi r^2 + 2\pi r h,$$

where $r > 0$ and $h > 0$. The quantity S to be minimized is expressed in terms of both r and h, but the formula $V = \pi r^2 h$ for the volume of a right circular cylinder will make possible the elimination of one of these variables.

$$V = 16\pi = \pi r^2 h$$

or

$$h = \frac{16}{r^2}.$$

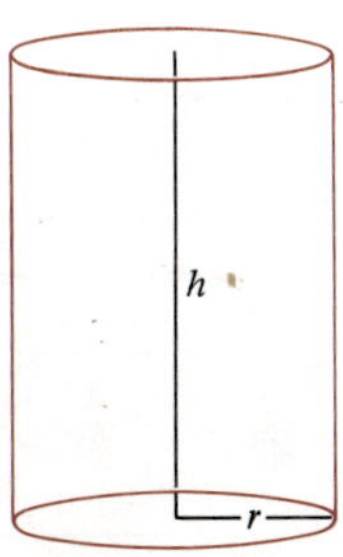

Figure 3.16

Therefore,

$$S = 2\pi r^2 + 2\pi r \frac{16}{r^2}$$

$$= 2\pi r^2 + \frac{32\pi}{r}$$

$$= \frac{2\pi r^3 + 32\pi}{r}.$$

Now S is expressed as a function of a single variable r; the domain of this function is $(0, \infty)$, since the radius of the can must be a positive number.

$$\frac{dS}{dr} = \frac{r(6\pi r^2) - (2\pi r^3 + 32\pi)(1)}{r^2}$$

$$= \frac{4\pi r^3 - 32\pi}{r^2}$$

$$= 4\pi \frac{r^3 - 8}{r^2}$$

$$= 4\pi \frac{(r - 2)(r^2 + 2r + 4)}{r^2}$$

$$= \frac{(4\pi)(r - 2)[(r + 1)^2 + 3]}{r^2}.$$

The only critical number is

$$r = 2 \text{ inches},$$

obtained by setting dS/dr equal to zero. If $r = 2$, then

$$h = \frac{16}{(2)^2} = 4 \text{ inches}$$

and

$$S = \frac{2\pi(2)^3 + 32\pi}{2} = 24\pi \text{ square inches.}$$

Next we construct Table 3.12. The solution to the problem is to make the radius of each can 2 inches and the height 4 inches, which gives each can the least possible surface area under the restriction that the volume be 16π cubic inches. ▮

Table 3.12

Interval	Calculation of sign of $\frac{dS}{dr}$	Result	Conclusion
$(0, 2)$	$\frac{(+)(-)(+)}{(+)}$	$-$; S decreasing	
			$(2, 24\pi)$ is minimum point
$(2, \infty)$	$\frac{(+)(+)(+)}{(+)}$	$+$; S increasing	

APPLICATIONS

Example 5

Suppose that a company sells a quantity q of a particular item each year. The company buys the item and stores it before selling it, even though money is spent to store the item. Assume that adequate storage space is available. It also costs money to reorder the item, but the item can be reordered frequently if necessary. Assume that sales are distributed *evenly* over the year and that all costs remain unchanged. Suppose that the company buys x items each time it reorders, and that this can be done in such a manner that a shipment arrives just as the previous inventory is depleted—thus since sales are evenly distributed, the average inventory is $x/2$. The problem is to find x so as to minimize the total annual cost. If x is too large, the storage cost (also known as carrying cost) is too great; but if x is too small, the reorder costs are too great.

Let p be the unit price of each item, r be the reorder cost associated with each shipment, and s be the storage cost of one item for a year. For 1 year the basic cost of the items is pq. The number of shipments each year is q/x, and hence the total annual reorder cost is rq/x. The annual storage cost is $(s)(x/2)$, since $x/2$ is the average inventory. The total cost is

$$C(x) = pq + \frac{rq}{x} + \frac{sx}{2},$$

where x must be positive and p, q, r, and s are positive constants.

$$\begin{aligned} C'(x) &= -\frac{rq}{x^2} + \frac{s}{2} \\ &= \frac{-2rq + sx^2}{2x^2} \\ &= \frac{s(x^2 - 2rq/s)}{2x^2} \\ &= \frac{(s)(x + \sqrt{2rq/s})(x - \sqrt{2rq/s})}{(2x)(x)}. \end{aligned}$$

If $C'(x) = 0$, then $x = \pm\sqrt{2rq/s}$. The number $x = \sqrt{2rq/s}$ is the only critical number (the negative of this number is not in the domain of C). If $x = 0$, the denominator of $C'(x)$ is zero, but $x = 0$ is not a dividing point of C' because it is not within the domain $(0, \infty)$ of C. The only candidate for a maximum or minimum point is $(\sqrt{2rq/s}, C(\sqrt{2rq/s}))$. See Table 3.13. From the table the total annual cost is minimized if the size of each shipment is $x = \sqrt{2rq/s}$; this formula is often called the **optimal lot-size formula,** and it does not depend upon the price p of each item. ∎

*Example 6 The Mode of a Measurement

When a measurement (whose domain is a sample space) is defined, it is useful to locate numbers that in some manner are representative of the

Table 3.13

Interval	Calculation of sign of $C'(x)$	Result	Conclusion
$\left(0, \sqrt{\frac{2rq}{s}}\right)$	$\frac{(+)(+)(-)}{(+)(+)}$	$-$; C decreasing	$\left(\sqrt{\frac{2rq}{s}}, C\left(\sqrt{\frac{2rq}{s}}\right)\right)$ is minimum point
$\left(\sqrt{\frac{2rq}{s}}, \infty\right)$	$\frac{(+)(+)(+)}{(+)(+)}$	$+$; C increasing	

values of the measurement. We shall define three such representatives, which are normally referred to as **measures of central tendency.** These measures, the mode, the mean (average), and the median, are often used as references to which the other values are compared. In this example we shall be concerned only with the mode of a measurement; the mean and the median will be defined later. For a *discrete* measurement the **mode** is defined to be the measurement value (or values) that has the largest associated weight. For example, consider a sample space U of 10 students who take a test. A measurement, with domain U, is defined by assigning to each student his test grade. Suppose that the grade of 75 is assigned to 5 students, 80 to 3 students, and 65 to 2 students. The weighting function w that has domain $\{65, 75, 80\}$ can be defined as follows: $w(65) = \frac{2}{10}$, $w(75) = \frac{5}{10}$, $w(80) = \frac{3}{10}$. Because $\frac{5}{10}$ is the largest weight, 75 is the mode.

For a *continuous* measurement the **mode** is defined to be the measurement value (or values) for which the associated probability density function is a maximum. As an illustration consider a measurement whose values measure the life span (in decades) of a certain type of machine. Suppose that the probability density function of this measurement is given by

$$p(x) = \begin{cases} \frac{3}{2}x - \frac{3}{4}x^2 & \text{if } 0 \leq x \leq 2, \\ 0 & \text{otherwise.} \end{cases}$$

By the methods of this chapter one can find that p has a maximum point at $x = 1$, and hence one decade is the mode life span; that is, a life span of one decade has a greater density of probability than that of any other unit of time in the immediate vicinity of $x = 1$. See Figure 3.17.

In both the discrete and continuous cases, there can be more than one mode. ▮

EXERCISES

Suggested minimum assignment: Exercises 1, 5, 9, 13, and 15.

1. The owner of an apple orchard estimates that if he begins picking apples today, he will have a crop of 1000 bushels and can make a profit

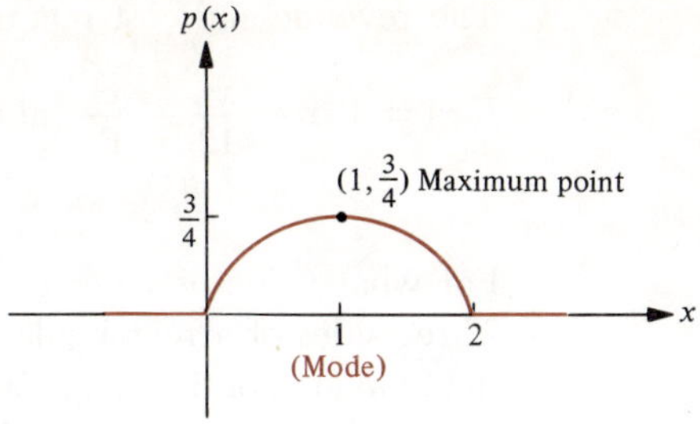

Figure 3.17. *Graph of probability density function p.*

of \$1 per bushel. He could wait as long as 25 days in order to begin picking; for each day he waits he estimates that he will have 50 extra bushels, but his profit per bushel will decrease by 2 cents for each day he waits. How many days should he wait to begin picking his crop so that he can earn a maximum profit? [*Hint:* Let x be the number of days that the owner waits; $P(x) =$ (profit per bushel)(number of bushels).]

2. Suppose that each apple tree in an orchard has an annual yield of 20 bushels, provided that there are no more than 30 trees per acre. For each additional tree per acre above 30, the annual yield decreases by $\frac{2}{5}$ bushel per tree. How many trees per acre will give the maximum annual yield?
3. An alumni group wants to charter a bus for a football game. The bus company offers a bus that seats 100 persons and insists that at least 40 persons make the trip. The company will charge \$15 per person if exactly 40 members of the alumni group make the trip, but the company will decrease the price by 10 cents per person for each person above 40 (for example, if 50 persons make the trip, the charge is \$14 per person). How many persons would have to make the trip in order to maximize the revenue received by the bus company?
4. In Example 2 $P(x) = R(x) - C(x) = (14x - \frac{1}{3}x^3) - (5x + 2)$, provided that $0 \le x \le 5$. Make a table (like Table 3.10) to find all maximum and minimum points of P. (Be sure that your maximum point agrees with Example 2.)
5. For a certain item the equation of the demand curve is $y = f(x) = 70 - \dfrac{3x}{2} - \dfrac{x^2}{15}$, and the cost function is given by $C(x) = 50x + 5$. Assume that $0 \le x \le 15$. The profit function is given by $P(x) = xf(x) - C(x)$. For what value of x is the profit largest?
6. Suppose that a profit function P is given by

$$y = P(x) = -x^2 + 8x - 5, \qquad 0 < x < 8.$$

Show that maximum profit is attained at $x = 4$ and find the maximum profit. (See Example 5 of Section 2.1.)

7. The revenue and cost functions for an item are given by

$$R(x) = 12x - \frac{x^2}{12} - \frac{x^3}{6} \quad \text{and} \quad C(x) = 2 + 4x - \frac{x^2}{12},$$

where $0 \leq x < 6$.

For what value of x is the profit greatest?

8. Three sides of a rectangular plot of ground are to be fenced so that the total area of the plot is 5000 square feet. Find the length and width of the plot so that the number of feet of fencing is minimized.
9. Rework Example 3 if the corporation is willing to spend \$1800.
10. The space within a 440-yard racetrack for track meets is to consist of a rectangle with a semicircle at each end. Find the length and width of the rectangle so that the area of the rectangle is maximized.
11. Rework Example 4 if the volume of each can is 54π cubic inches.
12. Rework Example 4 if the volume of each can is 16 cubic inches.
13. Find the radius and height of an open (no top) right circular cylinder with minimum surface area if the volume of the cylinder is to be 8π cubic feet. [*Hint:* Surface area $S = \pi r^2 + 2\pi rh$.]
14. (*a*) Find two positive numbers x and y, if possible, whose product is 100, such that the sum of the two numbers is minimized.
    (*b*) Find two positive numbers x and y, if possible, whose product is 100, such that the sum of the two numbers is maximized.
15. A rectangular-shaped advertisement sign with total area of 50 square feet is planned. The sign is to have margins of 6 inches at the top and at the bottom with margins of 1 foot at the left and right. Find the dimensions of the sign so that the area in the center for printed matter is maximum.
16. Suppose that the concentration y of a drug in the blood x hours after an injection is given by $y = \dfrac{x}{5x^2 + 80}$. How many hours after the injection does the concentration reach its maximum value?
17. Suppose that the reaction r of the body to a dose d of a drug can be represented by

$$r = f(d) = d^2\left(\frac{c}{2} - \frac{d}{3}\right),$$

where $c > 0$ is the greatest amount of the drug that may be given. The rate of change of r with respect to d is given by

$$f'(d) = cd - d^2.$$

Find the value of d for which $f'(d)$ is greatest. The answer will tell when the rate of change in r is greatest, and it is sometimes useful in determining the exact dose of a drug to be given. For more discussion, see pp. 216–218 of Thrall et al. [23].

18. An open box is manufactured from a piece of cardboard 24 inches × 24 inches by cutting equal squares out of the corners and turning up

the sides. How large a square should be cut out of each corner to obtain a box with maximum volume?

19. An average cost function (see Example 6 of Section 2.4) is given by

$$f(x) = \frac{C(x)}{x} = \frac{\frac{1}{10}x^2 + 7x + 10}{x}, \qquad 0 < x \leq 30.$$

For what value of x is the average cost minimized?

20. A company sells 3000 of an item each year, and the conditions of Example 5 are met. The reorder cost of each shipment is \$125, and the storage cost of one item for a year is \$3. How many of the item should the company order at one time? How often should it reorder this particular item?

21. (*a*) In Example 6 of Section 1.6 a discrete measurement and the weighting function were given for the experiment of flipping a coin until a head appears. What is the mode of the measurement?

(*b*) In Example 7 of Section 2.2 a continuous measurement and the probability density function were given for a machine problem. What is the mode of the measurement?

3.5 Concavity, Second Derivative Test

The population of a particular type of object over a certain time interval is graphed in Figure 3.18. From time 0 to time 2 the population is increasing, and the slope, which represents the rate of change of population, is also increasing. For $2 < t < 4$, however, the slope starts decreasing, even though the population continues to rise; in other words, the population continues to rise but at a decreasing rate. Eventually, at $t = 4$ the continuing decreasing slope (rate) causes the population itself to start decreasing. Then at $t = 5$ something else happens to cause the slope (rate) to start increasing again; this causes the population to decrease at a slower rate and eventually (at $t = 6$) even begin to increase. It is important to distinguish between the case of decreasing slope (rate) and the case of increasing

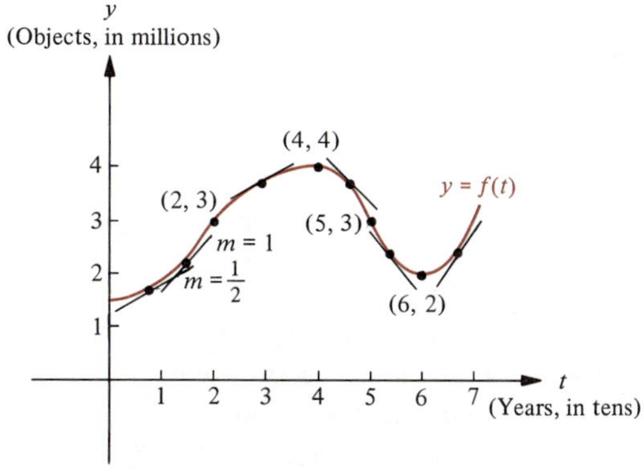

Figure 3.18

slope (rate). When the slope is increasing on the intervals (0, 2) and (5, 7), notice that a tangent line will be below the curve; it is customary to say that the curve is **concave upward** in this case. When the slope is decreasing on the interval (2, 5), notice that a tangent line will be above the curve; it is customary to say that the curve is **concave downward** in this case.

Definition 4

Suppose that a function f is differentiable at each point of an interval. Then

(*i*) f is said to be **concave upward on the interval** if the tangent line at each point of the interval is below the graph of f.

(*ii*) f is said to be **concave downward on the interval** if the tangent line at each point of the interval is above the graph of f.

The interval referred to in Definition 4 may be open, closed, or half-open and may or may not be an infinite interval; we shall be especially interested in the case in which it is an open interval.

Concave upward and concave downward are the two types of **concavity.** The following theorem, whose proof is omitted, gives a practical method of determining on which open intervals a function possesses a particular type of concavity.

Theorem 4

(*i*) If $f''(x) > 0$ at each point of an open interval, then f is concave upward on the interval.

(*ii*) If $f''(x) < 0$ at each point of an open interval, then f is concave downward on the interval. ◀

Example 1

Suppose that a function f is defined by

$$y = f(x) = \tfrac{1}{16}(-x^3 + 12x^2 - 32).$$

Use Theorem 4 to find open intervals on which f is concave upward and open intervals on which f is concave downward.

$$\begin{aligned} f'(x) &= \tfrac{1}{16}(-3x^2 + 24x), \\ f''(x) &= \tfrac{1}{16}(-6x + 24) \\ &= (-\tfrac{3}{8})(x - 4). \end{aligned}$$

According to Theorem 4 the problem is solved by calculating the sign of $f''(x)$ on various open intervals. The only dividing point of f'' is 4, at which a change in sign of $f''(x)$ may occur. The work is shown in Table 3.14—let it be emphasized that in determining the concavity, it is necessary to calculate the *sign of* $f''(x)$. The graph of f is shown in Figure 3.19—observe that the curve is drawn so that the concavity agrees with Table 3.14. The maximum and minimum points indicated in Figure 3.19 are found by the methods of Section 3.3. ▮

Table 3.14

Interval	Calculation of sign of $f''(x)$	Result	Conclusion
$(-\infty, 4)$	$(-)(-)$	$+$	f concave upward
$(4, \infty)$	$(-)(+)$	$-$	f concave downward

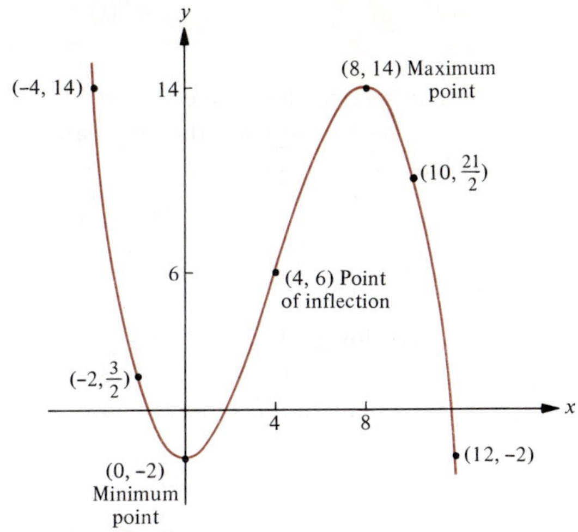

Figure 3.19. $y = \frac{1}{16}(-x^3 + 12x^2 - 32)$.

The function f in Example 1 is differentiable at $x = 4$, and the point $(4, 6)$ separates a portion of the curve that is concave upward from a portion that is concave downward—such a point is called a **point of inflection** of the function. If the tangent line to the curve at the point of inflection $(4, 6)$ were drawn, it would cross the curve rather than be either entirely above or entirely below the curve in the immediate vicinity of $(4, 6)$.

A point of inflection of a function must occur at an interior point of the domain of the function. It cannot occur at an endpoint because a function is defined on only one side (left or right) of an endpoint; therefore, an endpoint cannot separate a portion of the curve that is concave upward from a portion that is concave downward. In Example 1 there is a point of inflection at $x = 4$ and $f''(4) = 0$. Theorem 5 says that for a rational function the value of the second derivative at a point of inflection *must* be zero. The significance of this statement is that the only *candidates* for points of inflection of a rational function are those in the interior of the domain at which the value of the second derivative is zero.

▶ **Theorem 5**

If $(x_1, f(x_1))$ is a point of inflection of a rational function f, then $f''(x_1) = 0$.

A hint for the proof of Theorem 5 will be given in Exercise 18. ◀

Example 2
Find all points of inflection of the function defined by

$$y = f(x) = x^5 + \tfrac{5}{2}x^4 - 2.$$

We begin by finding $f''(x)$ in factored form.

$$\begin{aligned} f'(x) &= 5x^4 + 10x^3, \\ f''(x) &= 20x^3 + 30x^2 \\ &= (20x)(x)(x + \tfrac{3}{2}). \end{aligned}$$

The second derivative is zero if $x = 0$ or if $x = -\frac{3}{2}$. Therefore, $(0, f(0))$ and $(-\frac{3}{2}, f(-\frac{3}{2}))$ are the only candidates for points of inflection. We calculate

$$f(0) = 0 + 0 - 2 = -2$$

and

$$f(-\tfrac{3}{2}) = -\tfrac{243}{32} + \tfrac{5}{2}(\tfrac{81}{16}) - 2 = \tfrac{49}{16}.$$

According to Table 3.15, $(-\frac{3}{2}, \frac{49}{16})$ is a point of inflection because it separates portions of the curve with different concavity. The candidate $(0, -2)$ is not a point of inflection because the concavity does not change at this point. See Figure 3.20. ∎

Table 3.15

Interval	Calculation of sign of $f''(x)$	Result	Conclusion
$(-\infty, -\frac{3}{2})$	$(-)(-)(-)$	$-$; concave downward	
			$(-\frac{3}{2}, \frac{49}{16})$ is point of inflection
$(-\frac{3}{2}, 0)$	$(-)(-)(+)$	$+$; concave upward	
			$(0, -2)$ is *not* point of inflection
$(0, \infty)$	$(+)(+)(+)$	$+$; concave upward	

Let us return to the problem of determining maximum and minimum points of a rational function f. Suppose that x_1 is an interior point of the domain of f, and suppose that $f'(x_1) = 0$. The first derivative test, Theorem 3 of Section 3.3, is effective in deciding whether the critical point $(x_1, f(x_1))$ is a maximum point, a minimum point, or neither; Theorem 3 *always* yields a definite conclusion. There is another test that is often easier to apply that *usually* gives a definite conclusion about a critical point. This test is called the **second derivative test** and is stated in Theorem 6.

▶ Theorem 6

Suppose that $(x_1, f(x_1))$ is a critical point of a rational function f.
(*i*) $(x_1, f(x_1))$ is a minimum point if $f''(x_1) > 0$.
(*ii*) $(x_1, f(x_1))$ is a maximum point if $f''(x_1) < 0$.
[If $f''(x_1) = 0$, no conclusion can be made; use Theorem 3 in this case.]

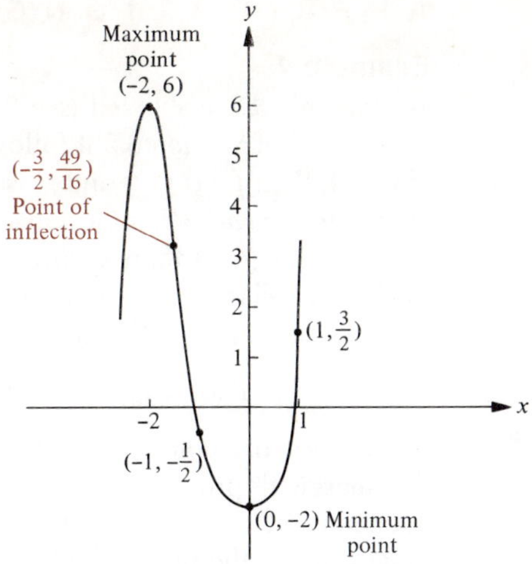

Figure 3.20. $y = x^5 + \frac{5}{2}x^4 - 2$.

Proof: We shall prove (*i*) and leave the similar proof of (*ii*) as Exercise 19. Since x_1 is an interior point of the domain of the *rational* function f, the function f'' is continuous at x_1. Because f'' is continuous at x_1 and $f''(x_1) > 0$, there is an open interval containing x_1 throughout which $f''(x) > 0$ by Exercise 14 of Section 1.9. This interval can be chosen so that x_1 is the only critical number in the interval. By Theorem 1, f' is increasing on this open interval. Since $f'(x_1) = 0$, $f'(x)$ changes from negative to zero to positive on this interval. By Theorem 3(*ii*), $(x_1, f(x_1))$ is a minimum point. ◀

Example 3

Find all maximum and minimum points of f if

$$y = f(x) = x^5 + \tfrac{5}{2}x^4 - 2.$$

This is the same function as given in Example 2.

$$f'(x) = 5x^4 + 10x^3 = 5x^3(x + 2).$$

It follows that $f'(-2) = 0$ and $f'(0) = 0$. The only candidates for maximum or minimum points are $(-2, 6)$ and $(0, -2)$. Since $f''(x) = 20x^3 + 30x^2$,

$$f''(-2) = 20(-8) + 30(4) = -40 < 0,$$

and hence $(-2, 6)$ is a maximum point by Theorem 6. Also

$$f''(0) = 20(0) + 30(0) = 0,$$

and in this case Theorem 6 does not apply. In Theorem 6, if $f''(x_1) = 0$, then no conclusion can be made—in other words, if $f''(x_1) = 0$, then $(x_1, f(x_1))$ may be either a maximum point, a minimum point, or neither. In this problem $(0, -2)$ can be shown to be a minimum point by applying Theorem 3. Our conclusions agree with Figure 3.20. ▮

APPLICATIONS

Example 4

In Example 1 we showed (see Table 3.14) that $f''(x) > 0$ if $x < 4$. From Theorem 1 of Section 3.2 it follows that f' is increasing if $x < 4$. However, if $x > 4$, then $f''(x) < 0$ and f' is decreasing. Hence, at the inflection point (4, 6), the *derivative* must attain a maximum value—in Figure 3.19 the slope is greater at (4, 6) than at any nearby point. At a point of inflection the slope is a (relative) maximum or minimum.

Suppose that a profit function P is given by

$$P(x) = \tfrac{1}{16}(-x^3 + 12x^2 - 32), \qquad 0 \leq x \leq 10.$$

(Except for the domain, P is the same function as in Example 1.) $P'(x)$ is the marginal profit. From our discussion above, the marginal profit is maximum at $x = 4$, since P changes from concave upward to concave downward at the point of inflection (4, 6). Therefore, the additional profit obtained by production of one more item is greatest at $x = 4$. ▮

Example 5

A book that contains a wealth of applications of calculus in economics is *Quantitative Economic Theory* by Brems [5]. For example, on pp. 37–41, Brems discusses the problem of finding the optimal useful life of consumer durables. If an item such as a television set, washing machine, or car is replaced frequently, repair costs are not very great but the cost of the new item is incurred too often; if one waits too long to replace the item, the repair costs become excessive. Brems makes certain reasonable assumptions, and then finds the useful life that will minimize cost; he uses the second derivative test (Theorem 6) to prove that the cost has been *minimized*. The optimal useful life can also be used to predict demand. A glance at Brems's book should convince the student of the importance of calculus in economics, although many of his examples depend upon material that will be presented in Chapters 6 and 7 of this text. ▮

EXERCISES

Suggested minimum assignment: Exercises 1, 5, 9, 11, 14(*b*), and 15.

In Exercises 1–4 use Theorem 4 to find open intervals on which f is concave upward and open intervals on which f is concave downward. Make a table like Table 3.14.

1. $y = f(x) = \dfrac{x^5}{20} - \dfrac{x^3}{6}$.
2. $y = f(x) = x^4 + 2x^3 - 12x^2 + x + 3$.
3. $y = f(x) = \dfrac{x}{x^2 + 3}$.
4. $y = f(x) = x^4 + 1$.

In Exercises 5–8 find all points of inflection. Make a table like Table 3.15.

5. $y = f(x) = x^3 + 6x^2 + 4x - 6$.
6. $y = f(x) = x^3$.
7. $y = f(x) = x^4 - 24x^2$.
8. $y = f(x) = \dfrac{x}{x^2 + 3}$. (See Exercise 3.)

9. Given $y = f(x) = \dfrac{1}{x}$. On what open intervals is f concave upward and on what open intervals is f concave downward? Why is there no point of inflection at $x = 0$?

In Exercises 10–12 find all maximum and minimum points. Use Theorem 6 if possible—otherwise use Theorem 3. Also sketch the graph of f.

10. $y = f(x) = \frac{2}{3}x^3 + \frac{3}{2}x^2 + x - 1$.
11. $y = f(x) = x^2 - x^3$.
12. $y = f(x) = 1 - (x - 1)^6$.

13. For each of the following profit functions, find the value of x for which the profit is maximum.
    (*a*) $P(x) = -x^3 + 3x^2 + 9x - 4,\ 0 \le x \le 4$.
    (*b*) $P(x) = -\frac{1}{3}x^3 + 2x^2 + 96x - 10,\ 0 \le x \le 16$.
14. Sketch a graph of each of the given functions after doing the following: find the open intervals where f is increasing and the open intervals where f is decreasing; find all maximum and minimum points; find the open intervals where f is concave upward and the open intervals where f is concave downward; and find all points of inflection.
    (*a*) $y = f(x) = \frac{1}{3}x^3 - x^2 - 8x + 2$.
    (*b*) $y = f(x) = x^4 - 4x^3$.
15. Let revenue and cost functions be given by

$$R(x) = -\tfrac{1}{6}x^3 + 4x^2 + 110x, \qquad 0 \le x \le 14,$$

and

$$C(x) = \tfrac{1}{6}x^3 + 2x^2 + 14x + 10, \qquad 0 \le x \le 14.$$

Find the value x_1 at which the profit is maximum. Show that $P''(x_1) < 0$. Also verify that marginal revenue equals marginal cost at the point of maximum profit.

16. Repeat Exercise 15 for

$$R(x) = 2x^2 - \tfrac{1}{32}x^3, \qquad 0 \le x \le 10,$$

and

$$C(x) = \tfrac{1}{32}x^3 + \tfrac{5}{4}x^2 + 2, \qquad 0 \le x \le 10.$$

Also explain the relationship of this exercise to Figure 3.19.

17. The rate y of a certain autocatalytic reaction is given by

$$y = kx(a - x), \qquad 0 \le x \le a,$$

where $a > 0$ and $k > 0$. Use Theorem 6 to find the value of x at which the rate y is greatest. (In an autocatalytic reaction an amount a of one substance is converted into a new substance, called the product, in such a way that the product catalyzes its own formation. The amount of the product is denoted by x; $x = 0$ when the reaction begins. The equation $y = kx(a - x)$ says that the rate of the reaction is proportional to the quantity $x(a - x)$, where $k > 0$ is the constant of proportionality. For more discussion, see pp. 33–34 of Thrall et al. [23].)

18. Prove Theorem 5. [*Hint:* Show that $f''(x_1) > 0$ and $f''(x_1) < 0$ are impossible—use Exercise 14 of Section 1.9, Theorem 4, and the definition of a point of inflection.]
19. Prove Theorem 6(*ii*).
20. At what point does the function f in Exercise 5 have minimum slope? What is the slope at this point?
21. For each of the profit functions in Exercise 13, find the value of x for which the marginal profit is maximum.

3.6 Curve Sketching

Pictures are often worth many words in mathematics as in other areas of life. In this section we shall recall several ideas from previous sections that are helpful in sketching the graph of an equation. We shall be aided by the new concepts of **symmetry** and **asymptotes.**

First we discuss symmetry—*the discussion of symmetry applies to equations in general and is not restricted to an equation of a rational function.* In Figure 3.21 the graph of the equation $x = 9 - y^2$ is shown. For any point (x, y) on the graph of the equation, there exists a point $(x, -y)$ directly across the x-axis which also satisfies the equation, and therefore we say that the graph is **symmetric to the x-axis.** To show that this is true, one can start with the original equation $x = 9 - y^2$ [we are assuming that (x, y) satisfies this equation when we write the equality] and determine whether $(x, -y)$ also satisfies the equation:

$$x = 9 - y^2 \qquad \text{(original equation).}$$
$$x = 9 - (-y)^2 \qquad \text{[replace } (x, y) \text{ with } (x, -y)\text{].}$$
$$x = 9 - y^2 \qquad \text{(simplify).}$$

Since the original equation is obtained again, we have shown that $(x, -y)$ satisfies the equation whenever (x, y) does, and therefore the symmetry to the x-axis has been proved. Whenever symmetry to the x-axis can be detected at the outset, it is an aid in sketching the curve. Incidentally, the graph of $x = ay^2 + by + c$, $a \neq 0$, is called a parabola and opens to the left if $a < 0$, and to the right if $a > 0$ (parabolas that open upward or downward were mentioned in Section 1.6).

In a similar way the graph of an equation can be shown to be **symmetric to the y-axis** if replacement of (x, y) by the point $(-x, y)$ yields the original equation again. For example, the graph of $y = x^4 - 2x^2 - 1$ is symmetric

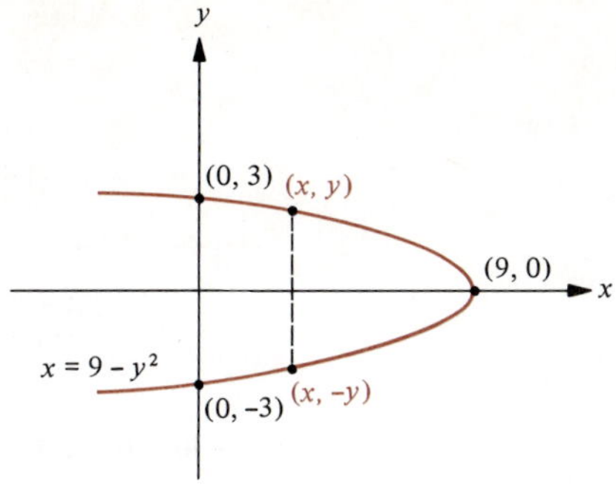

Figure 3.21. *Curve symmetric to x-axis.*

to the y-axis because $y = (-x)^4 - (2)(-x)^2 - 1$ reduces to $y = x^4 - 2x^2 - 1$. Observe in Figure 3.22 that the portion of the curve on one side of the y-axis is a "reflection" across the y-axis of the portion on the other side. The symmetry of a curve to any line can be defined by this type of reflection property.

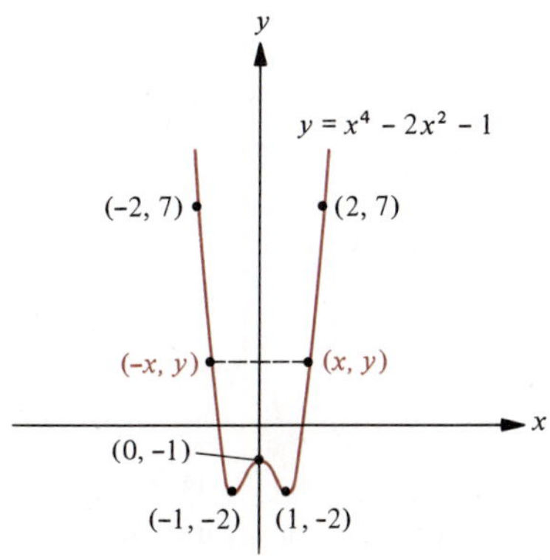

Figure 3.22. *Curve symmetric to y-axis.*

In addition to symmetry about a line such as the x-axis or y-axis, there is another type of symmetry that we shall discuss briefly. We are referring to symmetry to a point, and, in particular, symmetry to the origin. The graph of an equation is said to be **symmetric to the origin** if for any point (x, y) that satisfies the equation, the point $(-x, -y)$ also satisfies the equation. For example, the graph of $x = \frac{1}{2}y^3$ is symmetric to the origin because $-x = \frac{1}{2}(-y)^3$ can be simplified to yield the original equation. See Figure 3.23.

Example 1

The circle $x^2 + y^2 = 9$ is symmetric to the x-axis, the y-axis, and the origin. The parabola $y = x^2 + 4$ is symmetric to the y-axis but is not symmetric to the x-axis or to the origin. The graph of the rational function defined by $y = x/(x^2 + 1)$ is symmetric to the origin but is not symmetric to the

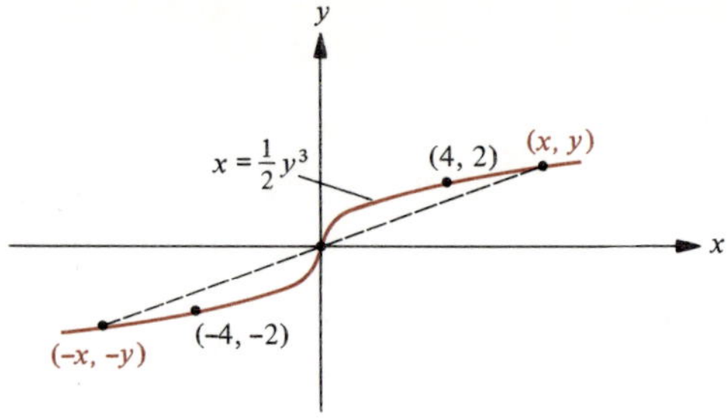

Figure 3.23. *Curve symmetric to origin.*

x-axis or y-axis. The line $y = 2x + 1$ is not symmetric to the x-axis, the y-axis, or the origin. ▮

For the remainder of this section we shall consider only equations that define rational functions—in other words, we shall discuss equations of the form

$$y = \frac{g(x)}{h(x)},$$

where g and h are polynomial functions. We shall introduce the ideas of **vertical asymptotes** and **horizontal asymptotes** in Example 2.

Example 2

Consider the function f defined by

$$y = f(x) = \frac{g(x)}{h(x)} = \frac{2x}{x - 3}.$$

Notice that $f(3)$ is undefined. However, for values of x that are close to 3, the corresponding values of y are large in absolute value as seen in Table 3.16. For values of x even closer to 3 than those shown in Table 3.16, the corresponding y values are still larger in absolute value, and the points on the curve are very close to the vertical line $x = 3$. In this case, the line $x = 3$ is called a **vertical asymptote** (see Figure 3.24). Vertical asymptotes can be identified according to the following rule: If $h(x_1) = 0$ and $g(x_1) \neq 0$, then the line $x = x_1$ is a vertical asymptote of $y = g(x)/h(x)$. The rule is a result of the fact that $|y| \to \infty$ as $x \to x_1$.

Next, calculate

$$\lim_{x\to\infty} \frac{2x}{x - 3} = \lim_{x\to\infty} \frac{2}{1 - (3/x)} = \frac{2}{1 - 0} = 2.$$

Table 3.16

x	2.9	2.99	2.999	3.1	3.01	3.001
y	−58	−598	−5998	62	602	6002

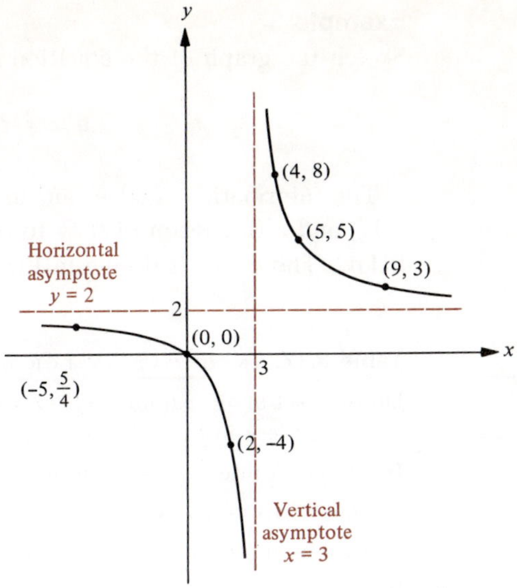

Figure 3.24. $y = \dfrac{2x}{x-3}$.

Hence $y \to 2$ as $x \to \infty$, and the horizontal line $y = 2$ is a **horizontal asymptote** in this example. For large values of x, the curve approaches the line $y = 2$ (see Figure 3.24). Similarly, it can be shown that

$$\lim_{x \to -\infty} \frac{2x}{x-3} = 2,$$

and therefore the curve also approaches the line $y = 2$ as $x \to -\infty$. If either $\lim_{x \to \infty} [g(x)/h(x)] = k$ or $\lim_{x \to -\infty} [g(x)/h(x)] = k$, where k is a real number, then the line $y = k$ is a horizontal asymptote of $y = g(x)/h(x)$. ▮

To draw a sufficiently accurate sketch of the graph of a rational function f, there are several things we have studied that one may attempt to find; in a given problem it may not be practical or necessary to find all of them.

1. Domain and range (Section 1.4).
2. Intercepts: x-intercepts and y-intercepts (Section 1.3).
3. Symmetry (Section 3.6).
4. Vertical and horizontal asymptotes (Section 3.6).
5. Intervals where f is increasing or decreasing (Section 3.2).
6. Maximum and minimum points (Sections 3.3 and 3.5).
7. Intervals where f is concave upward or concave downward (Section 3.5).
8. Points of inflection (Section 3.5).
9. Plot points other than those found in items 2, 6, and 8 (Section 1.3).
10. Slope at key points (Chapter 2).

Example 3
Sketch the graph of the equation

$$y = f(x) = \frac{1}{x^2 - 4}.$$

The information used to aid in drawing the sketch is shown in Table 3.17, and a discussion of how to obtain some of the information is given below. The curve is drawn in Figure 3.25.

Table 3.17. INFORMATION FOR $y = 1/(x^2 - 4)$.

Domain: set of all real numbers except -2 and 2. Range: set of all real numbers except those in the interval $(-\frac{1}{4}, 0]$.

Intercepts: x-intercepts—none; y-intercepts—$-\frac{1}{4}$.

Symmetric to y-axis. Not symmetric to x-axis or origin.

Vertical asymptotes: $x = -2$, $x = 2$. Horizontal asymptotes: $y = 0$.

Increasing: $x < -2$, $-2 < x < 0$. Decreasing: $0 < x < 2$, $x > 2$.

Maximum points: $(0, -\frac{1}{4})$. Minimum points: none.

Concave upward: $x < -2$, $x > 2$. Concave downward: $-2 < x < 2$.

Points of inflection: none.

Other points: $(-3, \frac{1}{5})$, $(-1, -\frac{1}{3})$, $(\frac{7}{4}, -\frac{16}{15})$, $(\frac{21}{10}, \frac{100}{41})$.

Slope at $(-3, \frac{1}{5})$: $\frac{6}{25}$.

The remainder of this example is devoted to a discussion of how some of the information in Table 3.17 is obtained. The denominator is zero at $x = -2$ and $x = 2$. Hence the domain consists of all real numbers except ± 2, and the only points of discontinuity are at $x = \pm 2$. Since the numerator is not zero at these points, the lines $x = \pm 2$ are vertical asymptotes. If the equation is solved for x, the result is

$$x = \pm\sqrt{\frac{1 + 4y}{y}}.$$

If $-\frac{1}{4} < y < 0$, we are led to the square root of a negative number, and therefore numbers in $(-\frac{1}{4}, 0)$ must be excluded from the range. Also, because division by zero is undefined, y cannot equal zero. Because

$$\lim_{x\to\infty} \frac{1}{x^2 - 4} = 0,$$

the line $y = 0$ is a horizontal asymptote. Note that $\lim_{x\to-\infty} [1/(x^2 - 4)] = 0$ also. After differentiating and simplifying, one can obtain

$$f'(x) = \frac{(-2)(x)}{(x^2 - 4)^2} = \frac{(-2)(x)}{(x + 2)^2(x - 2)^2}$$

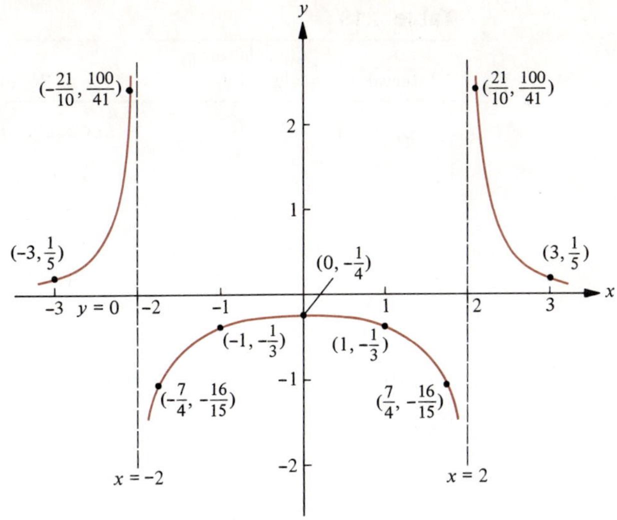

Figure 3.25. $y = \dfrac{1}{x^2 - 4}$.

and

$$f''(x) = \frac{6x^2 + 8}{(x^2 - 4)^3} = \frac{6x^2 + 8}{(x + 2)^3(x - 2)^3},$$

which lead to Tables 3.18 and 3.19. ▮

Table 3.18

Interval	Calculation of sign of $f'(x)$	Result	Conclusion
$(-\infty, -2)$	$\frac{(-)(-)}{(-)^2(-)^2}$	$+$; f increasing	
			No maximum or minimum point at $x = -2$, since -2 not in domain
$(-2, 0)$	$\frac{(-)(-)}{(+)^2(-)^2}$	$+$; f increasing	
			$(0, -\frac{1}{4})$ is maximum point
$(0, 2)$	$\frac{(-)(+)}{(+)^2(-)^2}$	$-$; f decreasing	
			No maximum or minimum point at $x = 2$, since 2 not in domain
$(2, \infty)$	$\frac{(-)(+)}{(+)^2(+)^2}$	$-$; f decreasing	

Table 3.19

Interval	Calculation of sign of $f''(x)$	Result	Conclusion
$(-\infty, -2)$	$\dfrac{+}{(-)^3(-)^3}$	$+$; concave upward	
			No point of inflection at $x = -2$, since -2 not in domain
$(-2, 2)$	$\dfrac{+}{(+)^3(-)^3}$	$-$; concave downward	
			No point of inflection at $x = 2$, since 2 not in domain
$(2, \infty)$	$\dfrac{+}{(+)^3(+)^3}$	$+$; concave upward	

APPLICATIONS

Example 4

In Example 6 of Section 2.4 we defined the average cost function f for an item by

$$f(x) = \frac{C(x)}{x},$$

where C is the cost function for the item. If C is a polynomial function, then f is a rational function. If the fixed cost $C(0) \neq 0$, the line $x = 0$ is a vertical asymptote of f.

Suppose, as an example, that the cost function for an item is given by

$$C(x) = x^2 + 10x + 36, \qquad 0 \leq x \leq 12.$$

The equation of the average cost function is

$$y = f(x) = \frac{C(x)}{x} = \frac{x^2 + 10x + 36}{x}, \qquad 0 < x \leq 12.$$

Let us sketch the graph of f after some preliminary analysis. The function f has $x = 0$ as a vertical asymptote, which is typical for an average cost function, as we suggested above. There are no horizontal asymptotes, since the domain is $(0, 12]$; it must be possible to let $x \to \infty$ or $x \to -\infty$ before there is a chance of a horizontal asymptote. It can be shown by the quotient rule that

$$f'(x) = \frac{x^2 - 36}{x^2} = \frac{(x + 6)(x - 6)}{x^2}.$$

The only critical number in the domain of f is $x = 6$. It is easily seen that f is decreasing for $0 < x < 6$ and increasing for $6 < x < 12$, which means that by Theorem 3 there is a minimum point at $x = 6$. The fact that the average cost is minimized at $x = 6$ is probably the most significant feature in this example. The quotient rule can be applied again to yield

$$f''(x) = \frac{72}{x^3}.$$

Since $f''(x) > 0$ for $0 < x < 12$, f is concave upward on the interval (0, 12). Also note that $f''(6) > 0$, which shows by Theorem 6 that f has a minimum point at $x = 6$. Complete information is given in Table 3.20, and the graph of the average cost function is revealed in Figure 3.26. ∎

Table 3.20. INFORMATION FOR

$$y = \frac{x^2 + 10x + 36}{x}, \qquad 0 < x \leq 12.$$

Domain: $\{x:0 < x \leq 12\}$. Range: $\{y:y \geq 22\}$.
Intercepts: x-intercepts—none; y-intercepts—none.
Not symmetric to x-axis, y-axis, or origin.
Vertical asymptotes: $x = 0$. Horizontal asymptotes: none.
Increasing: $6 < x < 12$. Decreasing: $0 < x < 6$.
Minimum point: (6, 22). Endpoint maximum: (12, 25).
Concave upward: $0 < x < 12$.
Points of inflection: none.
Other points: (1, 47), (2, 30), (3, 25), (9, 23).
Slope at (2, 30): -8.

EXERCISES

Suggested minimum assignment: Exercises 1(*b*) (*e*) (*g*) (*i*), 2(*c*), 3(*a*), 4(*c*), and 4(*g*).

1. Determine whether the graph of each of the following equations is symmetric to (1) the x-axis, (2) the y-axis, and (3) the origin.

   (*a*) $y = \dfrac{x^2 - 4}{x^2 - 9}$. (*b*) $y^2 = \dfrac{x^2 - 4}{x^2 - 9}$.

   (*c*) $x^2 - y^2 = 16$. (*d*) $y = x^6 + x^4 + x^2$.
   (*e*) $y = x^3$. (*f*) $x^2 + 4y^2 = 5$.
   (*g*) $y = |x|$. (*h*) $xy = 8$.
   (*i*) $x^3 = y^2$. (*j*) $y = x^2 + 3x$.
2. Find the equations of all vertical asymptotes.

   (*a*) $y = \dfrac{1}{x^2 - 4}$. (*b*) $y = \dfrac{x - 2}{x^2 - 4}$.

   (*c*) $y = \dfrac{x + 1}{x^2 - 3x}$. (*d*) $y = \dfrac{x^2 - 1}{x^2 + 1}$.

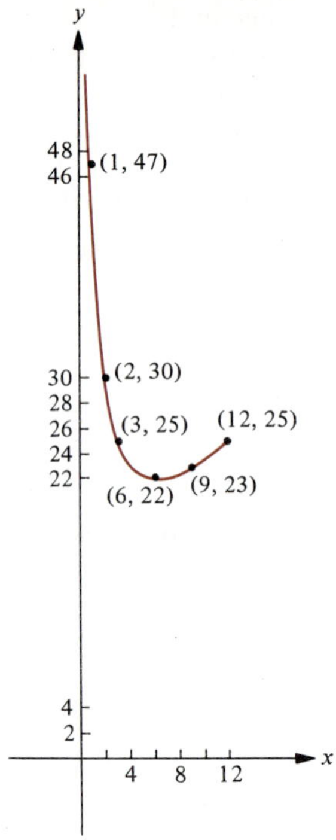

Figure 3.26. $y = \dfrac{x^2 + 10x + 36}{x}$, $0 < x \leq 12$.

3. Find the equations of all horizontal asymptotes.

   (*a*) $y = \dfrac{2x^2}{3x^2 - 12}$. (*b*) $y = \dfrac{x}{5x - 10}$.

   (*c*) $y = \dfrac{2x}{x^2 - 1}$. (*d*) $y = \dfrac{x^2 + 4}{x - 1}$.

4. Construct a table similar to Table 3.17 for the graph of each of the following equations. Then make a sketch.

   (*a*) $y = \dfrac{3}{x - 1}$. (*b*) $y = \dfrac{x - 2}{x + 2}$.

   (*c*) $y = \dfrac{5}{x^2 + 3}$. (*d*) $y = \dfrac{5}{x^2 - 1}$.

   (*e*) $y = \dfrac{4x}{x^2 + 9}$. (*f*) $y = \dfrac{4x}{x^2 - 9}$.

   (*g*) $y = \dfrac{2x^2}{x^2 - 4}$. (*h*) $y = \dfrac{x^2 - 9}{x^2 - 4}$.

5. Sketch the function in Example 3 of Section 3.3.
6. Suppose that the concentration y of a drug in the blood after x hours

is given by

$$y = f(x) = \frac{x}{20x^2 + 50x + 80}, \qquad x \geq 0.$$

Sketch the graph of f. (See Example 8 of Section 1.9 and Example 4 of Section 3.2.)

7. Suppose that the equation of a demand curve is

$$y = f(x) = \frac{100}{x + 10}, \qquad 0 \leq x \leq 40.$$

Sketch the graph of f.

8. Suppose that the equation of a supply curve is

$$y = f(x) = \frac{2x + 400}{x + 800}, \qquad 0 \leq x \leq 1200.$$

Sketch the graph of f.

9. Sketch the graph of the average cost function given by

$$y = f(x) = \frac{C(x)}{x} = \frac{\frac{1}{10}x^2 + 7x + 10}{x}, \qquad 0 < x \leq 30.$$

(See Exercise 19 of Section 3.4 and Example 4 of this section.)

3.7 Mean Value Theorem for Derivatives (Optional)

Earlier in this chapter (Theorem 1) we stated the fact that if $f'(x) > 0$ at each point of an open interval (a, b), then f is increasing on (a, b). This result will be proved in this section. In order to accomplish the proof of Theorem 1, however, we need the **mean value theorem for derivatives** that will be stated in Theorem 8. The mean value theorem for derivatives is also needed to prove Theorem 1 of Chapter 4. This section will be more theoretical in nature than other sections in this text; if one is willing to accept without proof the results referred to above, then this section can be omitted. *The validity of the results of this section is not restricted to rational functions.*

In order to prove the mean value theorem for derivatives, we shall use Theorem 7, which is known as **Rolle's theorem.** The geometric interpretation in Figure 3.27 will aid in understanding Theorem 7 and its proof.

▶ Theorem 7

If a function f is continuous on $[a, b]$, differentiable on (a, b), and if $f(a) = f(b)$, then there exists at least one number c, where $a < c < b$, such that $f'(c) = 0$.

Outline of Proof: If f is a constant function on $[a, b]$, then $f'(c) = 0$ for any number c in (a, b), and the theorem is proved. If, however, f is

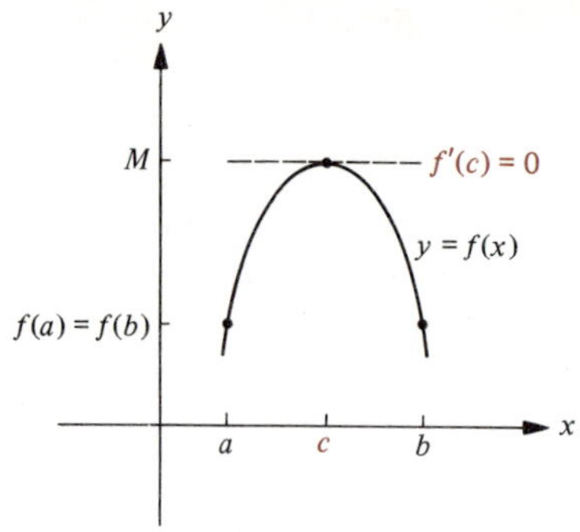

Figure 3.27

not a constant function on $[a, b]$, then, by the statement of Exercise 13(a) of Section 1.9, $f(x)$ must take on a least value m and a greatest value M on $[a, b]$. Either $M > f(a)$ or $m < f(a)$. Suppose that f attains the maximum value M at $x = c$, where $a < c < b$. Since f is differentiable on (a, b), $f'(c)$ exists. If f is a rational function, then $f'(c) = 0$ by Theorem 2 of Section 3.3. The proof of Theorem 2 is also valid for functions other than rational functions if $f'(c)$ is known to exist. Hence $f'(c) = 0$ is valid whether or not f is a rational function. ◀

If f is not a constant function on $[a, b]$, then under the hypotheses of Rolle's theorem there must be a maximum or minimum point between a and b at which the slope is zero (see Figure 3.27).

A geometric interpretation of the next theorem, known as the **mean value theorem for derivatives,** can be obtained from Figure 3.28. In that figure we notice that the secant line (see Section 2.1) between $(a, f(a))$ and $(b, f(b))$ is parallel to the tangent line at $x = c$. The slope of the secant line is $[f(b) - f(a)]/(b - a)$, and the slope of the tangent line is $f'(c)$. It seems evident geometrically that if the function f satisfies certain conditions, then there is at least one point c between a and b at which the slope of the tangent line is equal to the slope $[f(b) - f(a)]/(b - a)$ of the secant line.

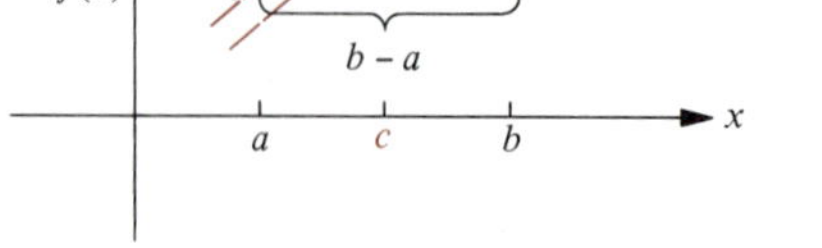

Figure 3.28

▶ **Theorem 8**

If a function f is continuous on $[a, b]$ and differentiable on (a, b), then there exists at least one number c, where $a < c < b$, such that

$$\frac{f(b) - f(a)}{b - a} = f'(c).$$

Outline of Proof: Create a function F with domain $[a, b]$ such that

$$F(x) = f(x) - f(a) - \frac{f(b) - f(a)}{b - a}(x - a).$$

It can be shown that F satisfies the hypotheses of Rolle's theorem (Theorem 7). Therefore, there exists a number c, where $a < c < b$, such that $F'(c) = 0$. Since we can calculate

$$F'(x) = f'(x) - \frac{f(b) - f(a)}{b - a},$$

the equation $F'(c) = 0$ becomes

$$f'(c) - \frac{f(b) - f(a)}{b - a} = 0$$

or

$$\frac{f(b) - f(a)}{b - a} = f'(c). \blacktriangleleft$$

Example 1

If $f(x) = x^3 - x + 2$, $a = 0$, and $b = 3$, find one number that can serve as c in Theorem 8.

Because f is a polynomial function, the hypotheses of Theorem 8 are satisfied. Since

$$f(b) = f(3) = 27 - 3 + 2 = 26$$

and

$$f(a) = f(0) = 0 - 0 + 2 = 2,$$

then

$$\frac{f(b) - f(a)}{b - a} = \frac{26 - 2}{3 - 0} = 8.$$

Also $f'(x) = 3x^2 - 1$, and therefore

$$f'(c) = 3c^2 - 1.$$

Theorem 8 guarantees that the equation $f'(c) = [f(b) - f(a)]/(b - a)$, which in this problem is

$$3c^2 - 1 = 8,$$

is satisfied by a number c in $(0, 3)$. When we solve the last equation for c, we find that this number is

$$c = \sqrt{3}. \blacksquare$$

Theorem 1 of Section 3.2 is restated below. Its proof can be given now that the mean value theorem for derivatives is available.

▶ **Theorem 1**

(*i*) If $f'(x) > 0$ at each point of (a, b), then f is increasing on (a, b).
(*ii*) If $f'(x) < 0$ at each point of (a, b), then f is decreasing on (a, b).

Proof: For part (*i*), assume that $f'(x) > 0$ at each point of (a, b). Let x_1 and x_2 be arbitrary numbers such that $a < x_1 < x_2 < b$. According to Definition 1 of Section 3.2, we need to show that $f(x_1) < f(x_2)$. Since f is differentiable on (a, b), f is continuous on (a, b). Therefore, the hypotheses of the mean value theorem for derivatives are satisfied on $[x_1, x_2]$, and hence by Theorem 8 there exists a number c, where $x_1 < c < x_2$, such that

$$\frac{f(x_2) - f(x_1)}{x_2 - x_1} = f'(c).$$

Since $f'(c)$ is positive by hypothesis and $x_2 - x_1 > 0$, it follows that

$$f(x_2) - f(x_1) > 0,$$

and therefore

$$f(x_1) < f(x_2).$$

The proof of part (*ii*) requires only minor modifications of the proof above and is left as Exercise 5. ◀

Theorem 9 is another consequence of Theorem 8. Theorem 9 will be used to prove Theorem 10, which in turn leads immediately to the first theorem of Chapter 4.

▶ **Theorem 9**

If $f'(x) = 0$ for all values of x in (a, b), then f is a constant function on (a, b).

Proof: Let x_1 and x_2 be arbitrary numbers such that $a < x_1 < x_2 < b$. To show that f is a constant function on (a, b), we must prove that $f(x_1) = f(x_2)$. Theorem 8 applies on $[x_1, x_2]$; therefore, there exists a number c, where $x_1 < c < x_2$, such that

$$\frac{f(x_2) - f(x_1)}{x_2 - x_1} = f'(c).$$

Since $f'(c) = 0$ by hypothesis and $x_1 \neq x_2$, it follows that $f(x_1) = f(x_2)$. ◀

▶ **Theorem 10**

If g and h are differentiable functions that have equal derivatives on (a, b), then g and h differ by a constant on (a, b).

Proof: We are given that

$$\frac{d}{dx}(g(x)) = \frac{d}{dx}(h(x)) \text{ on } (a, b).$$

Therefore,

$$\frac{d}{dx}(g(x) - h(x)) = 0 \text{ on } (a, b).$$

By Theorem 9,

$$g(x) - h(x) = C \text{ on } (a, b),$$

where C is a constant. ◀

If continuity on $[a, b]$ is assumed for the functions in Theorems 1, 9, and 10, then the conclusions of those theorems are valid on $[a, b]$ instead of (a, b).

APPLICATIONS

The primary applications of the material in this section are indirect and numerous. The theorems stated and proved here occupy a prominent position in the theory of calculus; hence many applications that follow directly from other sections can be considered as rooted in the results of this section. One direct application, however, is given in the following example.

Example 2

Suppose that a cost function C satisfies the hypotheses of the mean value theorem for derivatives (Theorem 8) on the interval $[0, 100]$. Also assume that the fixed cost is zero—in other words, $C(0) = 0$. From Theorem 8 it can be concluded that there is at least one number c, where $0 < c < 100$, such that

$$C'(c) = \frac{C(100) - C(0)}{100 - 0} = \frac{C(100)}{100}. \tag{3.1}$$

If 100 items are produced, then $C(100)/100$ is called the average cost per item (see Example 6 of Section 2.4). Therefore, (3.1) may be interpreted as meaning that there is at least one level of production between 0 and 100 at which the marginal cost equals the average cost of producing 100 items. The result above might not hold if $C(0) > 0$ [instead of $C(0) = 0$]; in that case the marginal cost might always remain less than $C(100)/100$. ∎

EXERCISES

Suggested minimum assignment: Exercises 1(*c*), 3(*b*), 4(*a*), and 7.

1. By finding the x-coordinates of the maximum and minimum points of f, find all numbers that can serve as c in Theorem 7. [Observe that $f(a) = f(b)$.]

(*a*) $f(x) = x^2 - 4,\ a = -2,\ b = 2.$
(*b*) $f(x) = x^3 + x^2,\ a = -1,\ b = 0.$
(*c*) $f(x) = \dfrac{x^4}{4} - 2x^2,\ a = -3,\ b = 3.$

2. In each of the following there is no number c, where $a < c < b$, such that $f'(c) = 0$. Explain in each case why Theorem 7 is not contradicted.
   (*a*) $f(x) = x^2 + 3,\ a = 1,\ b = 2.$
   (*b*) $f(x) = |x|,\ a = -1,\ b = 1.$
3. Find all numbers that can serve as c in Theorem 8.
   (*a*) $f(x) = x^2 + 5x + 2,\ a = 0,\ b = 2.$
   (*b*) $f(x) = x^3 - x,\ a = -1,\ b = 2.$
4. Find the x-coordinates of all points between a and b at which the tangent line is parallel to the secant line joining $(a, f(a))$ and $(b, f(b))$. Make a sketch.
   (*a*) $y = f(x) = (x - 1)^2,\ a = 0,\ b = 3.$
   (*b*) $y = f(x) = 2x^3 - 6x,\ a = -2,\ b = 2.$
5. Prove Theorem 1(*ii*).
6. If $f'(x) = 0$ on the interval $(2, 7)$, what can be said about f?
7. Suppose that F is a function with the property that $F'(x) = f(x) = 2x$ on an interval (a, b). Let G be any function defined on (a, b) for which $G'(x) = 2x$. Show, using Theorem 10, that $G(x)$ is of the form $F(x) + C$ on (a, b), where C is constant. (This result illustrates Theorem 1 of Chapter 4.) What is C if $F(x) = x^2 + 3$ and $G(x) = x^2 + 7$?
8. Suppose that a cost function C is given by $C(x) = 200x - x^2$, provided that $0 \leq x \leq 100$. Find a value of x between 0 and 100 at which the marginal cost equals the average cost of producing 100 items. (See Example 2.)

NEW VOCABULARY

finite interval 3.1
infinite interval 3.1
sign change of a factor 3.1
dividing point 3.1
demand curve 3.2
supply curve 3.2
function increasing (decreasing) on an interval 3.2
maxima, minima 3.3
extreme point 3.3
(relative) maximum point 3.3
(relative) minimum point 3.3
endpoint (relative) maximum 3.3
endpoint (relative) minimum 3.3
critical number 3.3
critical point 3.3
first derivative test 3.3
concave upward (downward) on an interval 3.5
concavity 3.5
point of inflection 3.5
second derivative test 3.5
symmetric to the x-axis 3.6
symmetric to the y-axis 3.6
symmetric to the origin 3.6
vertical asymptote 3.6
horizontal asymptote 3.6
Rolle's theorem 3.7
mean value theorem for derivatives 3.7

4 Antiderivatives

As stated earlier, the study of calculus is ordinarily divided into two parts—the differential calculus and the integral calculus. The last two chapters have been concerned with the differential calculus. The purpose of Chapters 4 and 5 is to introduce the other part—the integral calculus. In the differential calculus, problems were encountered where functions were known and instantaneous rates of change of the functions were sought. Illustrations in this chapter, however, will show that in many applied situations the inverse problem is encountered; the instantaneous rates of change are known and functions relating the variables are sought. The operation that determines a function f whose derivative f' is known is called integration. Many historians credit Isaac Barrow (1630–1677), Newton's teacher, with the exploitation of the inverse relationship between the operations of differentiation and integration.

Prerequisites: Chapter 1, 2.1–2.6, 3.1, 3.2
Suggested Sections for Moderate Emphasis: 4.1, 4.2
Suggested Sections for Minimum Emphasis: 4.1, 4.2

4.1 Introduction

Suppose that we are given the function F defined by

$$y = F(x) = \tfrac{1}{3}x^3,$$

where the domain of F is the set of all real numbers. In Chapter 2 we learned that the derivative of F is given by

$$y' = F'(x) = x^2,$$

and that the differential dy is given by

$$dy = F'(x)\,dx = x^2\,dx.$$

Observe that functions that differ from F by only a constant have the same derivative and differential that F has; for example, if $y = \tfrac{1}{3}x^3 + C$, where C is any real number, then $y' = x^2$ as before and $dy = x^2\,dx$ as before.

In this section we shall be concerned with the inverse problem in which either $F'(x)$ or $F'(x)\,dx$ is known, and the object is to find $F(x)$. If we use our experience with differentiation and work backward, it may not be too difficult to find by trial and error that if $F'(x) = x^2$, then $F(x) = \tfrac{1}{3}x^3 + C$, where C is any real number. If $F'(x) = f(x) = x^2$ and $F(x) = \frac{1}{3}x^3 + 5$, then F is called an **antiderivative** of f according to Definition 1.

Definition 1

If F is a function whose derivative is the function f, then F is called an **antiderivative** of f.

It is understood in the definition above that F is differentiable and that $F'(x) = f(x)$ holds for all x in a suitable interval of the real numbers, although this interval will not be specifically stated in each problem.

Theorem 1

If $F'(x) = f(x)$, then every antiderivative of f is of the form $F(x) + C$, where C is a real number.

A hint for the proof of Theorem 1 is given in Exercise 26. ◀

Example 1

Suppose that the slope of the tangent line to a curve at (x, y) is given by $y' = F'(x) = x^2$. The equation of the curve must be of the form $y = F(x) = \frac{1}{3}x^3 + C$. After one antiderivative F such as the one given by $F(x) = \frac{1}{3}x^3$ is found, we know by Theorem 1 that no further search for antiderivatives is necessary because *every* possible antiderivative is given by an expression of the form $\frac{1}{3}x^3 + C$. If marginal revenue is given by $R'(x) = x^2$, then the revenue function is given by an expression of the form $R(x) = \frac{1}{3}x^3 + C$; if more information is known, such as $R(0) = 0$, then $0 = 0 + C$ and C can be determined. (Discussion of the determination of C in this manner will be postponed until Section 4.3.) ▮

The notation most commonly used to indicate that *every* antiderivative of f is an expression of the form $F(x) + C$ is

$$\int f(x)\,dx = F(x) + C.$$

The symbol $\int$ is called an **integral sign** and when placed before $f(x)\,dx$, *it indicates that the most general antiderivative of f is to be found;* the process of finding the most general antiderivative of f is often called **integration.** In the notation above, $f(x)$ is called the **integrand** and C is called the **constant of integration.** If $\int x^2\,dx$ is written, we say that we are required to **integrate** the given function, and we complete this operation by writing

$$\int x^2\,dx = \tfrac{1}{3}x^3 + C;$$

the result of integration can be checked by recalling that this process is the inverse of differentiation so that either

$$D_x(\tfrac{1}{3}x^3 + C) = x^2$$

or

$$d(\tfrac{1}{3}x^3 + C) = x^2\,dx$$

is a foolproof check. This type of check will aid in proving the various parts of Theorem 2, in which we state some formulas that can be used for the integration of certain functions.

Theorem 2

(*i*) $\int dx = x + C.$

(*ii*) $\int kf(x)\,dx = k\int f(x)\,dx$, where k is a real number.

(*iii*) $\int [f(x) + g(x)]\,dx = \int f(x)\,dx + \int g(x)\,dx.$

(*iv*) $\int x^n\,dx = \dfrac{x^{n+1}}{n+1} + C$, where n is a real number and $n \neq -1$.

The proof of Theorem 2 is left as Exercises 27 and 28. ◀

Example 2

$$\int \frac{dx}{x^2} = \int x^{-2}\,dx = \frac{x^{-1}}{-1} + C = -\frac{1}{x} + C.$$

Notice that the integrand $1/x^2$ was first rewritten in the form x^n (with $n = -2$) so that Theorem 2(*iv*) could be applied. The answer is checked as follows:

$$D_x\left(-\frac{1}{x} + C\right) = -D_x x^{-1} + 0 = -(-1)x^{-2} = \frac{1}{x^2},$$

which is the original integrand. A check of the answer in this manner is optional. ∎

Example 3

$$\begin{aligned}\int (2x^3 + x^5)\,dx &= 2\int x^3\,dx + \int x^5\,dx && \text{by Theorem 2}(ii)\text{ and }(iii)\\ &= 2\left(\frac{x^4}{4} + C_1\right) + \frac{x^6}{6} + C_2 && \text{by Theorem 2}(iv)\\ &= \frac{x^4}{2} + \frac{x^6}{6} + (2C_1 + C_2)\\ &= \frac{x^4}{2} + \frac{x^6}{6} + C.\end{aligned}$$

With practice, the first step above can be omitted; also the use of constants C_1 and C_2 is unnecessary as they can be combined, so the addition of one constant C at the end is all that is required. In the future we shall write

$$\int (2x^3 + x^5)\,dx = \frac{2x^4}{4} + \frac{x^6}{6} + C$$
$$= \frac{x^4}{2} + \frac{x^6}{6} + C. \blacksquare$$

Example 4

$$\int (2x + 1)(x - 2)\,dx = \int (2x^2 - 3x - 2)\,dx$$
$$= \frac{2x^3}{3} - \frac{3x^2}{2} - 2x + C. \blacksquare$$

Theorem 2(*iii*) is easily extended to handle addition (or subtraction) of three or more terms, as we assumed above. Observe that part (*i*) was used in integrating the third term in Example 4. It was necessary to multiply $(2x + 1)$ and $(x - 2)$ together before integrating because in general it can be shown that

$$\int f(x)g(x)\,dx \neq \left(\int f(x)\,dx\right)\left(\int g(x)\,dx\right).$$

Example 5

$$\int \frac{\sqrt[3]{t} + \sqrt{t^5} + t^{-2/3}}{t}\,dt = \int \left(\frac{t^{1/3}}{t} + \frac{t^{5/2}}{t} + \frac{t^{-2/3}}{t}\right) dt$$
$$= \int (t^{-2/3} + t^{3/2} + t^{-5/3})\,dt$$
$$= \frac{t^{1/3}}{\frac{1}{3}} + \frac{t^{5/2}}{\frac{5}{2}} + \frac{t^{-2/3}}{-\frac{2}{3}} + C$$
$$= 3t^{1/3} + \frac{2}{5}t^{5/2} - \frac{3}{2}t^{-2/3} + C. \blacksquare$$

APPLICATIONS

The material in Sections 4.1 and 4.2 will be applied in Section 4.3 and in Chapters 5 and 6.

EXERCISES

Suggested minimum assignment: Exercises 3, 8, 9, 11, 16, 18, 21, and 23.

In Exercises 1–10 perform the indicated integration.

1. $\int x^3\,dx.$

2. $\int x^{-3}\,dx.$

3. $\int (x^2 - 3x)\,dx.$

4. $\int (1 + 2x^4)\,dx.$

5. $\int \frac{1}{t^2}\,dt.$

6. $\int \sqrt[3]{t}\,dt.$

7. $\int (x^{\pi} - x^2 + 3x + 8)\,dx.$

8. $\int \frac{1 + x^{\sqrt{3}}}{2x^2}\,dx.$

9. $\int (x - 1)(2x - 9)\,dx.$

10. $\int x(1 + \sqrt{x})\,dx.$

In Exercises 11–14 perform the indicated integration and check your answer.

11. $\int (5x^6 - 2x^2)\,dx.$

12. $\int \sqrt{x}(1 + 2x^2)\,dx.$

13. $\int \frac{(x^2 + 1)^2}{x^2}\,dx.$

14. $\int (5x - 1)(x)(7x - 3)\,dx.$

15. If $g(x) = \frac{x^5}{15} + 3$ and $f(x) = \frac{x^4}{3}$, is g an antiderivative of f? Why?

16. If $F(x) = x^3$ and $f(x) = 3x^2 + 1$, is F an antiderivative of f? Why?

17. If $F(x) = 4x^3$ and $f(x) = x^4$, is F an antiderivative of f? Why?

In Exercises 18–25 perform the indicated integration.

18. $\int x\sqrt[3]{x}\,dx.$

19. $\int \frac{\sqrt{x} + \sqrt[3]{x}}{x^2}\,dx.$

20. $\int (2 + \sqrt{x})^2\,dx.$

21. $\int \left(2 + \frac{1}{\sqrt{x^3}}\right)^2 dx.$

22. $\int \frac{dx}{x\sqrt{3x}}.$

23. $\int x^{-5}(2x - 1)^2\,dx.$

24. $\int \sqrt[7]{y^6}\,dy.$

25. $\int (2y - 3)^3\,dy.$

26. Prove Theorem 1. [*Hint:* Let G be any antiderivative of f. Use Theorem 10 of Section 3.7.]

27. (*a*) Prove Theorem 2(*i*). (*b*) Prove Theorem 2(*ii*).
[*Hint:* Show that the derivative of the right side of each formula is equal to the integrand of the left side.]

28. (*a*) Prove Theorem 2(*iii*). (*b*) Prove Theorem 2(*iv*).

4.2 An Important Integration Formula

In Section 2.3 we presented the important power rule for differentiation, and in this section we study the following corresponding **power rule for integration:**

$$\int [f(x)]^n f'(x)\,dx = \frac{[f(x)]^{n+1}}{n + 1} + C, \tag{4.1}$$

provided that n is a real number and $n \neq -1$. The validity of (4.1) for $n \neq -1$ can be checked by differentiation:

$$D_x\left(\frac{[f(x)]^{n+1}}{n+1} + C\right) = \frac{1}{n+1}(n+1)[f(x)]^{(n+1)-1}f'(x) + 0$$
$$= [f(x)]^n f'(x).$$

Notice that the power rule for differentiation was used in making the above check, and of course we assumed that f was differentiable. If we substitute

$$u = f(x),$$

then

$$du = f'(x)\,dx,$$

and we can state Theorem 3.

▶ Theorem 3

If f is differentiable and $u = f(x)$, then

$$\int u^n\,du = \frac{u^{n+1}}{n+1} + C,$$

provided that n is a real number and that $n \neq -1$. ◀

Theorem 3 is the form of the important **power rule for integration** to which we shall refer frequently, but it is a somewhat difficult formula to apply correctly at first, and therefore we illustrate with some examples.

Example 1

$$\int (2x^2 + 1)^8(4x)\,dx.$$

With $u = 2x^2 + 1$ and $n = 8$, it follows that $du = 4x\,dx$ and that the given problem fits exactly the left side of the formula in Theorem 3. Therefore,

$$\int (2x^2 + 1)^8(4x\,dx) = \int u^8\,du = \frac{u^9}{9} + C$$
$$= \frac{1}{9}(2x^2 + 1)^9 + C. \blacksquare$$

Example 2

$$\int (2x^2 + 1)^8 x\,dx.$$

This example differs from Example 1 in that the number 4 is missing from the integrand. Since the number 4 is needed as part of the du (see

Example 1), we simply insert the needed 4 and counteract this operation by also multiplying by $\frac{1}{4}$ so as not to change the integrand. Hence we have

$$\int (2x^2 + 1)^8(\tfrac{1}{4})(4x\,dx).$$

By Theorem 2(*ii*) of Section 4.1, the constant $\frac{1}{4}$ can be brought across the integral sign to yield $\frac{1}{4}\int (2x^2 + 1)^8(4x\,dx)$. Except for the extra $\frac{1}{4}$ the problem is the same as Example 1, so that

$$\begin{aligned}\int (2x^2 + 1)^8 x\,dx &= \frac{1}{4}\int (2x^2 + 1)^8(4x\,dx)\\ &= \frac{1}{4}\int u^8\,du = \frac{1}{4}\frac{u^9}{9} + C = \frac{1}{36}(2x^2 + 1)^9 + C. \ \blacksquare\end{aligned}$$

Examples 1 and 2 can be checked by differentiation. Example 2 does not check unless the extra $\frac{1}{4}$ is obtained in front of the integral sign as shown above. The insertion of a nonzero constant to make the exact required du appear, together with the corresponding insertion of the reciprocal of that constant in front of the integral sign, is a legitimate procedure that is frequently needed. The insertion of a *variable* in the integrand and the reciprocal of that *variable* in front of the integral sign, however, is *not* a legitimate procedure—*only constants can be brought across the integral sign*. For the problem

$$\int (2x^2 + 1)^8\,dx,$$

it might be tempting to put $4x$ in the integrand and $1/4x$ in front of the integral sign, but this leads to the wrong answer—as can be checked by differentiation. In order to solve the integration problem

$$\int (2x^2 + 1)^8\,dx,$$

it is necessary to expand $(2x^2 + 1)^8$ and then integrate each term.

Example 3

Suppose that it is desired to find the equation $y = F(x)$ of a demand curve (see Example 1 of Section 3.2). Assume that it is known that the rate of change of price y with respect to demand x is given by

$$\frac{dy}{dx} = F'(x) = -\frac{1}{(\frac{1}{5}x + 1)^2},$$

provided that $0 \leq x \leq 100$. (Observe that since $dy/dx < 0$ throughout the interval $[0, 100]$, it follows that the price decreases as the demand increases.)

Then $F(x)$ can be found by using the power rule for integration as follows:

$$\begin{aligned}
F(x) = \int F'(x)\,dx &= \int \frac{(-1)\,dx}{(\frac{1}{5}x + 1)^2} \\
&= (-1)\int \left(\frac{1}{5}x + 1\right)^{-2} dx \\
&= (-1)(5)\int \left(\frac{1}{5}x + 1\right)^{-2}\left(\frac{1}{5}\,dx\right) \\
&= -5\int u^{-2}\,du \\
&= -5\frac{u^{-1}}{-1} + C \\
&= \frac{5}{u} + C \\
&= \frac{5}{\frac{1}{5}x + 1} + C.
\end{aligned}$$

Hence the equation of the demand curve (for $0 \le x \le 100$) is

$$y = \frac{5}{\frac{1}{5}x + 1} + C.$$

More information is needed in order to find a specific value for the constant C; for example, if it were known that $y = 2$ when $x = 10$, then it could be determined that $C = \frac{1}{3}$. ▮

Example 4

$$\begin{aligned}
\int \frac{(4x + 80)\,dx}{\sqrt{x^2 + 40x + 1}} &= \int (x^2 + 40x + 1)^{-1/2}(4x + 80)\,dx \\
&= 2\int (x^2 + 40x + 1)^{-1/2}(2x + 40)\,dx \\
&= 2\int u^{-1/2}\,du = 2\,\frac{u^{1/2}}{\frac{1}{2}} + C \\
&= 4\sqrt{u} + C = 4\sqrt{x^2 + 40x + 1} + C. \ ▮
\end{aligned}$$

Notice that if $u = x$, then $du = dx$, and the power rule for integration,

$$\int u^n\,du = \frac{u^{n+1}}{n + 1} + C, \qquad \text{provided that } n \neq -1,$$

reduces to

$$\int x^n\,dx = \frac{x^{n+1}}{n + 1} + C,$$

which is Theorem 2(*iv*) of Section 4.1.

EXERCISES

Suggested minimum assignment: Exercises 1, 5, 11, 12, 14, 18, 20, and 21.

Perform the indicated integration in Exercises 1–16.

1. $\int (3x^2 + 1)^7(6x)\, dx.$
2. $\int x(3x^2 + 1)^7\, dx.$
3. $\int \sqrt{4x + 3}\, dx.$
4. $\int \dfrac{dx}{3x^5}.$
5. $\int x\sqrt{x^2 + 3}\, dx.$
6. $\int \dfrac{t\, dt}{(5t^2 + 1)^2}.$
7. $\int \dfrac{x\, dx}{\sqrt{2x^2 + 5}}.$
8. $\int \dfrac{(x + 1)\, dx}{\sqrt{100x^2 + 200x + 77}}.$
9. $\int \dfrac{(1 + \sqrt{x})^5}{\sqrt{x}}\, dx.$
10. $\int \sqrt[3]{8y + 27}\, dy.$
11. $\int (x^2 + \sqrt{3x})\, dx.$
12. $\int (2x + 1)^{10}\, dx.$
13. $\int x\sqrt{x}\, dx.$
14. $\int (7x + 3)^{8/3}\, dx.$
15. $\int \dfrac{dx}{(3x + 1)^4}.$
16. $\int \sqrt{x^3 + 7x - 3}(3x^2 + 7)\, dx.$
17. Check your answer to Exercise 7.
18. Check your answer to Exercise 14.

In Exercises 19 and 20 select the only one of the three parts that can be done by use of Theorem 3. Then work out the one that you select.

19. (*a*) $\int x^2\sqrt{9 + x^2}\, dx.$ (*b*) $\int x\sqrt{9 + x^2}\, dx.$ (*c*) $\int \sqrt{9 + x^2}\, dx.$
20. (*a*) $\int \dfrac{dx}{\sqrt{x^2 + 4x + 1}}.$ (*b*) $\int \dfrac{(x + 1)\, dx}{\sqrt{x^2 + 4x + 1}}.$ (*c*) $\int \dfrac{(x + 2)\, dx}{\sqrt{x^2 + 4x + 1}}.$
21. Find the equation $y = F(x)$ of a demand curve if it is known that the rate of change of price y with respect to demand x is given by $dy/dx = F'(x) = -100/(3x + 1)^3$, provided that $0 \leq x \leq 5$.
22. Find the equation $y = C(x)$ of a cost function if the marginal cost $C'(x)$ is given by $C'(x) = (1 + x/3)^3$, provided that $0 \leq x \leq 6$.

4.3 Differential Equations*

In this section we shall give the reader additional practice in using the integration techniques studied in Sections 4.1 and 4.2. Determination of the constant of integration, provided that certain **initial conditions** are

* This section is a prerequisite to the discussion of exponential growth and decay in Section 6.5; if Section 6.5 is to be omitted, this section may be considered optional.

given, will be stressed. **Differential equations,** a subject of importance in many applications of calculus, will be introduced next.

A **differential equation** is an equation that involves derivatives of an unknown function. For example,

$$\frac{d^2y}{dx^2} - \frac{dy}{dx} + 2y = 8x^2, \tag{4.2}$$

is a differential equation in which y is an unknown function of x. Usually, the object is to find a function f, as given by $y = f(x)$, which satisfies the equation; such a function is called a **solution** of the differential equation.

Example 1

Let us demonstrate that the function f given by

$$y = f(x) = 4x^2 + 4x - 2$$

is a solution of (4.2).

$$\frac{dy}{dx} = 8x + 4,$$

$$\frac{d^2y}{dx^2} = 8.$$

Substitution in (4.2) yields

$$8 - (8x + 4) + 2(4x^2 + 4x - 2) = 8x^2,$$

which simplifies to the identity

$$8x^2 = 8x^2. \blacksquare$$

We have verified in Example 1 that f is a solution of the differential equation (4.2), although we do not claim to have explained how f was found in the first place. Instead of saying that "the function f given by $y = f(x) = 4x^2 + 4x - 2$ is a solution of (4.2)," we shall find it convenient to say "$y = 4x^2 + 4x - 2$ is a solution of (4.2)."

The **order** of a differential equation is the order of the highest-ordered derivative appearing in the equation. The order of equation (4.2) is 2 because d^2y/dx^2 is the highest-ordered derivative that appears. The equation

$$\frac{dy}{dx} = -\frac{x}{y} \tag{4.3}$$

has order 1. Equation (4.3) can be written as

$$y\,dy = -x\,dx,$$

in which the x's and y's are placed on opposite sides of the equation; when this can be done, we say that it is possible to **separate the variables.** In the equation $dy/dx = x + y$, it is *not* possible to separate the variables. The only type of first-order differential equation that we shall learn to solve is the type for which it is possible to separate the variables. After separating the variables, the next step is to integrate each side of the equation; we demonstrate this procedure in the next example by solving (4.3).

Example 2

Solve $dy/dx = -x/y$.

Separate the variables to obtain

$$y\,dy = -x\,dx,$$

and integrate as follows:

$$\int y\,dy = \int -x\,dx,$$

$$\frac{y^2}{2} + C_1 = -\frac{x^2}{2} + C_2,$$

$$\frac{x^2}{2} + \frac{y^2}{2} = C_2 - C_1.$$

The last equation can be written

$$x^2 + y^2 = C,$$

where $C = 2(C_2 - C_1)$. Either

$$y = \sqrt{C - x^2}$$

or

$$y = -\sqrt{C - x^2}$$

can be shown to satisfy the original equation, provided that $C > 0$ and $-C < x < C$. ∎

In Example 3 enough information is given to determine a specific value for C.

Example 3

Find the solution of

$$\frac{dy}{dx} = x^2$$

which also satisfies the condition that $y = \frac{14}{3}$ when $x = 2$.

We separate the variables to obtain

$$dy = x^2\,dx$$

and integrate to get

$$y = \tfrac{1}{3}x^3 + C.$$

By substituting $y = \frac{14}{3}$ and $x = 2$, C can be determined as follows:

$$\tfrac{14}{3} = \tfrac{1}{3}(2)^3 + C,$$

$$C = 2.$$

The desired solution is

$$y = \tfrac{1}{3}x^3 + 2. \blacksquare$$

A condition such as $y = \frac{14}{3}$ when $x = 2$, in Example 3, is called an **initial condition.** Without the initial condition we stop with the **general solution** $y = \frac{1}{3}x^3 + C$; an appropriate initial condition makes it possible to determine a special case $y = \frac{1}{3}x^3 + 2$ of the general solution, and such a special case is called a **particular solution.** Geometrically, the general solution $y = \frac{1}{3}x^3 + C$ represents a family of curves (one for each real number C). See Figure 4.1. The particular solution is the one curve (shown in color) that passes through the point $(2, \frac{14}{3})$.

Example 4 illustrates again the use of an initial condition to determine a constant of integration.

Example 4

Suppose that for the production of a certain item, a firm has determined that its marginal cost is given by

$$C'(x) = 50 + 4x - \frac{x^2}{20}, \qquad \text{provided that } 0 \le x \le 60.$$

Also it is known that the fixed cost is 100 [this means that $C(0) = 100$]. Find the cost of producing 30 items.

$$\begin{aligned} C'(x) &= \frac{dC}{dx} = 50 + 4x - \frac{x^2}{20}, \\ dC &= \left(50 + 4x - \frac{x^2}{20}\right) dx, \\ C(x) &= \int \left(50 + 4x - \frac{x^2}{20}\right) dx \\ &= 50x + 4\frac{x^2}{2} - \frac{1}{20}\frac{x^3}{3} + B, \end{aligned}$$

where B is the constant of integration. Since $C(0) = 100$, we can find B as follows:

$$100 = 0 + 0 - 0 + B$$

and

$$B = 100.$$

Hence the cost function is given by

$$C(x) = 50x + 2x^2 - \frac{x^3}{60} + 100.$$

To compute the cost of producing 30 items, let $x = 30$ to obtain

$$C(30) = 50(30) + 2(30)^2 - \frac{(30)^3}{60} + 100 = 2950. \blacksquare$$

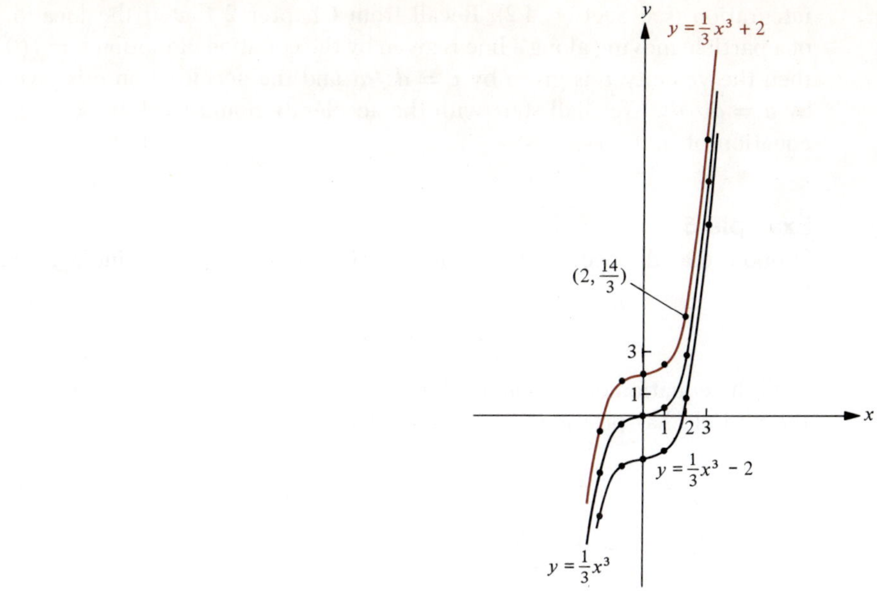

Figure 4.1. *Family of curves* $y = \frac{1}{3}x^3 + C$.

The only other type of differential equation that we shall learn to solve is a second-order equation of the form $d^2y/dx^2 = f(x)$. Consider, as an example,

$$\frac{d^2y}{dx^2} = x^5.$$

This can be written

$$\frac{dy'}{dx} = x^5$$

or

$$dy' = x^5\,dx.$$

Then

$$y' = \frac{x^6}{6} + C_1$$

so that

$$dy = \left(\frac{x^6}{6} + C_1\right) dx.$$

Therefore,

$$y = \frac{x^7}{42} + C_1x + C_2.$$

The method of solution consists of separating the variables and integrating, and then repeating the process.

The next example involves the same ideas as above, but initial conditions will be given that make possible the determination of both C_1 and C_2. Also the reader should be alert in Example 5 for the use of the power rule for integration (see Section 4.2). Recall from Chapter 2 that if the location of a particle moving along a line is given by the equation of motion $s = f(t)$, then the velocity v is given by $v = ds/dt$ and the acceleration a is given by $a = dv/dt$. We shall start with the acceleration and work back to the equation of motion.

Example 5

Suppose that the acceleration a of a particle moving along a line is given by

$$a = \sqrt{2t + 1}.$$

Find the equation of motion for the particle if it is known that when $t = 4$ the particle has position $s = \frac{81}{5}$ and velocity $v = 15$.

$$\frac{dv}{dt} = \sqrt{2t + 1},$$

$$dv = \sqrt{2t + 1}\, dt,$$

$$v = \int \sqrt{2t + 1}\, dt = \frac{1}{2}\int (2t + 1)^{1/2} 2\, dt$$

$$= \frac{1}{2}\int u^{1/2}\, du = \frac{1}{2}\frac{u^{3/2}}{\frac{3}{2}} + C_1 = \frac{1}{3}(2t + 1)^{3/2} + C_1.$$

Since $v = 15$ when $t = 4$,

$$15 = \tfrac{1}{3}(9)^{3/2} + C_1,$$

$$= \tfrac{1}{3}(27) + C_1,$$

and hence

$$C_1 = 6.$$

Therefore,

$$v = \frac{ds}{dt} = \frac{1}{3}(2t + 1)^{3/2} + 6,$$

$$ds = \left[\frac{1}{3}(2t + 1)^{3/2} + 6\right] dt,$$

$$s = \frac{1}{3}\frac{1}{2}\int (2t + 1)^{3/2}(2\, dt) + \int 6\, dt,$$

$$= \frac{1}{6}\frac{(2t + 1)^{5/2}}{\frac{5}{2}} + 6t + C_2.$$

Next, substitute $s = \frac{81}{5}$ and $t = 4$ to determine C_2 as follows:

$$\tfrac{81}{5} = \tfrac{1}{15}(9)^{5/2} + 24 + C_2,$$

$$\tfrac{81}{5} = \tfrac{1}{15}(243) + 24 + C_2,$$

$$C_2 = -24.$$

Therefore, the required equation of motion is

$$s = \tfrac{1}{15}(2t + 1)^{5/2} + 6t - 24,$$

and the position s of the particle can be calculated for a given value of t. ▮

APPLICATIONS

Example 6

A certain publishing company would like to price a new book so as to maximize its annual profit from sales of the book. The company considers it reasonable to charge any price y between \$8 and \$25. From experience with a similar book it is believed that the annual demand x would be 3900 if $y = \$10$. Also, from a study of the demand curve of the similar book, it is expected that the rate of change dy/dx of price with respect to demand can be approximated by $dy/dx = -y^2/40{,}000$. The revenue from sales of the book will be the price y per book times the demand x. The annual cost is given by $(8100 + x)$. Find the price y that will maximize the annual profit P from sales of the new book.

In the differential equation

$$\frac{dy}{dx} = -\frac{y^2}{40{,}000},$$

the variables can be separated to obtain

$$dx = -40{,}000y^{-2}\,dy.$$

Integrating both sides, we get

$$\begin{aligned}\int dx &= -40{,}000 \int y^{-2}\,dy,\\ x &= -40{,}000\frac{y^{-1}}{-1} + C\\ &= \frac{40{,}000}{y} + C.\end{aligned}$$

Since $x = 3900$ if $y = 10$, it follows that $C = -100$ and that the equation of the demand curve is given by

$$x = \frac{40{,}000}{y} - 100.$$

The profit equals the revenue minus the cost, and hence

$$\begin{aligned}P &= yx - (8100 + x)\\ &= y\left(\frac{40{,}000}{y} - 100\right) - \left[8100 + \left(\frac{40{,}000}{y} - 100\right)\right]\\ &= 32{,}000 - 100y - \frac{40{,}000}{y}.\end{aligned}$$

By the methods of Chapter 3 it can be shown that maximum profit occurs at $y = \$20$. ▮

EXERCISES

Suggested minimum assignment: Exercises 1, 2(*b*), 3(*a*), 4, 5, 7, and 9.

1. State the order of each of the following differential equations.

   (*a*) $\dfrac{dy}{dx} = x^3 + 8x + 3.$ (*b*) $x\dfrac{d^2y}{dx^2} + \dfrac{dy}{dx} - 8y = 0.$

   (*c*) $\dfrac{d^6y}{dx^6} + \left(\dfrac{dy}{dx}\right)^7 = 3.$ (*d*) $\dfrac{d^2y}{dx^2} = x + y.$

2. Solve each of the following differential equations.

   (*a*) $\dfrac{dy}{dx} = 2x.$ (*b*) $\dfrac{dy}{dx} = x^2 + 8x + 3.$

   (*c*) $\dfrac{d^2y}{dx^2} = x^2 + 2x.$ (*d*) $(y + 1)\dfrac{dy}{dx} = \sqrt{x}.$

   (*e*) $\dfrac{dy}{dx} = \sqrt{xy}.$ (*f*) $\dfrac{d^2y}{dx^2} = 4x^3.$

3. Find the solution of each of the following differential equations that passes through the point $(1, 3)$.

   (*a*) $\dfrac{dy}{dx} = 4x.$ (*b*) $\dfrac{dy}{dx} = 3x^2 + 4.$

4. Sketch at least three members of the family of curves that arises in solving Exercise 3(*a*), including the one that passes through $(1, 3)$.
5. Find the equation of the curve that satisfies $d^2y/dx^2 = 2x$ if it is known that the curve has slope 2 at the point $(1, 4)$.
6. Find the equation of the curve that satisfies $d^2y/dx^2 = \sqrt{2x + 3}$ if it is known that the curve has slope 4 at the point $(3, 0)$.
7. The velocity v of a particle moving along a line is given by

   $$v = \frac{ds}{dt} = t\sqrt[3]{t^2 + 4}.$$

   Find the equation of motion for the particle if it is known that $s = 8$ when $t = 2$.
8. The acceleration a of a particle moving along a line is given by

   $$a = 6t + 3.$$

   Find the equation of motion for the particle if it is known that when $t = 3$ the particle has position $s = 8$ and velocity $v = 20$. Also find s when $t = 2$.
9. Suppose that marginal cost is given by

   $$C'(x) = 1000 - 2x.$$

   The fixed cost is 2000. Find $C(x)$. Find $C(10)$.
10. Suppose that marginal revenue is given by

$$R'(x) = \frac{x^2}{10} + \frac{x}{5} + 100,$$

and that $R(0) = 0$. Find $R(x)$.

11. An object is shot vertically upward from the ground. Let s be the height in feet above the ground after t seconds. The acceleration due to gravity is given by $a = -32$ feet per second per second. If $s = 0$ when $t = 0$ and velocity $v = v_0$ when $t = 0$, find a formula that expresses s in terms of t.
12. Suppose that the rate of change of price with respect to weekly demand for an item is given by

$$\frac{dy}{dx} = -\frac{y^3}{1000},$$

provided that $1.5 \leq y \leq 3$. If the weekly demand is $x = 100$ when the price is $y = 2$, find the equation of the demand curve.

13. In Example 6 show that maximum profit occurs at $y = \$20$.

NEW VOCABULARY

antiderivative 4.1
integral sign 4.1
integration 4.1
integrand 4.1
constant of integration 4.1
power rule for integration 4.2
differential equation 4.3
solution of a differential equation 4.3
order of a differential equation 4.3
separate the variables 4.3
initial condition 4.3
general solution 4.3
particular solution 4.3

5 The Definite Integral

Integral calculus includes two basic operations—integration, and evaluation of definite integrals. The purpose of this chapter is to define the definite integral and to explore the very important relationship between the antiderivative and the definite integral. The vehicle for showing this relationship is known as the Fundamental Theorem of Integral Calculus.

Actually, the basic ideas leading to the definite integral were known for centuries before the development of the differential calculus in the 17th century. Eudoxus (408–355 B.C.*) and Archimedes (287–212* B.C.*) contributed methods whereby formulas for areas and volumes could be constructed. Little was done to develop the subject further, however, until the 17th century.*

An indication of the importance of the definite integral in probability and statistics will be given in the applications of Sections 5.2, 5.3, 5.5, and 5.6.

Prerequisites: Chapter 1, 2.1–2.5, 4.1, 4.2; the first part of 3.6 is needed for 5.4.
Suggested Sections for Moderate Emphasis: 5.1–5.4
Suggested Sections for Minimum Emphasis: 5.1–5.3

5.1 Definition

An interesting problem that can often be solved by methods of calculus is the problem of finding the area of a region surrounded by curves whose equations are known. In statistics such an area might represent the probability of an event, as we shall learn in Example 4 of Section 5.2. In economics such an area might represent a quantity called *consumers' surplus* or a quantity called *producers' surplus,* as we shall illustrate in Example 4 of Section 5.4. In this section we first learn how to approximate the area of a region in order to motivate the definition of the **definite integral.**

Consider the problem of finding the area of the region bounded by $y = f(x) = 4x - \frac{1}{2}x^2$, the x-axis, and the vertical lines $x = 1$ and $x = 6$ (see Figure 5.1). Using the fact that the area of a rectangle is the product of its length and width, we shall approximate the desired area by adding the areas of the four rectangles shown in Figure 5.1. We begin by deciding arbitrarily to divide the closed interval $[1, 6]$ into four subintervals by means of the **division points** 1, 2, 4, 5, and 6. This set $\{1, 2, 4, 5, 6\}$ of division points is called a **partition** of $[1, 6]$. The widths of the approximating

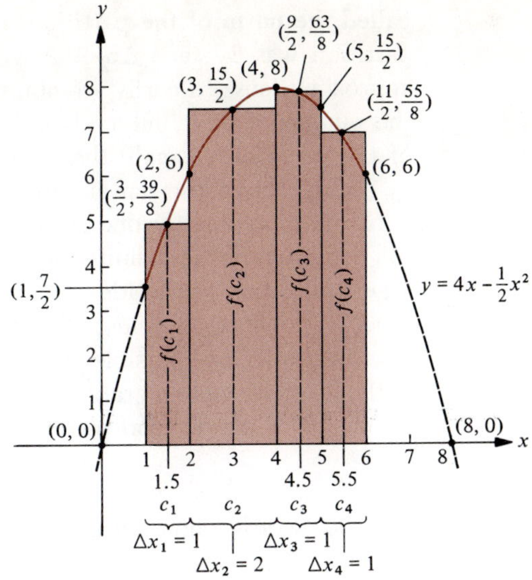

Figure 5.1. *Approximation of area with rectangles.*

rectangles shown in the figure are $\Delta x_1 = 2 - 1 = 1$, $\Delta x_2 = 4 - 2 = 2$, $\Delta x_3 = 5 - 4 = 1$, and $\Delta x_4 = 6 - 5 = 1$. We next select point c_1 in, or at an end of, the first subinterval and draw a vertical line segment from this point to the curve $y = f(x)$; we let the length $f(c_1)$ of this segment be the length (height) of the first approximating rectangle. Arbitrarily choose c_1 to be the x-coordinate of the midpoint (the point equidistant from the endpoints) of the first subinterval; hence $c_1 = \frac{3}{2}$. The area of the first rectangle is

$$[f(c_1)][\Delta x_1] = [f(\tfrac{3}{2})][1] = f(\tfrac{3}{2}) = 4(\tfrac{3}{2}) - \tfrac{1}{2}(\tfrac{3}{2})^2 = \tfrac{39}{8}\,.$$

Similarly, suppose that we decide to let c_2, c_3, and c_4 be the x-coordinates of the midpoints of the other subintervals, respectively; hence $c_2 = 3$, $c_3 = \frac{9}{2}$, $c_4 = \frac{11}{2}$, and the lengths of the other rectangles are $f(3)$, $f(\frac{9}{2})$, and $f(\frac{11}{2})$, respectively. The sum S_4 of the areas of the four rectangles is

$$\begin{aligned} S_4 &= f(c_1)\,\Delta x_1 + f(c_2)\,\Delta x_2 + f(c_3)\,\Delta x_3 + f(c_4)\,\Delta x_4 \\ &= [f(\tfrac{3}{2})][1] + [f(3)][2] + [f(\tfrac{9}{2})][1] + [f(\tfrac{11}{2})][1] \\ &= (\tfrac{39}{8})(1) + (\tfrac{15}{2})(2) + (\tfrac{63}{8})(1) + (\tfrac{55}{8})(1) \\ &= \tfrac{277}{8} = 34.625 \text{ square units.} \end{aligned}$$

By methods we shall learn in the next section it can be shown that the exact area under the curve and above the x-axis between $x = 1$ and $x = 6$ is $34\,\frac{1}{6}$ square units. Notice that the actual area and the approximated area differ by less than $\frac{1}{2}$ square unit. Ordinarily, a closer approximation can be obtained by using more subintervals. A very wide subinterval tends to yield more error in the approximation; the width of the largest subinterval is called the **norm** of the partition. In the example above the norm of the partition was 2, since $\Delta x_2 = 2$ was larger than Δx_1, Δx_3, or Δx_4. Good approximations ordinarily are obtained by choosing division points so that the number n of subintervals is large and in such a way that the norm of the partition is small; the choice of location of points such as c_1, c_2, c_3, and c_4 above also has an effect on the accuracy of an approximation.

Next we prepare for one of the most important definitions in calculus—the definition of the definite integral. The discussion that follows presents a generalization of the ideas given in the illustration above. Start with a function f defined on a closed interval $[a, b]$. The reader may refer to Figure 5.2 to visualize one possible graph of f, although what follows can be done without reference to a graph. Divide $[a, b]$ into a number n of subintervals by means of the division points $x_0, x_1, x_2, \ldots, x_{n-1}, x_n$, where

$$a = x_0 < x_1 < x_2 < \cdots < x_{n-1} < x_n = b.$$

In other words, choose a partition $\{x_0, x_1, x_2, \ldots, x_{n-1}, x_n\}$ of $[a, b]$. Let

$$\begin{aligned} \Delta x_1 &= x_1 - x_0, \\ \Delta x_2 &= x_2 - x_1, \\ &\vdots \\ \Delta x_n &= x_n - x_{n-1}. \end{aligned}$$

Let δ_n, called the norm of the partition, be the largest of the numbers Δx_1, $\Delta x_2, \ldots, \Delta x_n$. Choose numbers $c_1, c_2, \ldots, c_n$ to satisfy

$$\begin{aligned} x_0 &\le c_1 \le x_1, \\ x_1 &\le c_2 \le x_2, \\ &\vdots \\ x_{n-1} &\le c_n \le x_n. \end{aligned}$$

Form the sum

$$S_n = f(c_1)\,\Delta x_1 + f(c_2)\,\Delta x_2 + \cdots + f(c_n)\,\Delta x_n.$$

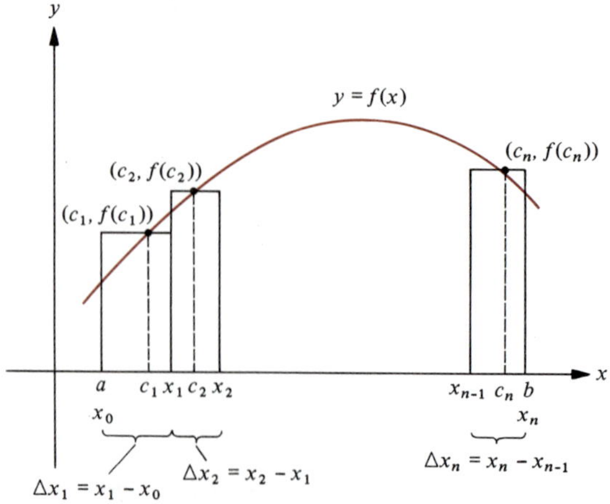

Figure 5.2

Next we think of repeating the above process, letting n get larger and larger (we require that $\delta_n \to 0$ as $n \to \infty$ in order to make sure that $\Delta x_1, \Delta x_2, \ldots, \Delta x_n$ *all* become very small as n gets large). This suggests the idea of a limit and leads to the conjecture that S_n may approach some finite number as $n \to \infty$ and $\delta_n \to 0$. In other words, we are interested in finding

$$\lim_{\substack{n\to\infty \\ \delta_n\to 0}} S_n$$

if this limit exists.

Definition 1

Let a function f be defined on $[a, b]$. If

$$\lim_{\substack{n\to\infty \\ \delta_n\to 0}} S_n$$

exists, then it is called the **definite integral** of f from a to b and will be denoted temporarily by $I_a^b(f)$.

Several comments about Definition 1 need to be made. First, in Figure 5.2, f is continuous on $[a, b]$ and $f(x)$ is *nonnegative* on $[a, b]$; in this case S_n is the sum of the areas of the n rectangles and represents an *approximation* of the area above the x-axis and below the curve $y = f(x)$ between $x = a$ and $x = b$. The exact area is defined to be the limit of S_n, namely, $I_a^b(f)$. Second, in Definition 1 when we say that

$$\lim_{\substack{n\to\infty \\ \delta_n\to 0}} S_n$$

exists, we mean that this limit does not depend upon the choice of the points $x_1, x_2, \ldots, x_{n-1}$ or $c_1, c_2, \ldots, c_n$; at any stage of the limiting process S_n *does* depend upon the selection of these points, but when we say that $I_a^b(f)$ exists, the *limit* of S_n does *not* depend upon their selection. Third, it is *not* correct in general to write $I_a^b(f) = \lim_{n\to\infty} S_n$ because n could become large without δ_n becoming small; this would happen if the first subinterval is never made smaller as $n \to \infty$, because δ_n would always be as large as Δx_1. If, however, all subintervals are *equal* in length at every stage of the limiting process, then δ_n does approach 0 as $n \to \infty$; this idea is used in optional Section 5.5, where definite integrals are evaluated using Definition 1.

We began this section with an example in which we computed a sum S_4 as an approximation of an area—that sum was an approximation of $I_1^6(f)$, where $f(x) = 4x - \frac{1}{2}x^2$. Then we considered the sum S_n, and finally we defined the definite integral of a function f from a to b (where $a < b$) by

$$I_a^b(f) = \lim_{\substack{n\to\infty \\ \delta_n\to 0}} S_n,$$

provided that this limit exists. Whenever this limit exists, f is said to be **integrable** on $[a, b]$. One can see from the definition why people often speak of the "limit of a sum" when discussing a definite integral.

APPLICATIONS

Example 1

The method of approximating an area by adding the areas of appropriate rectangles has been learned in this section. It has perhaps been implied that an equation $y = f(x)$ is always known in this type of problem. In applications it is frequently true that data are observed in some type of experiment and points are plotted from the data, but the correct equation of the curve passing through these points is not known. In the illustration that follows, it is demonstrated how an area can be approximated without knowledge of the equation for the correct curve passing through the given points.

Each day the temperature (in degrees Fahrenheit) is recorded at various times. Suppose that during the hours of darkness (when temperature often does not vary greatly) the temperature is not recorded as often as during the daytime hours. The temperatures on a particular winter day are shown in Table 5.1. The data are transferred to a graph in the manner shown in Figure 5.3. The temperature y is a function f (that is not known exactly) of the number of hours x after midnight—the domain of f is $[0, 24]$. Suppose that we are asked to approximate the area under $y = f(x)$ and above the x-axis between $x = 0$ and $x = 24$ (in Example 4 of Section 5.3 a use for this area will be given). In other words, an approximation of $I_0^{24}(f)$ is desired. Since 18 temperature readings are given, the approximation is accomplished by adding the areas of the 18 rectangles shown in Figure 5.3. As the rectangles are constructed, our estimation of the exact area is slightly high when the temperature is decreasing and slightly low when the temperature is increasing—these errors should almost cancel out over a period of 1 day. Hence the sum S_{18} gives a rather accurate approximation of $I_0^{24}(f)$:

$$\begin{aligned} S_{18} &= f(c_1)\,\Delta x_1 + f(c_2)\,\Delta x_2 + \cdots + f(c_{18})\,\Delta x_{18} \\ &= (16)(2) + (15)(2) + (13)(2) + (12)(2) + (15)(1) + (18)(1) \\ &\quad + (21)(1) + (24)(1) + (28)(1) + (32)(1) + (31)(1) + (30)(1) \\ &\quad + (28)(1) + (24)(1) + (20)(1) + (17)(1) + (15)(2) + (13)(2) \\ &= 456. \blacksquare \end{aligned}$$

Table 5.1. TEMPERATURES ON A WINTER DAY.

Time	mid-night	2	4	6	8	9	10	11	noon	1	2	3	4	5	6	7	8	10
Temperature	16	15	13	12	15	18	21	24	28	32	31	30	28	24	20	17	15	13

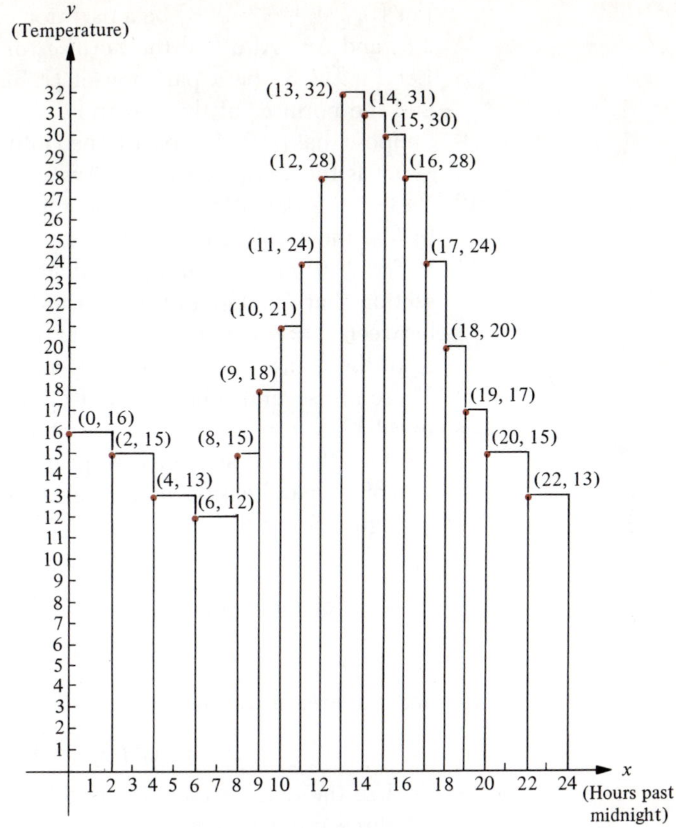

Figure 5.3. *Temperatures on a winter day.*

EXERCISES

Suggested minimum assignment: Exercises 1, 5, 7, 9, and 10.

In Exercises 1–6 sketch $y = f(x)$ between $x = a$ and $x = b$. From the given data, draw the appropriate rectangles as in Figure 5.1. Then compute S_n as an approximation of $I_a^b(f)$.

1. Given: $f(x) = x^2$, $a = 1$, $b = 3$; $n = 3$; division points are 1, 2, $\frac{5}{2}$, 3; and $c_1 = \frac{3}{2}$, $c_2 = \frac{9}{4}$, $c_3 = \frac{11}{4}$.
2. Same as Exercise 1, except that $c_1 = 1$, $c_2 = \frac{5}{2}$, $c_3 = \frac{11}{4}$.
3. Given: $f(x) = x^3$, $a = 0$, $b = 2$; $n = 4$; $\Delta x_1 = \Delta x_2 = \Delta x_3 = \Delta x_4 = \frac{1}{2}$; and $c_1 = \frac{1}{4}$, $c_2 = \frac{7}{8}$, $c_3 = \frac{9}{8}$, $c_4 = \frac{7}{4}$.
4. Same as Exercise 3, except that $c_1 = 0$, $c_2 = \frac{1}{2}$, $c_3 = \frac{5}{4}$, $c_4 = 2$.
5. Given: $f(x) = \frac{1}{2}x^2 + 1$, $a = -1$, $b = 2$; $n = 5$; the partition is $\{-1, -\frac{1}{2}, 0, \frac{1}{2}, \frac{3}{2}, 2\}$; and $c_1 = -1$, $c_2 = -\frac{1}{4}$, $c_3 = \frac{1}{4}$, $c_4 = 1$, $c_5 = 2$.
6. Same as Exercise 5, except that $c_1 = -1$, $c_2 = -\frac{1}{2}$, $c_3 = 0$, $c_4 = \frac{1}{2}$, $c_5 = \frac{3}{2}$.

7. Let $\{3, \frac{7}{2}, 4, \frac{17}{4}, \frac{19}{4}, \frac{11}{2}, 6\}$ be a partition of $[3, 6]$. Find Δx_1, Δx_2, Δx_3, Δx_4, Δx_5, and Δx_6. Also find the norm δ_6 of this partition.
8. Let $\{1, \frac{3}{2}, 2, 3\}$ be a partition of $[1, 3]$. Find Δx_1, Δx_2, and Δx_3. Also find the norm δ_3 of this partition.
9. Suppose that $[-2, 4]$ is partitioned into 12 subintervals of equal length. What is the value of each of $\Delta x_1, \ldots, \Delta x_{12}$?
10. Repeat the definition of the definite integral of a function f from a to b without reference to the text.
11. If $f(x) \geq 0$ on $[a, b]$ and f is integrable on $[a, b]$, we learned in this section that $I_a^b(f)$ gives the area under $y = f(x)$ and above the x-axis between $x = a$ and $x = b$. We also learned how to calculate a sum S_n of the areas of n rectangles as an approximation of the exact area. There are several other ways of approximating area, one of which is to calculate a sum T_n of the areas of n trapezoids (see Figure 5.4), using the fact that the area of a trapezoid is the product of one half the sum of the parallel sides and the altitude. The area of the first trapezoid in Figure 5.4 is

$$\tfrac{1}{2}\Delta x_1[f(x_0) + f(x_1)],$$

the area of the second trapezoid is

$$\tfrac{1}{2}\Delta x_2[f(x_1) + f(x_2)],$$

and the area of the last trapezoid is

$$\tfrac{1}{2}\Delta x_n[f(x_{n-1}) + f(x_n)].$$

(*a*) Use the data of Exercise 1 to find the sum T_3 of the appropriate three trapezoids.

(*b*) If $\Delta x_1, \Delta x_2, \ldots, \Delta x_n$ are all equal to the same number, Δx, find a general formula for T_n.

12. Three sides of a parcel of land are straight lines and the fourth side curves gently. A sketch of the land made to scale is bounded by $x = 0$, $x = 4$, $y = 0$, and $y = f(x)$. The equation $y = f(x)$ is unknown, except

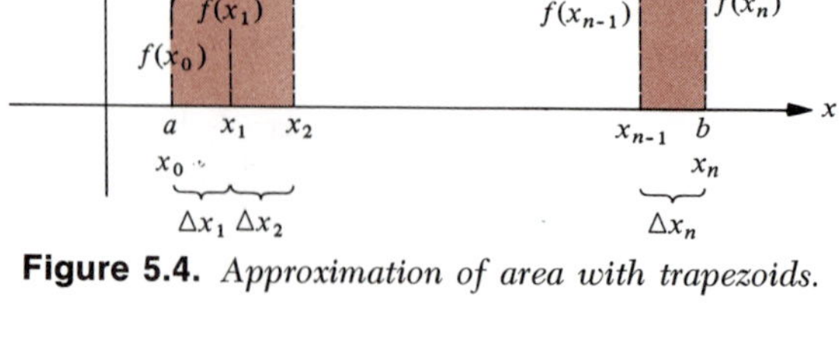

Figure 5.4. *Approximation of area with trapezoids.*

that it passes through (0, 1.52), (.5, 1.49), (1, 1.45), (1.5, 1.43), (2, 1.41), (2.5, 1.42), (3, 1.46), and (3.5, 1.50). The scale is fixed so that $I_0^4(f)$ equals the area in acres of the parcel of land. A buyer agrees to pay $5000 per acre for the land. Approximate $I_0^4(f)$ by the method of Example 1, and use the result to find the total price of the land.

13. The sum

$$5 + 10 + 15 + 20 + 25 + 30 + 35 + 40$$

can be written briefly as

$$\sum_{k=1}^{8} 5k,$$

which is an example of the **summation notation.** The Greek letter Σ (sigma) suggests the word *sum*. The notation $\sum_{k=1}^{8} 5k$ indicates that the values 1, 2, 3, 4, 5, 6, 7, and 8 are to be substituted successively for k in the expression $5k$ and that the results are to be added. The letter k is called the **index of summation,** but any other letter may be used; $\sum_{j=1}^{8} 5j$ represents the same sum. We did not use the summation notation when writing S_n in this section, but the instructor may wish to do so because it may be written conveniently as

$$S_n = \sum_{k=1}^{n} f(c_k)\,\Delta x_k.$$

To gain familiarity with summation notation, do each of the following:

(*a*) Write the sum represented by $\sum_{k=1}^{5} \frac{k}{k+1}$.

(*b*) Show that $\sum_{k=1}^{8} 5k = 5 \sum_{k=1}^{8} k$.

(*c*) Write $1^2 + 2^2 + 3^2 + \cdots + n^2$ using summation notation.
(*d*) Write $15 + 20 + 25 + 30 + 35 + 40$ using summation notation.
(*e*) Write $f(c_1)\,\Delta x_1 + f(c_2)\,\Delta x_2 + f(c_3)\,\Delta x_3$ using summation notation.

5.2 Fundamental Theorem of Integral Calculus

In Section 5.1 we defined the definite integral of a function f from a to b as a limit of a certain sum, provided that this limit exists. There is a condition that is sufficient to guarantee the existence of $I_a^b(f)$. The following theorem, whose proof is beyond the scope of this book, makes it clear that many functions that we shall encounter are integrable on an interval $[a, b]$.

▶ **Theorem 1**

If f is continuous on $[a, b]$, then $I_a^b(f)$ exists. ◀

There are functions that are not continuous on $[a, b]$ which are still integrable on $[a, b]$, but Theorem 1 guarantees that every continuous function on a closed interval is integrable on that interval.

The next theorem that we state relates the definite integral (defined in Section 5.1) to the idea of antiderivative (defined in Section 4.1), and also provides a practical method of evaluating many definite integrals. Because of its importance, Theorem 2 is frequently called the **Fundamental Theorem of Integral Calculus.**

▶ **Theorem 2**

If f is continuous on $[a, b]$ and a function F given by $F(x) = \int f(x)\,dx$ is an antiderivative of f on $[a, b]$, then

$$I_a^b(f) = F(b) - F(a).$$

A proof of Theorem 2 will be discussed in optional Section 5.6. ◀

The remarkable feature of the Fundamental Theorem is that it permits the evaluation of many definite integrals by using the seemingly unrelated concept of an antiderivative. We illustrate in Example 1.

Example 1

If $f(x) = x^2 + 1$, use Theorem 2 to evaluate the definite integral of f from $a = 2$ to $b = 3$. In other words, evaluate $I_2^3(f)$.

An antiderivative F of f on $[2, 3]$ is given by

$$F(x) = \int f(x)\,dx = \int (x^2 + 1)\,dx = \frac{x^3}{3} + x + C,$$

where C is any particular real number. Hence

$$\begin{aligned} I_2^3(f) &= F(3) - F(2) \\ &= \left(\frac{3^3}{3} + 3 + C\right) - \left(\frac{2^3}{3} + 2 + C\right) \\ &= (12 + C) - \left(\frac{14}{3} + C\right) \\ &= \frac{22}{3}. \quad ▮ \end{aligned}$$

In the type of problem illustrated in Example 1 the constant of integration C always cancels out, and hence we may as well simplify things by always selecting $C = 0$ when we use Theorem 2.

Instead of $I_a^b(f)$, a customary notation for the definite integral of f from a to b is $\int_a^b f(x)\,dx$. The notation $\int_a^b f(x)\,dx$ was not used in Section 5.1 because of the danger of confusing the definite integral with an antiderivative of f which was denoted by $\int f(x)\,dx$. The two concepts are quite different. One important thing about Theorem 2, however, is that it gives the relationship between these important concepts. In Example 1 we could write

$$\int_2^3 (x^2+1)\,dx = \left(\frac{x^3}{3}+x\right)\Big|_2^3$$
$$= \left(\frac{3^3}{3}+3\right) - \left(\frac{2^3}{3}+2\right)$$
$$= \frac{22}{3}.$$

The symbol $|_2^3$ indicates that 3 is to be substituted first; then from this result is subtracted the number obtained by substituting 2. (Notice that we selected $C = 0$.)

For the symbol $\int_a^b f(x)\,dx$, $f(x)$ is called the **integrand,** a is called the **lower limit,** b is called the **upper limit,** and x is called the **variable of integration.** Provided the hypothesis of Theorem 2 is met, the conclusion of the Fundamental Theorem of Integral Calculus can be written

$$\int_a^b f(x)\,dx = F(b) - F(a).$$

To evaluate the definite integral of f from a to b, find an antiderivative F of f on $[a, b]$, substitute the upper limit b to obtain $F(b)$, and then subtract $F(a)$, the result of substituting the lower limit a.

Example 2

Let $f(x) = 4x - \frac{1}{2}x^2$, $a = 1$, and $b = 6$. In Section 5.1 we approximated the area under $y = f(x)$ and above the x-axis between $x = a$ and $x = b$, and we learned that the exact area is given by $I_1^6(f) = \int_1^6 (4x - \frac{1}{2}x^2)\,dx$. Calculate this area using the Fundamental Theorem of Integral Calculus.

First find the antiderivative (with $C = 0$) of the function f.

$$F(x) = \int\left(4x - \frac{1}{2}x^2\right)dx = 4\frac{x^2}{2} - \frac{1}{2}\frac{x^3}{3} = 2x^2 - \frac{1}{6}x^3.$$

Then

$$\int_1^6 \left(4x - \frac{1}{2}x^2\right)dx = \left(2x^2 - \frac{1}{6}x^3\right)\Big|_1^6$$
$$= \left[2(6)^2 - \frac{1}{6}(6)^3\right] - \left[2(1)^2 - \frac{1}{6}(1)^3\right]$$
$$= (72-36) - \left(2 - \frac{1}{6}\right)$$
$$= 34\frac{1}{6}. \blacksquare$$

It follows from our discussion in this section and the previous section that if f is *continuous* and *nonnegative* on $[a, b]$, then the area bounded by $y = f(x)$, the x-axis, and the lines $x = a$ and $x = b$ is given by $\int_a^b f(x)\,dx$.

Example 3

Find the area bounded by $y = f(x) = 27/(2x + 3)^2$, the x-axis, and the lines $x = 0$ and $x = 3$.

The rational function f is continuous and nonnegative on $[0, 3]$. The desired area (see Figure 5.5) is given by $\int_0^3 27/(2x + 3)^2\,dx$. First find the antiderivative (with $C = 0$) of the function f.

$$\begin{aligned} F(x) &= \int \frac{27}{(2x+3)^2}\,dx = 27\int (2x+3)^{-2}\,dx \\ &= \frac{27}{2}\int (2x+3)^{-2}(2\,dx) = \frac{27}{2}\int u^{-2}\,du \\ &= \frac{27}{2}\,\frac{u^{-1}}{-1} = -\frac{27}{2}(2x+3)^{-1} \\ &= \frac{-27}{4x+6}. \end{aligned}$$

Therefore,

$$\begin{aligned} \int_0^3 \frac{27}{(2x+3)^2}\,dx &= \left.\frac{-27}{4x+6}\right|_0^3 \\ &= \frac{-27}{12+6} - \frac{-27}{0+6} \\ &= -\frac{3}{2} + \frac{9}{2} \\ &= 3. \ \blacksquare \end{aligned}$$

In optional Section 5.5 we shall demonstrate how to evaluate some rather simple definite integrals by using the *definition* of a definite integral from

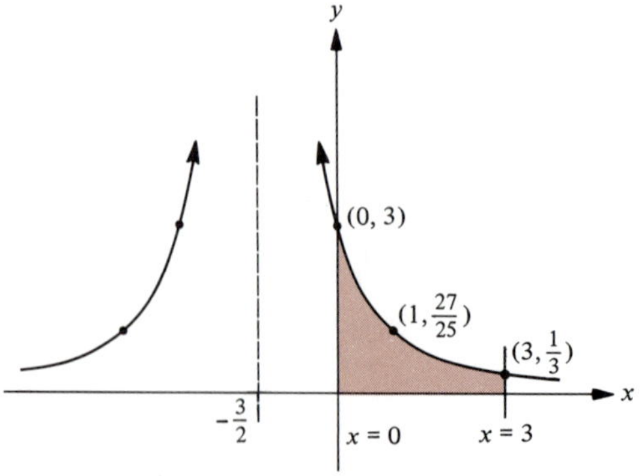

Figure 5.5. $y = \dfrac{27}{(2x + 3)^2}$.

Section 5.1 and carrying out a somewhat tedious limiting process; this demonstration will be helpful in gaining a better understanding of the definite integral. For most purposes, however, it is best to evaluate $\int_a^b f(x)\,dx$ by finding an antiderivative F of f on $[a, b]$, if possible, and then calculating $F(b) - F(a)$ according to the Fundamental Theorem of Integral Calculus. If, however, an antiderivative F cannot be found but the continuity requirement in Theorem 2 is satisfied, then $\int_a^b f(x)\,dx$ can be approximated by computing a sum S_n as explained in Section 5.1.

APPLICATIONS

*Example 4 Use of Probability Density Functions in Calculating Probability

For a discrete measurement, that is, a measurement having a finite or countably infinite number of values, the probability of an event can be determined by adding the weights of the outcomes comprising the event (see Example 6 of Section 1.6). For a continuous measurement, however, such a procedure is impossible; an event E of obtaining an outcome less than or equal to some real number b is an interval of real numbers and hence has an uncountable number of outcomes. The purpose of this example is to illustrate a method for finding probabilities of events corresponding to continuous measurements by means of a probability density function. In this section we shall limit consideration to continuous measurements whose ranges are finite closed intervals; we shall call such measurements **bounded continuous measurements.**

In Example 7 of Section 2.2 we explained how to obtain a probability density function p if the cumulative probability distribution function F were known. Below we define p in a manner that does not depend upon knowing F; the definition, of course, is consistent with our discussion in Example 7 of Section 2.2. A **probability density function** p of a *bounded* continuous measurement is a function, whose domain is the set of all real numbers, that possesses the following three properties:

1. $p(x) \geq 0$.
2. $\displaystyle\int_a^b p(x)\,dx = 1$, where $[a, b]$ is the range of the measurement.
3. $p(x) = 0$ if $x < a$ or $x > b$.

The probability of an event $E = \{s : s \leq c\}$ that the value s of a bounded continuous measurement is less than or equal to some real number c, where $a \leq c \leq b$, is denoted by $P(E)$. The connection between the definition of a probability density function p and the discussion in Example 7 of Section 2.2 is the definition $P(E) = \int_a^c p(x)\,dx$. Since $P(E)$ also equals $F(c)$, where F is the cumulative probability distribution function, we can write

$$P(E) = F(c) = \int_a^c p(x)\,dx.$$

As an illustration of the ideas above, suppose that an insurance company is interested in knowing at what age x in years the death of a man is likely to occur if the man is presently 40 years old. As a hypothetical case (that is not claimed to agree with mortality tables), suppose that x is in the interval [40, 100] and that the probability density function (see Figure 5.6) is given by

$$p(x) = \begin{cases} \frac{1}{36{,}000}(-x^2 + 140x - 4000) & \text{if } 40 \leq x \leq 100, \\ 0 & \text{otherwise.} \end{cases}$$

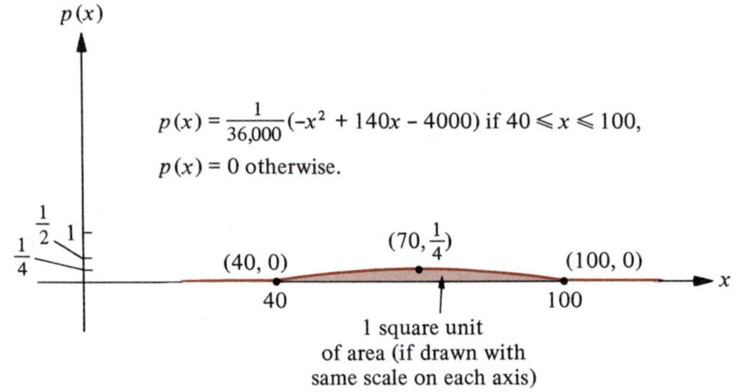

Figure 5.6. *Probability density function.*

Notice from the graph that $p(x) \geq 0$ whenever $40 \leq x \leq 100$ and $p(x) = 0$ otherwise; thus properties 1 and 3 of a probability density function are satisfied. Next we verify property 2. The function p is continuous on [40, 100], and hence we can use Theorem 2.

$$\int_{40}^{100} \frac{1}{36{,}000}(-x^2 + 140x - 4000)\, dx$$

$$= \frac{1}{36{,}000}\left(-\frac{x^3}{3} + 70x^2 - 4000x\right)\Bigg|_{40}^{100}$$

$$= \frac{1}{36{,}000}\left[\left(-\frac{1{,}000{,}000}{3} + 700{,}000 - 400{,}000\right)\right.$$

$$\left. - \left(-\frac{64{,}000}{3} + 112{,}000 - 160{,}000\right)\right]$$

$$= \frac{1}{36{,}000}\left(-\frac{936{,}000}{3} + 348{,}000\right)$$

$$= \frac{1}{36{,}000}(36{,}000) = 1.$$

This means that the company considers it a certainty that the man will die by the time he reaches age 100.

Now determine the probability of the event that the man will die on or before his 60th birthday. Hence $E = \{s : s \leq 60\}$, and

$$P(E) = \int_{40}^{60} \frac{1}{36{,}000}(-x^2 + 140x - 4000)\,dx$$
$$= \frac{1}{36{,}000}\left(-\frac{x^3}{3} + 70x^2 - 4000x\right)\Big|_{40}^{60}$$
$$= \frac{7}{27}.$$

(The arithmetic in the last step has been omitted.)

In summary, for a bounded continuous measurement whose range is $S = [a, b]$, the probability of an event $E = \{s : s \leq c\}$, where $a \leq c \leq b$, can be found by

$$P(E) = F(c),$$

where F is the cumulative probability distribution function, or by

$$P(E) = \int_a^c p(x)\,dx,$$

where p is the probability density function. Moreover, the two functions F and p are related according to

$$p(c) = F'(c),$$

provided that $F'(c)$ exists (see Example 7 of Section 2.2). ∎

★Example 5 Mean or Expected Value

A very important concept in the study of statistics is that of the mean or expected value of a measurement. It is one of three measures of central tendency discussed in this book. The others are the mode and the median.

For *discrete measurements* the **mean** (or **expected value**) is defined as

$$\mu = s_1 w(s_1) + s_2 w(s_2) + s_3 w(s_3) + \cdots,$$

where $S = \{s_1, s_2, s_3, \ldots\}$ is the sample space of all values of the discrete measurement and $w(s_1)$, $w(s_2)$, $w(s_3)$, ... are their respective weights. (The mean may not exist when S is countably infinite.)

As an illustration suppose that a man agrees to draw a single card from an ordinary well-shuffled deck with the following stipulations: If the man draws an ace he will collect \$5, if he draws a king he will collect \$4, and if he draws anything else he will pay \$1. The probabilities of drawing an ace, king, or other card are $\frac{1}{13}$, $\frac{1}{13}$, and $\frac{11}{13}$, respectively. We have $S = \{s_1, s_2, s_3\} = \{5, 4, -1\}$, $w(s_1) = \frac{1}{13}$, $w(s_2) = \frac{1}{13}$, and $w(s_3) = \frac{11}{13}$. The expected value then is

$$\mu = 5(\tfrac{1}{13}) + 4(\tfrac{1}{13}) + (-1)(\tfrac{11}{13}) = -\tfrac{2}{13}.$$

In other words, the man can expect to lose $\$\frac{2}{13}$ per draw in the long run. In *one* draw, however, it is impossible to lose $\$\frac{2}{13}$; rather, the man will win \$5, win \$4, or lose \$1.

For *bounded continuous measurements* the **mean** (or **expected value**) can be calculated by the formula

$$\mu = \int_a^b xp(x)\,dx,$$

where $[a, b]$ is the range of the measurement. A definition of μ and a derivation of this formula is given in Example 3 of Section 5.5. Using the illustration of Example 4 of this section, we calculate the expected time of death, or the mean death age, to be

$$\mu = \int_{40}^{100} \frac{1}{36{,}000} x(-x^2 + 140x - 4000)\,dx = 70. \blacksquare$$

★Example 6 Variance

Once the mean of a measurement is known, it is desirable to have some indication of how scattered the values of the measurement are. One such indicator, called the variance of the measurement, is denoted by σ^2.

For *discrete measurements* the **variance** is defined as

$$\sigma^2 = (s_1 - \mu)^2 w(s_1) + (s_2 - \mu)^2 w(s_2) + (s_3 - \mu)^2 w(s_3) + \cdots,$$

where $S = \{s_1, s_2, s_3, \ldots\}$ is the sample space of all values of the discrete measurement, $w(s_1)$, $w(s_2)$, $w(s_3)$, ... are their respective weights, and μ is the mean of the measurement. Notice that σ^2 can be viewed as the mean of the squares of the differences between the measurement values and measurement mean. The variance may not exist when S is countably infinite.

For *bounded continuous measurements* the **variance** can be calculated by the formula

$$\sigma^2 = \int_a^b (x - \mu)^2 p(x)\,dx,$$

where $[a, b]$ is the range of the measurement, μ its mean, and p the probability density function.

In both cases the positive square root σ of σ^2 is called the **standard deviation** of the measurement. $\blacksquare$

EXERCISES

Suggested minimum assignment: Exercises 1, 6, 7, 9, 11, 12, and 15.

In Exercises 1–8 evaluate the given definite integral.

1. $\int_1^3 x^2\,dx.$
2. $\int_{-4}^2 (2x + 3)\,dx.$
3. $\int_{-1}^1 (x + 1)(x - 1)\,dx.$
4. $\int_1^4 \sqrt{x}\,dx.$

5. $\int_0^1 (3x^2 + 4x + 7)\,dx.$

6. $\int_1^2 \frac{x^3 + 2}{x^2}\,dx.$

7. $\int_{-1}^0 (2x + 1)^8\,dx.$

8. $\int_0^3 x\sqrt{x^2 + 16}\,dx.$

9. Why is it not possible to use Theorem 2 to attempt to evaluate $\int_{-1}^1 \frac{dx}{x^2}$?

10. Is the function f defined by $f(x) = \frac{1}{(x + 1)^2}$ integrable on $[0, 5]$?

11. For the symbol $\int_3^4 x^2\,dx$ state the integrand, the upper limit, the lower limit, and the variable of integration.

In Exercises 12–16 find the area bounded by $y = f(x)$, the x-axis, and the lines $x = a$ and $x = b$.

12. $f(x) = x + 3$, $a = -1$, $b = 5$.
13. $f(x) = x^2$, $a = 3$, $b = 4$.
14. $f(x) = x^3 + 2x$, $a = 1$, $b = 2$.
15. $f(x) = \sqrt{4x + 5}$, $a = 1$, $b = 5$.
16. $f(x) = \frac{10}{(1 - x)^3}$, $a = -2$, $b = 0$.

17. In Example 4 find the probability that the man will die on or before his 50th birthday.

18. (*a*) Show that $\int_{-1}^1 \frac{3}{4}(1 - x^2)\,dx = 1.$

(*b*) Show that the function p defined by

$$p(x) = \begin{cases} \frac{3}{4}(1 - x^2) & \text{if } -1 \le x \le 1, \\ 0 & \text{otherwise,} \end{cases}$$

is a probability density function of a bounded continuous measurement.

(*c*) Show that

$$\int_{-1}^1 (x)(\tfrac{3}{4})(1 - x^2)\,dx = 0,$$

thus proving that the mean of the measurement is zero.

(*d*) Calculate the variance σ^2 of the measurement.

19. A man casts a pair of dice on a flat surface and observes the total that he throws. If the total is 7 or 11 he collects \$3, but for any other total he pays \$1. In the long run, what is the man's expected winning (or loss) per throw in this game?

5.3 Properties of Definite Integrals

In Section 5.1 we defined the definite integral of a function f from a to b as the limit of a certain sum, provided that this limit exists. In the last section we stated the Fundamental Theorem of Integral Calculus, which gives a practical method of evaluating many definite integrals. In this section we state and illustrate some basic properties of definite integrals.

▶ **Theorem 3**

Assume the existence of each definite integral; then the following properties are valid:

(*i*) $\int_a^b kf(x)\,dx = k\int_a^b f(x)\,dx$, where k is a real number.

(*ii*) $\int_a^b f(x)\,dx = \int_a^c f(x)\,dx + \int_c^b f(x)\,dx$, where $a < c < b$.

(*iii*) $\int_a^b [f(x) + g(x)]\,dx = \int_a^b f(x)\,dx + \int_a^b g(x)\,dx$.

(*iv*) $\int_a^b f(x)\,dx = \int_a^b f(t)\,dt$.

The proof of Theorem 3 depends upon the definition of the definite integral given in Section 5.1, and is left as Exercise 18. ◀

Example 1

(*a*) $\int_1^4 3x^2\,dx = 3\int_1^4 x^2\,dx$; by Theorem 3(*i*), the multiplier 3 may be moved across the integral sign—never move anything that involves the variable of integration in this manner, however.

(*b*) $\int_1^4 x^2\,dx = \int_1^3 x^2\,dx + \int_3^4 x^2\,dx$; by using Theorem 3(*ii*) again, we could write

$$\int_1^4 x^2\,dx = \int_1^2 x^2\,dx + \int_2^3 x^2\,dx + \int_3^4 x^2\,dx.$$

(*c*) $\int_1^4 (x^2 + x^3)\,dx = \int_1^4 x^2\,dx + \int_1^4 x^3\,dx$; Theorem 3(*iii*) also can be extended to handle sums of more than two functions.

(*d*) $\int_1^4 x^2\,dx = \int_1^4 t^2\,dt = \int_1^4 w^2\,dw$; by Theorem 3(*iv*), the particular letter used as the variable of integration is immaterial—hence it is sometimes called a "dummy variable." ▮

In Section 5.1 we defined $I_a^b(f)$ or $\int_a^b f(x)\,dx$ under the assumption that $a < b$. In order to give meaning to $\int_a^b f(x)\,dx$ in cases when $a \geq b$, it is customary to make the following definition.

Definition 2

(*i*) $\int_a^a f(x)\,dx = 0$, provided that $f(a)$ exists.

(*ii*) If $a > b$ and $\int_b^a f(x)\,dx$ exists, then

$$\int_a^b f(x)\,dx = -\int_b^a f(x)\,dx.$$

Example 2

(*a*) $\int_4^4 x^2\,dx = 0.$

(*b*) $\int_4^1 x^2\,dx = -\int_1^4 x^2\,dx.$ ∎

Theorem 3 is still valid if $a \geq b$, and in fact the restriction in part (*ii*) that c be between a and b also can be removed.

From our geometric interpretation of the definite integral in Sections 5.1 and 5.2 we know that if a region above the x-axis, below $y = f(x)$, and between $x = a$ and $x = b$ has an area of 5 square units and f is continuous on $[a, b]$, then $\int_a^b f(x)\,dx = 5$. See Figure 5.7. If the graph of $y = g(x)$ were below the x-axis between $x = a$ and $x = b$ (see Figure 5.8), then the numbers $g(c_1), g(c_2), \ldots, g(c_n)$ in S_n would be negative and this would lead to a negative value of the definite integral—that is, $\int_a^b g(x)\,dx = -5$. Suppose that the graph of $y = h(x)$ crosses the x-axis at a point c between a and b (see Figure 5.9). Then

$$\int_a^b h(x)\,dx = \int_a^c h(x)\,dx + \int_c^b h(x)\,dx$$
$$= -6 + 4 = -2.$$

The total shaded area, however, is $|-6| + |4| = 10$ square units.

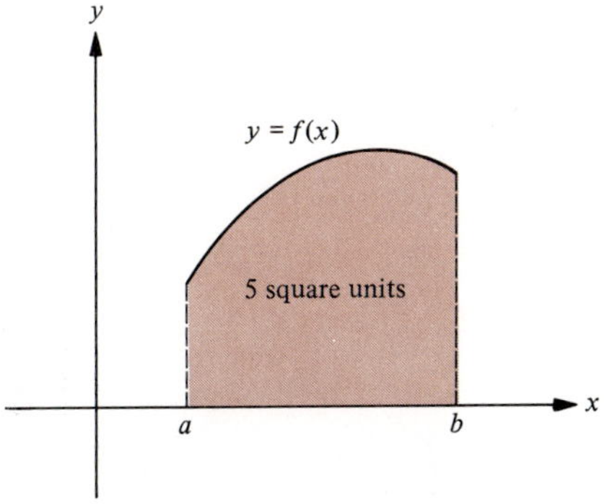

Figure 5.7. $\int_a^b f(x)\,dx = 5.$

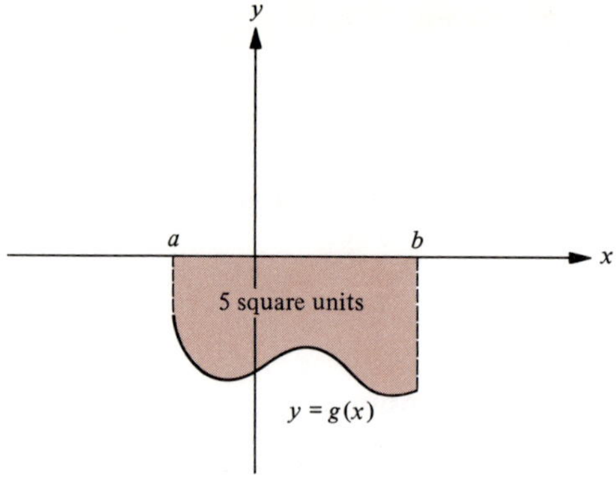

Figure 5.8. $\int_a^b g(x)\,dx = -5.$

The next property, which we state without proof, is often called the **Mean Value Theorem for Integrals.**

Theorem 4

If f is continuous on $[a, b]$, then there exists a number c, where $a \le c \le b$, such that

$$\int_a^b f(x)\,dx = (f(c))(b - a).$$

A geometric interpretation of Theorem 4 follows. In Figure 5.10 a continuous function f on a closed interval $[a, b]$ is shown (f is chosen intentionally to be positive throughout $[a, b]$ because the geometric interpretation of Theorem 4 is more readily understood in this case). In Figure 5.10 a rectangle $KLMN$ is constructed in such a way that its area, $(f(c))(b - a)$, is equal to the area under the curve, which is given by $\int_a^b f(x)\,dx$.

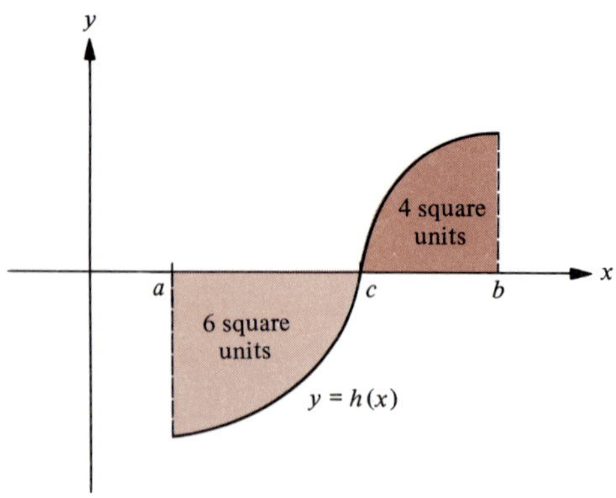

Figure 5.9. $\int_a^b h(x)\,dx = -2.$

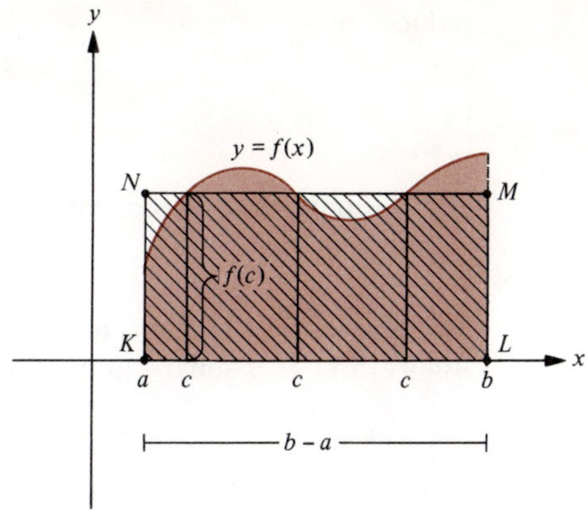

Figure 5.10. *Illustration of Theorem 4.*

There are three points in $[a, b]$ in Figure 5.10 at which c can be chosen so as to make the area under the curve, $\int_a^b f(x)\,dx$, equal the area $(f(c))(b - a)$ of rectangle $KLMN$; Theorem 4 guarantees the existence of at least one such number c in $[a, b]$.

Example 3

If $a = 0$, $b = 3$, and $f(x) = x^2$, find one value of the number c in Theorem 4.

In Figure 5.11 the portion of the parabola $y = x^2$ between $x = 0$ and $x = 3$ is shown.

$$\int_a^b f(x)\,dx = \int_0^3 x^2\,dx = \frac{x^3}{3}\Big|_0^3 = \frac{27}{3} - 0 = 9.$$

Hence the area between $x = 0$ and $x = 3$ which is above the x-axis and below $y = x^2$ is 9 square units. Thus, since $f(c) = c^2$ and $b - a = 3 - 0$,

$$\int_0^3 x^2\,dx = (f(c))(b - a)$$

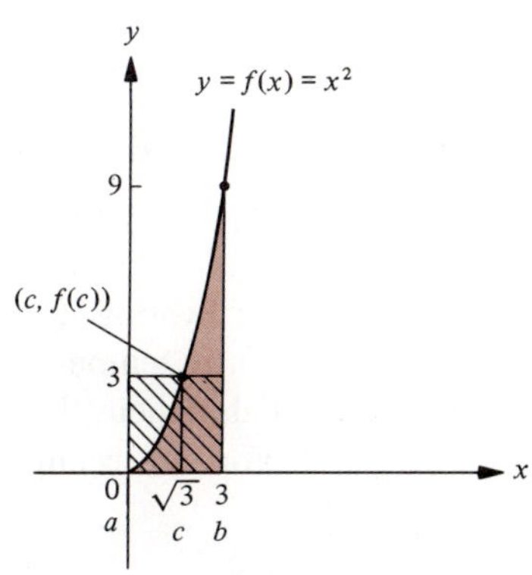

Figure 5.11

reduces to

$$9 = c^2(3 - 0)$$

or

$$9 = 3c^2,$$

and the only root of this equation in [0, 3] is

$$c = \sqrt{3}.$$

The answer can be checked by showing that the area of the rectangle in Figure 5.11 is 9 square units; the base is $b - a = 3$ and the height is

$$f(c) = f(\sqrt{3}) = (\sqrt{3})^2 = 3,$$

and hence the area of the rectangle is 9, which equals the area under the curve. ▮

APPLICATIONS

Example 4

Let a and b be real numbers, where $a < b$, and suppose that a function f is continuous on $[a, b]$. The **average value of f on $[a, b]$** is defined to be

$$\frac{1}{b - a}\int_a^b f(x)\,dx.$$

This quantity is equal to the number $f(c)$ in Theorem 4 of this section. From either Figure 5.10 or 5.11, it can be seen why $f(c)$ is considered to be the average y value on $[a, b]$ when $y = f(x)$ is plotted on $[a, b]$.

For a specific illustration let us return to Example 1 of Section 5.1. In that example a function f with domain [0, 24] was given in which temperatures for 1 day were expressed as a function of the hour of the day. The average value of this function on [0, 24] is called the **average daily temperature.** Using our approximation of 456 for $I_0^{24}(f) = \int_0^{24} f(x)\,dx$ from Example 1 of Section 5.1, we find that the average daily temperature in degrees Fahrenheit is given by

$$\frac{1}{24 - 0}\int_0^{24} f(x)\,dx \approx \frac{1}{24}(456) = 19.$$

If the average daily temperature is less than 65 degrees Fahrenheit, then this average can be subtracted from 65 to obtain what is defined to be the number of **heating degree days** for the particular day in question. If the average daily temperature for a given day is equal to or greater than 65 degrees Fahrenheit, then the number of heating degree days is defined to equal 0. For our example the number of heating degree days is

$$65 - 19 = 46.$$

The number of heating degree days is usually printed in a daily newspaper each day. It can be used, for example, by fuel suppliers to estimate the amount of fuel consumption and thus to determine when to make the automatic deliveries that are promised.

Observe from the data of Example 1 of Section 5.1 that the highest recorded temperature for the day is 32 and the low is 12. These can quickly be averaged to obtain 22, which can be used in place of 19 as an estimation of the average daily temperature. This shortcut method tends to be rather inaccurate, particularly under certain weather conditions. ∎

Example 5

Definite integrals have an important role in many applications. On pp. 194–200 of his *Mathematical Techniques for Physiology and Medicine* [20], Simon finds definite integrals useful in his discussion of the theory of blood flow measurement. In the indicator dilution method of blood flow measurement, a known amount of an indicator can be injected into the bloodstream and the flow rate determined by the rate of dilution of the indicator. Sometimes the indicator is injected into the bloodstream at a constant rate for enough time so that the concentration at some point downstream is approximately constant. The quantity of the indicator substance that passes the point downstream from time t_1 to time t_2 is given by

$$\int_{t_1}^{t_2} F(t)C(t)\,dt,$$

where $F(t)$ represents the flow rate at time t and $C(t)$ represents the concentration at time t. Simon also finds the concept of the average value of a function on an interval (see Example 4) to be helpful in his discussion of the theory of blood flow measurement. ∎

***Example 6**

In Example 4 of Section 5.2 we expressed the probability of the event $E = \{s:s \leq 60\}$ that a man would die on or before his 60th birthday as

$$P(E) = \int_{40}^{60} p(x)\,dx = \tfrac{7}{27}.$$

Suppose that for the same probability density function p, we wish to find the probability of the event $E_1 = \{s:50 < s \leq 60\}$ that the man will die between ages 50 and 60. Let $E_2 = \{s:s \leq 50\}$ so that

$$\begin{aligned} P(E_1) &= P(E) - P(E_2) \\ &= \int_{40}^{60} p(x)\,dx - \int_{40}^{50} p(x)\,dx. \end{aligned}$$

By Theorem 3(*ii*) of this section the latter expression is equal to

$$\int_{50}^{60} p(x)\,dx,$$

which can be calculated to be $\frac{5}{27}$. A valid geometric interpretation, which is suggested by the illustration above, is that the probability that a value of a bounded continuous measurement lies between numbers c and d is equal to the area under the graph of p and between $x = c$ and $x = d$. ▮

EXERCISES

Suggested minimum assignment: Exercises 2, 5, 7, 9, 11, and 14.

In Exercises 1–8 use Figure 5.12 to evaluate the given definite integral.

1. $\int_a^b f(x)\,dx.$
2. $\int_b^c f(x)\,dx.$
3. $\int_c^d f(t)\,dt.$
4. $\int_a^c f(x)\,dx.$
5. $\int_b^d 3f(x)\,dx.$
6. $\int_a^d f(x)\,dx.$
7. $\int_d^c f(x)\,dx.$
8. $\int_b^b f(x)\,dx.$

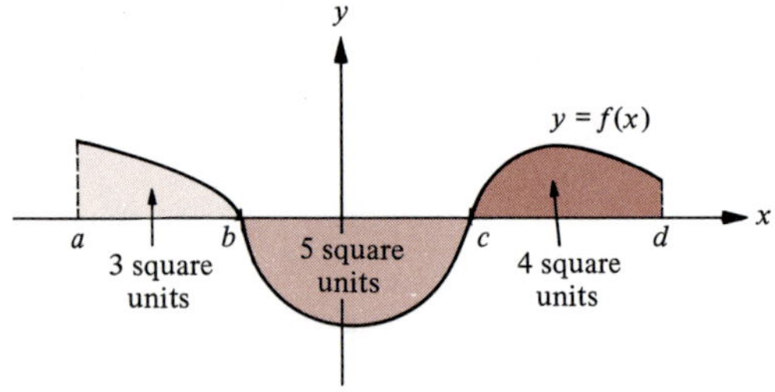

Figure 5.12. *Figure for Exercises 1–8.*

In Exercises 9–12 find one value of the number c in Theorem 4. Give a geometric interpretation.

9. $a = 1$, $b = 3$, $f(x) = x^2$.
10. $a = 0$, $b = 2$, $f(x) = x^2$.
11. $a = -1$, $b = 2$, $f(x) = 2x$.
12. $a = 0$, $b = 5$, $f(x) = x^2 + x$. (*Hint:* Use the quadratic formula.)

In Exercises 13–16 evaluate the given definite integral.

13. $\int_1^3 (x^3 + x^2)\,dx.$
14. $\int_{-2}^1 (3t + 2t^2)\,dt.$
15. $\int_0^4 (2x^2 - 5x + 3)\,dx.$
16. $\int_{-2}^3 |x|\,dx.$
17. Evaluate $\int_0^R \frac{P\pi}{2\eta l} r(R^2 - r^2)\,dr$, where R, P, η, l, and π are constants. (This definite integral occurs in the calculation of the volume of flow of blood per unit of time in an arteriole; see pp. 189–194 of Thrall et al. [23].)

18. Outline a proof of each of the following.
(*a*) Theorem 3(*i*). (*b*) Theorem 3(*ii*).
(*c*) Theorem 3(*iii*). (*d*) Theorem 3(*iv*).
19. Let $f(x) = x^3 - x^2 + 2$. Find the average value (see Example 4) of f on $[1, 3]$.
20. Refer to Example 6 and write (but do not calculate) the definite integral that gives the probability of death between ages 70 and 80. Give a geometric interpretation.

5.4 Area Between Curves

In many calculus books one can read about how to use definite integrals to calculate such things as areas, volumes, centroids, arc length, and work. The desired quantity is approximated by the type of sum which, after a limiting process, leads to a definite integral. In this section we show how to use definite integrals to calculate the area between curves whose equations are known.

First we illustrate one general type of situation that will occur; in Figure 5.13 it is desired to find the area from $x = a$ to $x = b$ between an upper curve $y = f(x)$ and a lower curve $y = g(x)$. We partition $[a, b]$ into n subintervals and choose arbitrary points $c_1, \ldots, c_n$ in these subintervals just as we did in Section 5.1. The desired area can be approximated if rectangles are drawn in such a way (the first rectangle and a typical kth rectangle are shown in Figure 5.13) that the sum S_n of the areas of these rectangles is given by

$$S_n = [f(c_1) - g(c_1)]\,\Delta x_1 + \cdots + [f(c_n) - g(c_n)]\,\Delta x_n.$$

To get the exact area, let $n \to \infty$ and let the norm δ_n of the partition approach zero; from the definition of a definite integral this limit yields

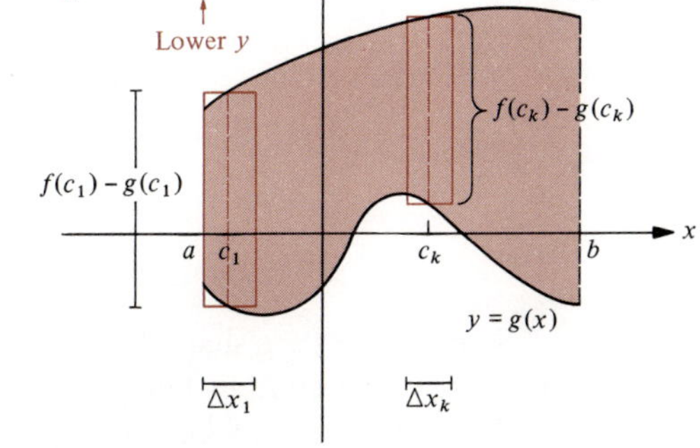

Figure 5.13. *Area using vertical strips.*

$\int_a^b [f(x) - g(x)]\,dx$, and hence the area A of the desired region is given by

$$A = \int_a^b [f(x) - g(x)]\,dx. \tag{5.1}$$

The upper curve $y = f(x)$ is always written first in (5.1) in order that $f(x) - g(x)$ will be nonnegative, and thereby assure that A is positive when $b > a$.

Example 1

Find the area of the region bounded by $y = \frac{1}{2}x^2$ and $y = 4\sqrt{x}$.

Always sketch the given curves accurately enough to be certain which area is desired and to find key points of intersection. The parabola $y = \frac{1}{2}x^2$ is symmetric to the y-axis and passes through (0, 0), (2, 2), and (4, 8). The equation $y = 4\sqrt{x}$ is the equation of the upper half of the parabola $y^2 = 16x$ and passes through (0, 0), (1, 4), and (4, 8). See Figure 5.14. When we compare our problem with the general case illustrated in Figure 5.13, it can be seen that in this example $a = 0$, $b = 4$, $f(x) = 4\sqrt{x}$, and $g(x) = \frac{1}{2}x^2$. Notice that $y = 4\sqrt{x}$ is the upper curve throughout *all* the interval [0, 4]; hence $f(x) - g(x) = 4\sqrt{x} - \frac{1}{2}x^2$ is nonnegative on [0, 4]. From (5.1) the area A of the shaded region in Figure 5.14 is given by

$$\begin{aligned} A &= \int_0^4 \left(4\sqrt{x} - \frac{1}{2}x^2\right) dx \\ &= \left(4\frac{x^{3/2}}{\frac{3}{2}} - \frac{1}{2}\frac{x^3}{3}\right)\Bigg|_0^4 \\ &= \left[\left(\frac{8}{3}\right)(8) - \left(\frac{1}{2}\right)\left(\frac{64}{3}\right)\right] - [0 - 0] = \frac{32}{3}. \quad \blacksquare \end{aligned}$$

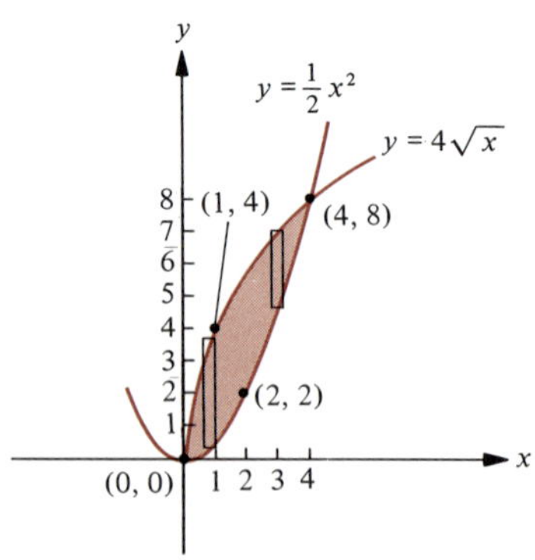

Figure 5.14

Some regions are shaped so that they do not have one curve as an upper boundary and another one as a lower boundary; instead, between $y = c$ and $y = d$, there may be one curve $x = f(y)$ which forms a right boundary and another curve $x = g(y)$ which serves as a left boundary (see Figure

5.15). It is convenient in this case to solve the equations for x in terms of y, if possible. After forming S_n from horizontal rectangles and passing to the limit, we obtain

$$A = \int_c^d [f(y) - g(y)]\, dy. \tag{5.2}$$

The curve $x = f(y)$ which serves as the right boundary is always written first in (5.2) in order that $f(y) - g(y)$ will be nonnegative, and thereby assure that A is positive when $d > c$.

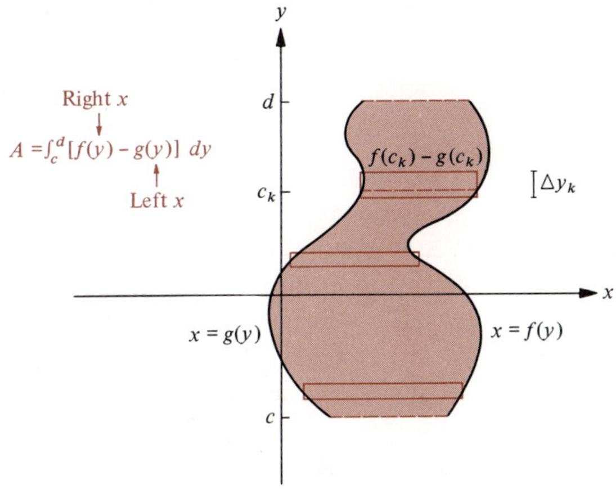

Figure 5.15. *Area using horizontal strips.*

Example 1 was selected so that it could also be worked using (5.2), and this is shown in Example 2.

Example 2

Rework Example 1 using (5.2). See Figure 5.16.

Notice that between $y = 0$ and $y = 8$ the curve $x = \sqrt{2y}$ is always on the right and $x = \frac{1}{16}y^2$ is always on the left so that, by (5.2), the area A

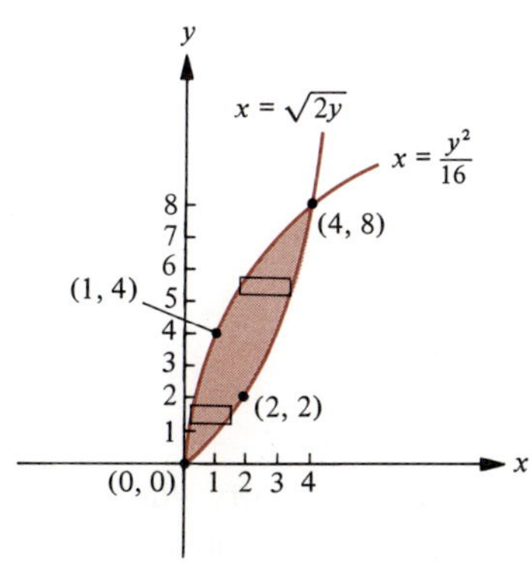

Figure 5.16

is given by

$$
\begin{aligned}
A &= \int_0^8 \left(\sqrt{2y} - \frac{1}{16}y^2\right) dy \\
&= \sqrt{2}\int_0^8 y^{1/2}\,dy - \frac{1}{16}\int_0^8 y^2\,dy \\
&= \left(\sqrt{2}\frac{y^{3/2}}{\frac{3}{2}} - \frac{1}{16}\frac{y^3}{3}\right)\Bigg|_0^8 \\
&= \left[\left(\frac{2\sqrt{2}}{3}\right)(8\sqrt{8}) - \frac{1}{16}\frac{(8)^3}{3}\right] - (0 - 0) \\
&= \left(\frac{2\sqrt{2}}{3}\right)(8)(2\sqrt{2}) - \left(\frac{1}{2}\right)\left(\frac{64}{3}\right) = \frac{64}{3} - \frac{32}{3} = \frac{32}{3}. \ \blacksquare
\end{aligned}
$$

Example 3

Find the area bounded by the parabola $x = y^2 - 1$ and the line $y = x - 1$. See Figure 5.17.

In this example it is preferable to solve the equations for x in terms of y and use formula (5.2) because, between $y = -1$ and $y = 2$, $x = f(y) = y + 1$ is always the right boundary of the region and $x = g(y) = y^2 - 1$ is always the left boundary.

$$
\begin{aligned}
A &= \int_{-1}^2 [(y + 1) - (y^2 - 1)]\,dy \\
&= \int_{-1}^2 (-y^2 + y + 2)\,dy = \left(-\frac{y^3}{3} + \frac{y^2}{2} + 2y\right)\Bigg|_{-1}^2 \\
&= \left(-\frac{8}{3} + 2 + 4\right) - \left(\frac{1}{3} + \frac{1}{2} - 2\right) = \frac{9}{2}. \ \blacksquare
\end{aligned}
$$

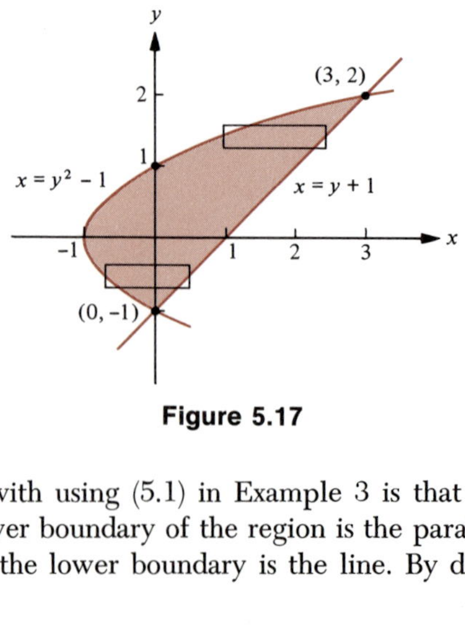

Figure 5.17

The difficulty with using (5.1) in Example 3 is that between $x = -1$ and $x = 0$ the lower boundary of the region is the parabola, but between $x = 0$ and $x = 3$ the lower boundary is the line. By dividing the region into two parts (a necessary technique for some problems but not recommended in Example 3), the area can be set up using (5.1) as follows:

$$A = \int_{-1}^{0} [\sqrt{x+1} - (-\sqrt{x+1})]\, dx + \int_{0}^{3} [\sqrt{x+1} - (x-1)]\, dx.$$

Now that the reader has seen some examples, we suggest a procedure to follow for finding the area between curves:

1. Draw a figure and mark clearly the area to be found. Label key points of intersection of the given curves (find these points in the process of plotting points or by solving the equations algebraically).

2. Decide whether to use (5.1), use (5.2), or divide the region into more than one part so that one of the formulas applies to each part. Show some approximating rectangles on your figure—vertical rectangles if you use (5.1) or horizontal rectangles if you use (5.2). Use (5.1) if the region has a single upper and a single lower boundary, or apply (5.2) if the region has a single right and a single left boundary.

3. Write the necessary definite integrals for the area. [Set them up so that the answer will be positive—in (5.1) let $a < b$ so that the Δx_k in an approximating sum S_n are positive, and in order that the $f(c_k) - g(c_k)$ are positive, let $y = f(x)$ be the upper curve and $y = g(x)$ the lower curve; in (5.2) let $c < d$, and let $x = f(y)$ be the bounding curve on the right and $x = g(y)$ the one on the left.] When you use (5.1), the upper and lower limits should be x limits and the integrand a function of x—for (5.2) the upper and lower limits should be y limits and the integrand a function of y.

4. Evaluate the definite integrals by using the Fundamental Theorem of Integral Calculus, and check the answer for gross errors by estimating the answer from the figure. (If you cannot evaluate the definite integrals, calculate an approximating sum as done in Section 5.1.)

APPLICATIONS

Example 4

Suppose that the equation of the demand curve for a certain style of record player is given by

$$y = 250 - \tfrac{5}{6}x.$$

In this equation y is the selling price in dollars and x is the number of units demanded by consumers each week. Suppose that the supply curve for the same item has the equation

$$y = \frac{x^2}{200} + 78,$$

where x is the number of record players that would be supplied weekly by producers for sale at y dollars each. The demand and supply curves are both shown in Figure 5.18.

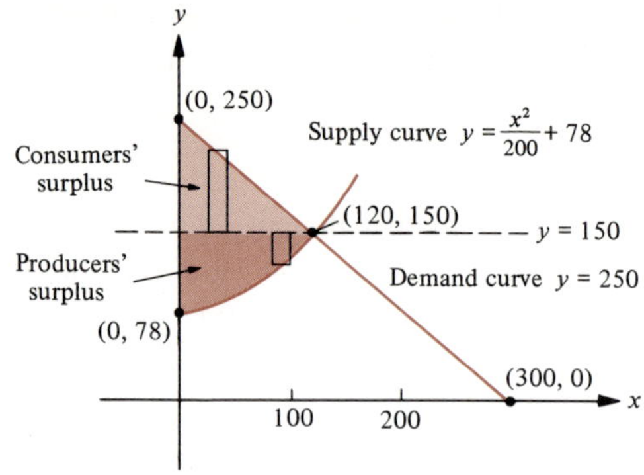

Figure 5.18. *Consumers' and producers' surpluses.*

It can be verified that the point (120, 150) satisfies each of the given equations; this point of intersection can be found by solving simultaneously the given equations of the line and the parabola. We shall assume that we are dealing with the situation of pure competition, in which case the actual selling price is determined by the intersection of the demand and supply curves. Hence 120 of our particular style of record player are sold each week for $150 apiece.

From Figure 5.18 it can be seen that some consumers are willing to pay up to $250 for one of the record players. When they find that the cost is only $150, they feel that they have saved money. The total amount saved in this way each week by the 120 consumers who make purchases is called **consumers' surplus** (CS) and is represented by the area above $y = 150$ and below $y = 250 - \frac{5}{6}x$. In this problem the area is a triangle and could be found without the use of definite integrals, but the use of (5.1) of this section is a more general approach for this type of problem; the calculation of consumers' surplus using (5.1) is as follows:

$$
\begin{aligned}
CS &= \int_0^{120} [(250 - \tfrac{5}{6}x) - (150)]\, dx \\
&= \int_0^{120} (100 - \tfrac{5}{6}x)\, dx \\
&= (100x - \tfrac{5}{12}x^2)\Big|_0^{120} \\
&= (12{,}000 - 6000) - (0 - 0) \\
&= \$6000.
\end{aligned}
$$

The producers of the record player also realize a gain, called the **producers' surplus,** in the sense that Figure 5.18 shows that some of them would have built the record players even if they had known that the selling price were below $150. The producers' surplus (PS) is represented by the area below $y = 150$ and above $y = \dfrac{x^2}{200} + 78$, and is computed using (5.1) as follows:

$$
\begin{aligned}
PS &= \int_0^{120} \left[150 - \left(\frac{x^2}{200} + 78\right)\right] dx \\
&= \int_0^{120} \left(72 - \frac{x^2}{200}\right) dx \\
&= \left(72x - \frac{x^3}{600}\right)\Bigg|_0^{120} \\
&= (8640 - 2880) - (0 - 0) \\
&= \$5760.
\end{aligned}
$$

In our computation of consumers' surplus and producers' surplus in this example, we have assumed that the equations of the demand and supply curves are valid at least in the interval [0, 120]. ▮

EXERCISES

Suggested minimum assignment: Exercises 3, 7, 8, and 13.

In each of Exercises 1–10 find the area of the region bounded by the given curves.

1. $y = x^2$, $y = 0$, $x = 1$, $x = 2$.
2. $y = x^2$, $y = 2x$.
3. $y = x^2$, $y = x + 2$.
4. $x = y^2$, $x = 4$.
5. $x = y^2$, $y = x - 2$.
6. $y = \sqrt{x}$, $x = 2y$. Work by two methods (see Examples 1 and 2).
7. $y = x^3$, $y = 8$, $x = 0$. Work by two methods (see Examples 1 and 2).
8. $x = y^2 - 4$, $x = -y^2$.
9. $y = x^2 - 2x - 3$, $y = 0$.
10. $y = x^2 - 2x - 3$, $y = 2x - 3$.

11. Find the total area between $x = -2$ and $x = 0$ that is bounded by the x-axis and the parabola $y = x^2 - 2x - 3$. (*Hint:* The parabola crosses the x-axis between $x = -2$ and $x = 0$.)
12. (*a*) Find the area bounded by $y = 2$, $x = 9$, and $x = 5 - 4y - y^2$.
    (*b*) How much of the area in part (*a*) lies in the first quadrant?
    (*c*) How much of the area in part (*a*) lies in the second quadrant?
    (*d*) How much of the area in part (*a*) lies in the fourth quadrant?
13. (*a*) Find the area bounded by $y = \sqrt{5x - 1}$, $y = 0$, and $x = 10$.
    (*b*) How much of the area in part (*a*) is above $y = 2$?
14. (*a*) Calculate the consumers' surplus in Example 4 using (5.2).
    (*b*) Calculate the producers' surplus in Example 4 using (5.2).
15. In Example 4 suppose that the equations of the demand and supply curves are, respectively, $y = 100 - \dfrac{x^2}{10}$ and $y = 50 + \dfrac{x^2}{40}$.
    (*a*) Compute the consumers' surplus.
    (*b*) Compute the producers' surplus.

5.5 More About Calculation of Definite Integrals (Optional)

In this section we shall try to develop a better understanding of definite integrals by showing how to evaluate some simple definite integrals by use of the long limit process of Definition 1 rather than the more efficient Fundamental Theorem of Integral Calculus (Theorem 2).

We start by assuming without proof two well-known formulas that are valid for each positive integer n:

$$1 + 2 + 3 + \cdots + n = \frac{n(n+1)}{2} \tag{5.3}$$

and

$$1^2 + 2^2 + 3^2 + \cdots + n^2 = \frac{n(n+1)(2n+1)}{6}. \tag{5.4}$$

Formula (5.3) is a formula for adding the first n positive integers; it may be familiar to anyone who has studied arithmetic progressions. We verify (5.3) for $n = 8$ as follows: the left side is

$$1 + 2 + 3 + 4 + 5 + 6 + 7 + 8 = 36,$$

and the right side is

$$\frac{8(8+1)}{2} = 36.$$

Formula (5.4) gives an expression for the sum of the squares of the first n positive integers and can be verified in a similar manner for any particular positive integer n. A general verification of these formulas involves the use of mathematical induction. Formula (5.3) will be used in Example 1, and (5.4) will be used in Example 2.

The evaluation of $I_a^b(f) = \int_a^b f(x)\,dx$ by Definition 1 can be very difficult, but in the examples that follow we let f be a polynomial function of degree 1 or 2 so that the procedure will not be unreasonably complicated. Polynomial functions are continuous on any interval $[a, b]$ and therefore integrable on $[a, b]$ by Theorem 1. Hence the division points for the n subintervals of $[a, b]$ may be chosen in any manner, and we elect to divide $[a, b]$ into n *equal* parts each of length

$$\Delta x = \frac{b-a}{n}.$$

Therefore, our division points are

$$\begin{aligned} x_0 &= a, \\ x_1 &= a + \Delta x, \\ x_2 &= a + 2\,\Delta x, \\ &\vdots \\ x_n &= b = a + n\,\Delta x. \end{aligned}$$

The points $c_1, \ldots, c_n$ in Section 5.1 may be chosen in any convenient manner, and we shall find that it works out well to let

$$\begin{aligned} c_1 &= x_1, \\ c_2 &= x_2, \\ &\vdots \\ c_n &= x_n. \end{aligned}$$

The norm δ_n of the partition $\{x_0, x_1, x_2, \ldots, x_n\}$ is $\delta_n = (b - a)/n$, since the length of each subinterval is $(b - a)/n$. Therefore,

$$\lim_{n\to\infty} \delta_n = \lim_{n\to\infty} \frac{b - a}{n} = 0.$$

This means that for our examples

$$\int_a^b f(x)\,dx = \lim_{\substack{n\to\infty \\ \delta_n\to 0}} S_n$$

can be simplified to

$$\int_a^b f(x)\,dx = \lim_{n\to\infty} S_n.$$

Example 1

Use Definition 1 to evaluate $\int_1^3 x\,dx$.

Note that $a = 1$, $b = 3$, $f(x) = x$, and

$$\Delta x = \frac{b - a}{n} = \frac{2}{n}.$$

We also have (see Figure 5.19)

$$\begin{aligned} c_1 &= x_1 = 1 + \frac{2}{n}, \\ c_2 &= x_2 = 1 + 2\left(\frac{2}{n}\right), \\ &\vdots \\ c_n &= x_n = 1 + n\left(\frac{2}{n}\right). \end{aligned}$$

Therefore,

$$\begin{aligned} S_n &= f(c_1)\,\Delta x_1 + f(c_2)\,\Delta x_2 + \cdots + f(c_n)\,\Delta x_n \\ &= f(c_1)\,\Delta x + f(c_2)\,\Delta x + \cdots + f(c_n)\,\Delta x \\ &= \frac{2}{n}[f(c_1) + f(c_2) + \cdots + f(c_n)]. \end{aligned}$$

Since $f(x) = x$, then $f(c_i) = c_i$ and S_n reduces to

$$\begin{aligned} S_n &= \frac{2}{n}[c_1 + c_2 + \cdots + c_n] \\ &= \frac{2}{n}\left[\left(1 + \frac{2}{n}\right) + \left(1 + 2\left(\frac{2}{n}\right)\right) + \cdots + \left(1 + n\left(\frac{2}{n}\right)\right)\right]. \end{aligned}$$

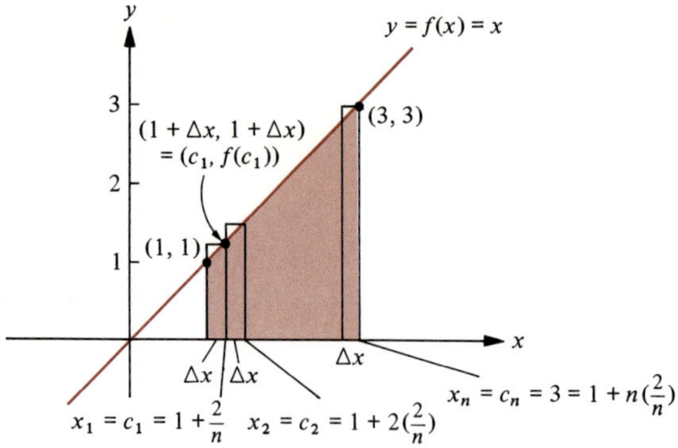

Figure 5.19

Using (5.3) and the fact that the sum of n ones is n, we obtain

$$\begin{aligned} S_n &= \frac{2}{n}\left[(1 + 1 + \cdots + 1) + \frac{2}{n}(1 + 2 + \cdots + n)\right] \\ &= \frac{2}{n}\left[n + \frac{2}{n}\frac{n(n + 1)}{2}\right] \\ &= \frac{2}{n}[n + (n + 1)] = \frac{4n + 2}{n}. \end{aligned}$$

Therefore,

$$\begin{aligned} I_1^3(f) = \int_1^3 x\,dx = \lim_{n\to\infty} S_n &= \lim_{n\to\infty} \frac{4n + 2}{n} \\ &= \lim_{n\to\infty} \frac{4 + 2/n}{1} = \frac{4 + 0}{1} = 4. \quad \blacksquare \end{aligned}$$

As a check on the answer to Example 1 we observe by the Fundamental Theorem of Integral Calculus that

$$I_1^3(f) = \int_1^3 x\,dx = \frac{x^2}{2}\bigg|_1^3 = \frac{9}{2} - \frac{1}{2} = 4.$$

Notice how much easier the latter method is. As another check we observe that geometrically $\int_1^3 x\,dx$ represents the area of the trapezoid bounded by $x = 1$, $x = 3$, $y = 0$, and $y = x$ (see Figure 5.19), and the area of this trapezoid can easily be seen to equal 4.

Example 2

Use Definition 1 to evaluate $\int_0^3 x^2\,dx$.

In this case $a = 0$, $b = 3$, $f(x) = x^2$, and $\Delta x = 3/n$. Next observe that

$$c_1 = x_1 = 0 + 1\left(\frac{3}{n}\right) = 1\left(\frac{3}{n}\right),$$

$$c_2 = x_2 = 0 + 2\left(\frac{3}{n}\right) = 2\left(\frac{3}{n}\right),$$

$$\vdots$$

$$c_n = x_n = 0 + n\left(\frac{3}{n}\right) = n\left(\frac{3}{n}\right).$$

Some of the approximating rectangles are shown in Figure 5.20; the fact that $f(x) = x^2$ means that

$$f(c_1) = c_1^2 = 1^2\left(\frac{3}{n}\right)^2,$$

$$f(c_2) = c_2^2 = 2^2\left(\frac{3}{n}\right)^2,$$

$$\vdots$$

$$f(c_n) = c_n^2 = n^2\left(\frac{3}{n}\right)^2,$$

and the appearance of $1^2, 2^2, \ldots, n^2$ leads to a need for (5.4) in the

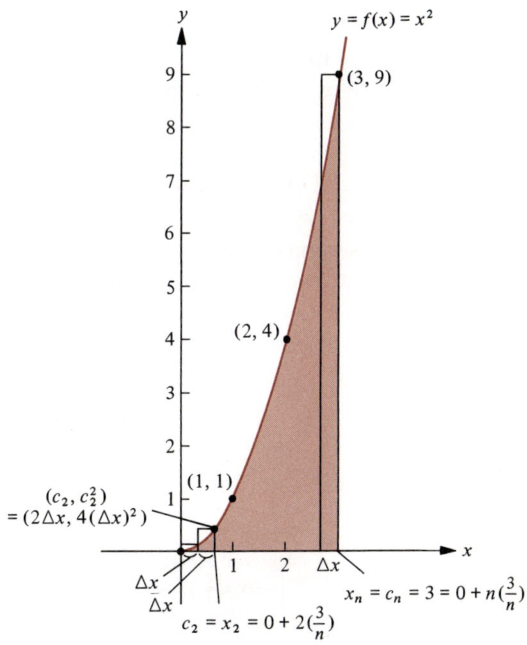

Figure 5.20

calculation that follows.

$$\begin{aligned} S_n &= f(c_1)\,\Delta x_1 + f(c_2)\,\Delta x_2 + \cdots + f(c_n)\,\Delta x_n \\ &= \frac{3}{n}[f(c_1) + f(c_2) + \cdots + f(c_n)] \\ &= \frac{3}{n}\left[1^2\left(\frac{3}{n}\right)^2 + 2^2\left(\frac{3}{n}\right)^2 + \cdots + n^2\left(\frac{3}{n}\right)^2\right] \\ &= \frac{3}{n}\left(\frac{3}{n}\right)^2[1^2 + 2^2 + \cdots + n^2] \\ &= \frac{27}{n^3}\left[\frac{n(n+1)(2n+1)}{6}\right] \\ &= \frac{9}{2}\left(\frac{2n^2 + 3n + 1}{n^2}\right). \end{aligned}$$

Therefore,

$$\begin{aligned} I_0^3(f) = \int_0^3 x^2\,dx &= \lim_{n\to\infty} \frac{9}{2}\left(\frac{2 + (3/n) + (1/n^2)}{1}\right) \\ &= \frac{9}{2}\left(\frac{2+0+0}{1}\right) \\ &= 9. \end{aligned}$$ ∎

APPLICATIONS

*Example 3

Let us continue our discussion of the mean or expected value of a measurement that was begun in Example 5 of Section 5.2. In this example we shall define the mean of a bounded continuous measurement, where $[a, b]$ is the range of the measurement.

First, let p be the probability density function for a bounded continuous measurement with range $[a, b]$. Divide the interval $[a, b]$ into n equal subintervals, each of length Δx, as shown in Figure 5.21. From Example 6 of Section 5.3 we know that the probability of the event E_k that a value of the measurement is a number in the kth subinterval is

$$P(E_k) = \int_{x_{k-1}}^{x_k} p(x)\,dx \qquad (k = 1, 2, \ldots, n).$$

By the Mean Value Theorem for Integrals (Theorem 4 of Section 5.3), we

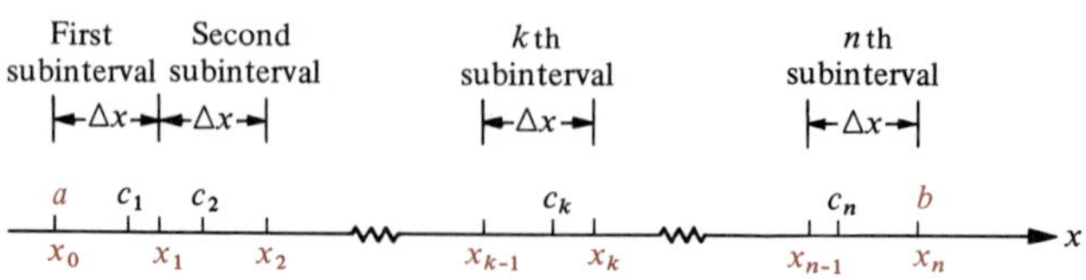

Figure 5.21

know that if p is continuous on $[a, b]$, there is some number c_k, where $x_{k-1} \le c_k \le x_k$, such that

$$\begin{aligned} P(E_k) &= p(c_k)(x_k - x_{k-1}) \\ &= p(c_k)\,\Delta x \qquad (k = 1, 2, \ldots, n). \end{aligned}$$

Second, consider the mean of a discrete measurement having values $c_1, c_2, \ldots, c_n$ with respective weights $w(c_1), w(c_2), \ldots, w(c_n)$; letting $w(c_k) = P(E_k)$, we have

$$w(c_k) = P(E_k) = p(c_k)\,\Delta x \qquad (k = 1, 2, \ldots, n).$$

Since there are a finite number of values of the discrete measurement, the mean is (see Example 5 of Section 5.2)

$$\begin{aligned} &c_1 w(c_1) + c_2 w(c_2) + \cdots + c_n w(c_n) \\ &\qquad = c_1[p(c_1)\,\Delta x] + c_2[p(c_2)\,\Delta x] + \cdots + c_n[p(c_n)\,\Delta x] \\ &\qquad = [c_1 p(c_1) + c_2 p(c_2) + \cdots + c_n p(c_n)]\,\Delta x. \end{aligned}$$

Third, the **expected value** or **mean** μ of a bounded continuous measurement with range $[a, b]$ is defined to be the limit of the last expression as the number of subintervals increases without bound. That is,

$$\mu = \lim_{n\to\infty} [c_1 p(c_1) + c_2 p(c_2) + \cdots + c_n p(c_n)]\,\Delta x,$$

which we know is the definite integral

$$\mu = \int_a^b xp(x)\,dx.$$

An illustration of the evaluation of an expected value using the above formula was given in Example 5 of Section 5.2. ▮

EXERCISES

Suggested minimum assignment: Exercises 2, 3, 4, and 11.

1. Verify that $1 + 2 + 3 + \cdots + n = \dfrac{n(n+1)}{2}$ is true for $n = 7$.
2. Verify that $1^2 + 2^2 + 3^2 + \cdots + n^2 = \dfrac{n(n+1)(2n+1)}{6}$ is true for $n = 6$.
3. If $[3, 8]$ is divided into seven subintervals of equal length, what is the length of each subinterval? Also answer the same question for n subintervals, where n is a positive integer.

In Exercises 4–11 evaluate the given definite integrals using the method of Examples 1 and 2 of this section.

4. $\displaystyle\int_0^4 x\,dx.$

5. $\displaystyle\int_2^4 x\,dx.$

6. $\int_{-1}^{2} 2x\,dx.$

7. $\int_{1}^{6} (x + 1)\,dx.$

8. $\int_{0}^{2} x^2\,dx.$

9. $\int_{0}^{4} (x^2 + 2)\,dx.$

10. $\int_{0}^{5} (x^2 + x)\,dx.$

11. $\int_{1}^{3} x^2\,dx.$

12. Define the variance of a bounded continuous measurement. (*Hint:* See Example 6 of Section 5.2 and Example 3 of this section.)

5.6 Discussion of Proof of Fundamental Theorem (Optional)

Our goal in this section is to discuss a proof of the Fundamental Theorem of Integral Calculus (Theorem 2); the proof depends upon other theorems whose proofs are beyond the scope of this text. In order to understand this discussion, it will be necessary to comprehend the concept of an **integral with a variable upper limit,** a concept that is also useful in applications related to statistics (see Examples 3 and 4).

Let f be continuous on $[a, b]$, and let x be a number such that $a \leq x \leq b$. In Theorem 5 of this section the **integral with a variable upper limit**

$$\int_{a}^{x} f(t)\,dt$$

will be encountered and so an explanation is needed. To get a geometric interpretation of this expression, visualize f as a function that is positive on $[a, b]$ and hence $\int_a^x f(t)\,dt$ equals the area from a to x under $y = f(t)$ and above the t-axis (see Figure 5.22). If the function f and the number a do not change during our discussion but x is allowed to vary within $[a, b]$,

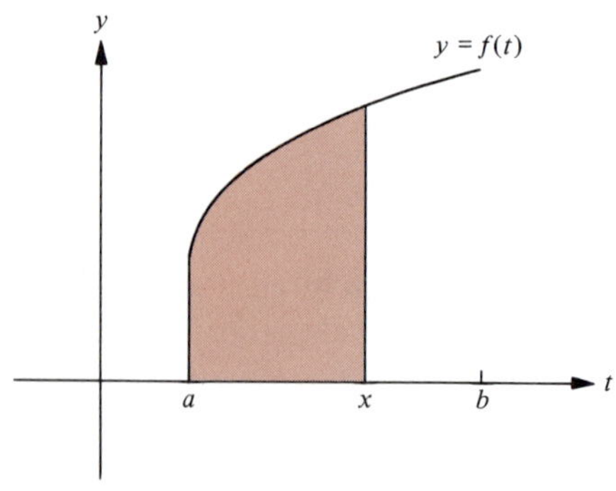

Figure 5.22. *Area* $= \int_{a}^{x} f(t)\,dt.$

then $\int_a^x f(t)\,dt$ depends only upon x and thus

$$F(x) = \int_a^x f(t)\,dt$$

defines a function F with domain $[a, b]$. If $f(t)$ is nonnegative on $[a, b]$, then $F(x)$ gives the area under the graph of f from a to x. Certainly, $F(x)$ does not depend upon the letter t that is used to label the horizontal axis; from Theorem 3(*iv*) we know that any letter could be used in place of t and, although x could be used, it is avoided because x is used as the variable upper limit.

Example 1

Consider $F(x) = \int_0^x 3t\,dt$, where $0 \le x \le 2$. Because $f(t) = 3t$ is nonnegative on $[0, 2]$, $F(x)$ equals the area of a triangle (see Figure 5.23) with base x and altitude $3x$. Therefore,

$$F(x) = \tfrac{1}{2}(x)(3x) = \tfrac{3}{2}x^2.$$

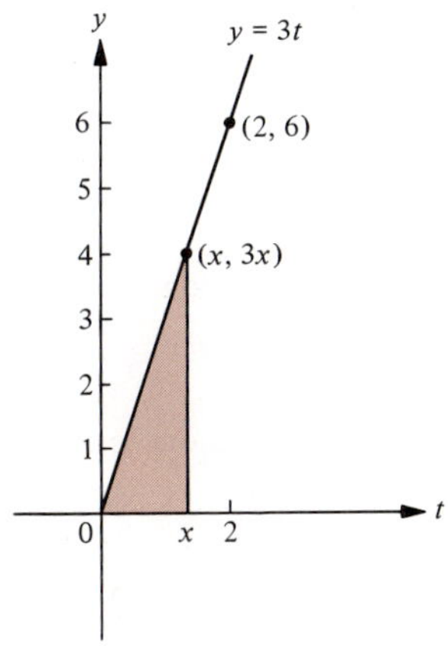

Figure 5.23

Next, notice that if F is differentiated, we get

$$F'(x) = \tfrac{3}{2}(2x) = 3x,$$

which is simply the integrand $3t$ evaluated at x. Theorem 5 guarantees that the latter result was no accident; that is, $F'(x) = f(x)$ on $[a, b]$ whenever f is continuous on $[a, b]$. ▮

In the preceding discussion we have defined a function F in terms of a function f by means of a definite integral with a variable upper limit, namely,

$$F(x) = \int_a^x f(t)\,dt.$$

Figures 5.22 and 5.23 were used to try to give some geometric interpretation to this type of function. Theorem 5 shows that f is the derivative of F, or that F is an antiderivative of f.

▶ Theorem 5

Let f be continuous on $[a, b]$, and let F with domain $[a, b]$ be defined by $F(x) = \int_a^x f(t)\,dt$. Then F is differentiable on $[a, b]$ and

$$F'(x) = f(x). \blacktriangleleft$$

The proof of Theorem 5 will not be given here. The proof depends strongly on the Mean Value Theorem for Integrals (Theorem 4) and can be found in many calculus texts. Some authors call Theorem 5 the Fundamental Theorem of Integral Calculus because they feel that the development of the result that F is an antiderivative of f is of primary importance and that Theorem 2 should be considered as a corollary of Theorem 5.

Example 2

Let $F(x) = \int_2^x \sqrt{1 + t^2}\,dt$, where $2 \le x \le 5$. Find $F'(x)$.

In this case $a = 2$, $b = 5$, and $f(t) = \sqrt{1 + t^2}$. Since $f(t) = \sqrt{1 + t^2}$, it follows that $f(x) = \sqrt{1 + x^2}$. By Theorem 5,

$$F'(x) = f(x) = \sqrt{1 + x^2}. \blacksquare$$

We now restate Theorem 2 and give its proof, which uses Theorem 5.

▶ Theorem 2

If f is continuous on $[a, b]$ and F is an antiderivative of f on $[a, b]$, then

$$I_a^b(f) = \int_a^b f(x)\,dx = F(b) - F(a).$$

Proof

STATEMENT	REASON
1. Let $G(x) = \int_a^x f(t)\,dt$, where $a \le x \le b$. Then G is an antiderivative of f on $[a, b]$; i.e., $G'(x) = f(x)$.	1. Theorem 5.
2. F is an antiderivative of f on $[a, b]$.	2. Hypothesis.
3. $F(x) = G(x) + C$, where C is a real number.	3. Statements 1 and 2, and Theorem 1 of Chapter 4.
4. $F(a) = G(a) + C = \int_a^a f(t)\,dt + C = 0 + C$.	4. Substitution of $x = a$ in Statements 3 and 1; Definition 2(*i*).

5. $F(b) = G(b) + C$ $= \int_a^b f(t)\,dt + F(a).$	5. Substitution of $x = b$ in Statements 3 and 1; by Statement 4, $C = F(a)$.
6. $\int_a^b f(x)\,dx = F(b) - F(a).$	6. Statement 5 and Theorem 3(*iv*). ◀

The result, $\int_a^b f(x)\,dx = F(b) - F(a)$, of Theorem 2 is also valid if $a \geq b$; this fact is a consequence of Definition 2 (see Exercises 7 and 8).

APPLICATIONS

***Example 3**

In Example 4 of Section 5.2 we defined a probability density function p for bounded continuous measurements. In this example we shall reconsider the related function F, called the cumulative probability distribution function, which was defined in Example 8 of Section 1.8 and discussed further in Example 7 of Section 2.2. We shall learn that for a bounded continuous measurement the function F can be expressed in terms of an integral with a variable upper limit.

Suppose that the value of a continuous measurement can be any real number s, where $a \leq s \leq b$. Suppose also that the probability density function p is continuous on $[a, b]$. Then the function F given by

$$F(x) = \begin{cases} 0 & \text{if } x < a, \\ \displaystyle\int_a^x p(s)\,ds & \text{if } a \leq x \leq b, \\ 1 & \text{if } x > b, \end{cases}$$

has the four properties required of a cumulative probability distribution function that are stated in Example 8 of Section 1.8. An outline of the proof of this statement follows.

Since $F(x) = 0$ if $x < a$ and $F(x) = 1$ if $x > b$, the properties

$$\lim_{x \to -\infty} F(x) = 0$$

and

$$\lim_{x \to \infty} F(x) = 1$$

are valid. Next observe that

$$F(a) = \int_a^a p(s)\,ds = 0.$$

Suppose that $a < x < b$. Then by Theorem 5 of this section F and p are related by the equation

$$F'(x) = p(x) \qquad \text{if } a < x < b.$$

Since $p(x) \geq 0$, $F'(x) \geq 0$ on (a, b), and hence F is a nondecreasing continuous function on (a, b). Also, observe that since p is a probability density function, $\int_a^b p(s)\, ds = 1$, and since $F(x) = \int_a^x p(s)\, ds$ if $a \leq x \leq b$, we have

$$F(b) = \int_a^b p(s)\, ds = 1.$$

From the known facts about p and F it now follows that F is a nondecreasing continuous function on $[a, b]$, and also that F has these same properties over the set of all real numbers; from this latter fact properties 3 and 4 of Example 8 of Section 1.8 are an immediate consequence. Therefore F is a cumulative probability distribution function, and hence the probability of an event $E = \{s : s \leq c\}$, where $a \leq c \leq b$, is given by

$$P(E) = F(c) = \int_a^c p(s)\, ds.$$

For a numerical example we shall use the probability density function p from Example 4 of Section 5.2; in that example p was given by

$$p(x) = \begin{cases} \frac{1}{36{,}000}(-x^2 + 140x - 4000) & \text{if } 40 \leq x \leq 100, \\ 0 & \text{otherwise.} \end{cases}$$

We determine the cumulative probability distribution function F when $40 \leq x \leq 100$ as follows:

$$\begin{aligned} F(x) &= \int_{40}^x \frac{1}{36{,}000}(-s^2 + 140s - 4000)\, ds \\ &= \frac{1}{36{,}000}\left(-\frac{s^3}{3} + 70s^2 - 4000s\right)\Bigg|_{40}^{x} \\ &= \frac{1}{36{,}000}\left[\left(-\frac{x^3}{3} + 70x^2 - 4000x\right) - \left(-\frac{64{,}000}{3} + 112{,}000 - 160{,}000\right)\right] \\ &= \frac{1}{36{,}000}\left(-\frac{x^3}{3} + 70x^2 - 4000x + \frac{208{,}000}{3}\right). \end{aligned}$$

The graph of F is shown in Figure 5.24. The height of the curve at 70 is $\frac{1}{2}$; this means that the probability of death by age 70 is $\frac{1}{2}$ and agrees with the fact that the area under the graph of $y = p(x)$ between 40 and 70 is $\frac{1}{2}$ (see Figure 5.6). In general, the height of the curve at x equals the area from 40 to x under the graph of $y = p(x)$ and gives the probability of death by age x. Note that

$$F(60) - F(50) = \tfrac{7}{27} - \tfrac{2}{27} = \tfrac{5}{27},$$

which agrees with our calculation in Example 6 of Section 5.3, which showed that the probability of death between ages 50 and 60 was $\frac{5}{27}$.

Another graphical observation follows from the fact that $F'(x) = p(x)$, namely, that the slope of F at any number x equals the height of p at the

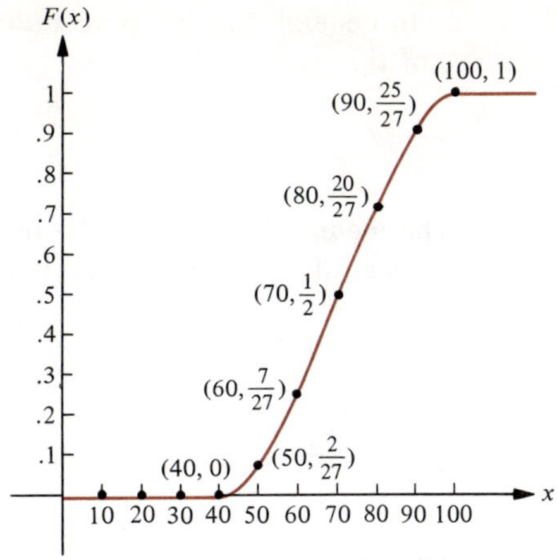

Figure 5.24. *Cumulative probability distribution function.*

same number x. Since $p(70) = \frac{1}{4}$ (see Figure 5.6), it follows that the slope of F at $(70, \frac{1}{2})$ is $\frac{1}{4}$; also since the maximum height of p is $\frac{1}{4}$, the maximum slope of F is $\frac{1}{4}$. (If the slope appears incorrect in Figure 5.24, it is because the same scale is not used on each axis.)

We showed above how to find the cumulative probability distribution function F from the probability density function p. It is even easier to find p if F is given because $F'(x) = p(x)$; in our example we have

$$F(x) = \frac{1}{36{,}000}\left(-\frac{x^3}{3} + 70x^2 - 4000x + \frac{208{,}000}{3}\right) \qquad \text{if } 40 \leq x \leq 100,$$

and by differentiation

$$p(x) = F'(x) = \frac{1}{36{,}000}(-x^2 + 140x - 4000) \qquad \text{if } 40 \leq x \leq 100. \; \blacksquare$$

***Example 4 Percentiles and Median of a Measurement**

A cumulative probability distribution function F is particularly useful in calculating **percentiles.** For a continuous measurement the **90th percentile** is the real number c for which 90 per cent of the measurement values are less than or equal to c. In other words, if $E = \{s : s \leq c\}$, the number c will satisfy the equation

$$P(E) = .9 \qquad \text{or} \qquad F(c) = .9.$$

For a bounded continuous measurement with range $[a, b]$, this is equivalent to solving for c in the equation

$$\int_a^c p(x)\,dx = .9,$$

where p is the probability density function. A geometric interpretation of the latter equation is that 90 per cent of the area under the curve $y = p(x)$ will be to the left of the line $x = c$.

In general, the **kth percentile,** where $0 < k < 100$, is the solution for c of the equation

$$F(c) = \frac{k}{100}.$$

The **median** is defined to be the 50th percentile.

As an illustration suppose that a measurement of the time required for the passage of legislation has the cumulative probability distribution function F and the probability density function p given by

$$F(x) = \begin{cases} 0 & \text{if } x < 0, \\ x^2 & \text{if } 0 \leq x \leq 1, \\ 1 & \text{if } x > 1, \end{cases} \qquad \text{and} \qquad p(x) = \begin{cases} 2x & \text{if } 0 \leq x \leq 1, \\ 0 & \text{otherwise.} \end{cases}$$

The 50th percentile or median time required is the (nonnegative) solution of the equation

$$F(c) = c^2 = .5,$$

or equivalently the (nonnegative) solution of

$$\int_0^c 2x\, dx = .5.$$

The result is $c = \sqrt{.5}$, which is approximately .707.

For a discrete measurement, to calculate the kth percentile, we consider two cases:

1. If there is a value s_i of the measurement such that $F(s_i) = k/100$, where $0 < k < 100$, then the kth percentile is $(s_i + s_{i+1})/2$, where s_{i+1} is the smallest value of the measurement that is larger than s_i. For example, consider four test scores of 60, 70, 80, and 90; then

$$F(x) = \begin{cases} 0 & \text{if } x < 60, \\ \frac{1}{4} & \text{if } 60 \leq x < 70, \\ \frac{1}{2} & \text{if } 70 \leq x < 80, \\ \frac{3}{4} & \text{if } 80 \leq x < 90, \\ 1 & \text{if } x \geq 90. \end{cases}$$

We see that $F(70) = \frac{1}{2}$. Hence the 50th percentile (median) is $(70 + 80)/2$, or 75.

2. If there is *no* value s_i of the measurement such that $F(s_i) = k/100$, where $0 < k < 100$, then the kth percentile is the smallest value of the measurement such that $F(s_i) > k/100$. For example, consider three test scores of 75, 85, and 95; then

$$F(x) = \begin{cases} 0 & \text{if } x < 75, \\ \frac{1}{3} & \text{if } 75 \leq x < 85, \\ \frac{2}{3} & \text{if } 85 \leq x < 95, \\ 1 & \text{if } x \geq 95. \end{cases}$$

We see that there is no value s_i of the measurement such that $F(s_i) = \frac{1}{2}$; therefore, the 50th percentile (median) is 85, since $F(85) > \frac{1}{2}$ and 85 is the smallest value for which this is the case. ▮

EXERCISES

Suggested minimum assignment: Exercises 1, 2, 3, and 4.

1. If $y = \int_1^w f(z)\,dz$, where $1 \le w \le 5$, then y is a function of what variable? What is the domain of this function?
2. Let $F(x) = \int_1^x (t + 1)\,dt$, $1 \le x \le 6$. Find $F(1)$. Find $F(6)$.

In Exercises 3–6 find $F'(x)$ by the use of Theorem 5.

3. $F(x) = \int_3^x t^2\,dt$, $3 \le x \le 4$.
4. $F(x) = \int_{-1}^x \sqrt{4 + t^3}\,dt$, $-1 \le x \le 1$.
5. $F(x) = \int_1^x t\sqrt{1 + t^2}\,dt$, $1 \le x \le 5$.
6. $F(x) = \int_2^x \sqrt{w}\,dw$, $2 \le x \le 4$.
7. Explain why $\int_a^b f(x)\,dx = F(b) - F(a)$ is valid if $a = b$.
8. Explain why $\int_a^b f(x)\,dx = F(b) - F(a)$ is valid if $a > b$.
9. In Example 3 calculate $F(80)$ and explain the meaning of the answer.
10. In this problem use the probability density function p in Exercise 18 of Section 5.2. Find $F(x)$ as described in Example 3 of this section. Then verify that $F(-1) = 0$, $F(1) = 1$, and $F'(x) = p(x)$.
11. In the first part of Example 4 we found the 50th percentile to be approximately .707. Find the 36th percentile.
12. A test is given to five students and the scores are 80, 85, 85, 90, and 95.
    (*a*) Find the 10th percentile.
    (*b*) Find the median.
    (*c*) Find the 80th percentile.

NEW VOCABULARY

division points 5.1
partition 5.1
norm of partition 5.1
definite integral 5.1

integrable 5.1
fundamental theorem of integral calculus 5.2
integrand 5.2
lower limit of definite integral 5.2
upper limit of definite integral 5.2
variable of integration 5.2
mean value theorem for integrals 5.3
integral with a variable upper limit 5.6

6 Exponential and Logarithmic Functions

Two particularly useful types of functions are the exponential and logarithmic functions. The purposes of this chapter are to define these functions, apply the techniques of differentiation and integration to them, and illustrate some of the many applications.

Historians generally credit John Napier (1550–1617) with the invention of logarithms. As a result of some of the properties of logarithmic functions, the invention was welcomed by astronomers as a significant time saver in performing various calculations; many of the great minds of the seventeenth century were thus freed from tedious calculations and were permitted more time for theoretical investigations. Today the logarithmic and exponential functions play an indispensable role in solving differential equations, which arise so often in the construction of mathematical models of real-world phenomena.

Prerequisites: For 6.1–6.3, the prerequisites are Chapter 1, 2.1–2.5, 3.1–3.3, and the first part of 3.6. For 6.4, additional prerequisites from previous chapters are 2.6, 4.1, 4.2, and 5.1–5.4. Section 4.3 is needed in 6.5. The reader may skip from 6.4 directly to either 6.6 or 6.7.
Suggested Sections for Moderate Emphasis: 6.1–6.4
Suggested Sections for Minimum Emphasis: 6.1–6.3

6.1 Exponential Functions

Suppose that a certain amount of a quantity is present at a given starting time ($t = 0$). Suppose also that, as time passes, the amount y which is present either always increases or always decreases. There are several practical situations in which it is reasonable to assume that the rate of change dy/dt of y with respect to time t is proportional to the amount y that is present. This relationship can be expressed as the differential equation

$$\frac{dy}{dt} = ky, \tag{6.1}$$

where k is the constant of proportionality. If $k > 0$, then y increases as t increases; if $k < 0$, then y decreases as t increases. The differential equation (6.1) might be used in a problem about population growth of humans or animals; it is applicable in problems on decay of radioactive material (useful in determining the approximate age of certain old substances). Later in this chapter we shall show that the solution of (6.1) is the important equation

$$y = Ae^{kt}, \tag{6.2}$$

where $e \approx 2.7$ is a number that will be introduced in this section and A is the value of y when $t = 0$. Equation (6.2) is useful for projecting the amount y expected to be present at future times. The type of function in (6.2) suggests that we begin this chapter by introducing **exponential functions.**

▶ Definition 1

Let b be any positive real number. The function f, whose domain is the set of all real numbers, determined by

$$f(x) = b^x,$$

is called the **exponential function with base *b*.**

The graph of an exponential function with base $b > 1$ has a characteristic appearance, as shown in the following example.

Example 1

Graph the exponential function f with base 2 determined by the equation

$$y = f(x) = 2^x.$$

Some corresponding values of x and y are displayed in Table 6.1, and the graph is shown in Figure 6.1. ▮

Table 6.1

x	-3	-2	-1	0	1	$\frac{3}{2}$	2	3
$y = 2^x$	$\frac{1}{8}$	$\frac{1}{4}$	$\frac{1}{2}$	1	2	$2\sqrt{2}$	4	8

In making Table 6.1, we assumed that the reader knew from algebra how to raise 2 to a rational power. Numbers with irrational exponents such as $2^{\sqrt{2}}$ can be defined by a limiting process—$2^{\sqrt{2}}$ may be thought of as the limit of the sequence

$$2^1, 2^{1.4}, 2^{1.41}, 2^{1.414}, \ldots,$$

where $1.414\ldots$ is the decimal representation of $\sqrt{2}$. As one may intuitively expect from Figure 6.1, an exponential function is continuous over the domain of all real numbers.

As Figure 6.1 suggests, an exponential function with base $b > 1$ is an increasing function; also, if $x \to -\infty$, then $b^x \to 0$, and if $x \to \infty$, then b^x increases without bound. If $0 < b < 1$, then the graph of $y = b^x$ is the graph of a decreasing function. For example, the graph of

$$y = (\tfrac{1}{2})^x = 2^{-x}$$

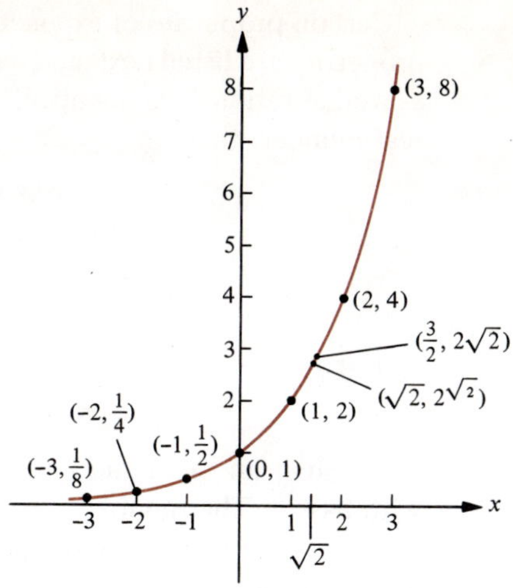

Figure 6.1. $y = f(x) = 2^x$.

is shown in Figure 6.2. Observe in this case that if $x \to \infty$, then $b^x \to 0$, and if $x \to -\infty$, then b^x increases without bound.

The graph of the exponential function with base $b = 1$ is a horizontal line because

$$f(x) = 1^x = 1$$

defines a constant function (see Section 1.6). If $b \neq 1$, the range of the exponential function with base b is the set of all positive numbers—see Figures 6.1 and 6.2. Hence equations such as

$$2^x = 0 \qquad \text{and} \qquad (\tfrac{1}{2})^x = -1$$

have no solutions.

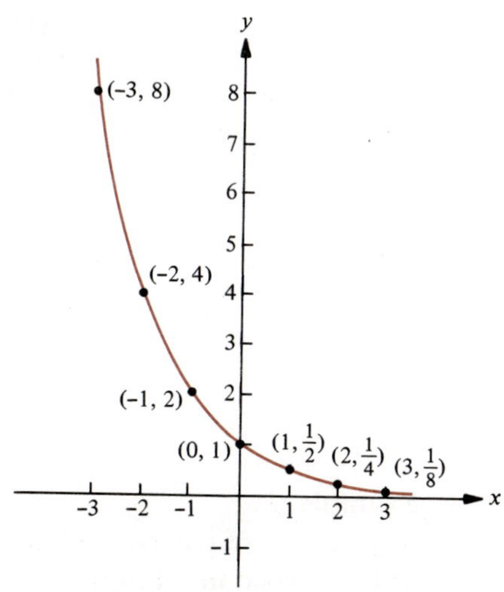

Figure 6.2. $y = (\frac{1}{2})^x = 2^{-x}$.

Certain properties of exponents are assumed in this book. Some of these properties are listed next, and we point out that they are valid for irrational as well as rational exponents. If x and y are real numbers and b is a positive real number, then

$$b^x b^y = b^{x+y},$$
$$(b^x)^y = b^{xy},$$
$$\frac{b^x}{b^y} = b^x b^{-y} = b^{x-y},$$

and, if $b \neq 1$,

$$b^x = b^y \qquad \text{implies that } x = y.$$

We shall find that the most important exponential function is the one with base e. The number e (named after the 18th-century mathematician Euler) is used frequently in many applications of mathematics and can be defined as follows:

Definition 2

$$e = \lim_{x \to 0} (1 + x)^{1/x}.$$

Table 6.2 suggests the existence of a limit for $(1 + x)^{1/x}$ as x approaches 0, but it is difficult to *prove* that the limit of Definition 2 exists. It does exist, however, and equals an irrational number whose value correct to five decimal places is

$$e = 2.71828.$$

Table 6.2

x	$(1 + x)^{1/x}$	x	$(1 + x)^{1/x}$
.1	$(1.1)^{10} = 2.59374$	$-.1$	$(.9)^{-10} = 2.86797$
.01	$(1.01)^{100} = 2.70481$	$-.01$	$(.99)^{-100} = 2.73200$
.001	$(1.001)^{1000} = 2.71692$	$-.001$	$(.999)^{-1000} = 2.71964$
.0001	$(1.0001)^{10,000} = 2.71815$	$-.0001$	$(.9999)^{-10,000} = 2.71842$
.00001	$(1.00001)^{100,000} = 2.71827$	$-.00001$	$(.99999)^{-100,000} = 2.71830$

In Example 2 we shall need an alternative formulation of the definition of e. Assume that $x > 0$ and let $z = 1/x$ in Definition 2. As $x \to 0$, $z \to \infty$ and we obtain

$$e = \lim_{z \to \infty} \left(1 + \frac{1}{z}\right)^z. \tag{6.3}$$

Example 2

A discussion of compound interest was begun at the outset of Section 1.6 and continued in Example 4 of Section 1.6 and Example 1 of Section 1.7. If a principal P is deposited at an annual interest rate r for t years and interest is compounded n times each year, then the amount A after t years is given by

$$A = P\left(1 + \frac{r}{n}\right)^{nt}.$$

Suppose that federal regulations limit a savings institution to a maximum annual interest rate r but that no restriction is placed on the number n of times per year that interest is compounded. To attract capital, the management decides to let $n \to \infty$; in this case we say that interest is **compounded continuously.** Let us calculate

$$\lim_{n\to\infty} P\left(1 + \frac{r}{n}\right)^{nt} = P \lim_{n\to\infty} \left[\left(1 + \frac{1}{n/r}\right)^{n/r}\right]^{rt}.$$

If $z = n/r$, then $z \to \infty$ as $n \to \infty$, and we shall assume that it is valid to write

$$P\left[\lim_{z\to\infty} \left(1 + \frac{1}{z}\right)^{z}\right]^{rt},$$

which equals, upon application of (6.3),

$$Pe^{rt}.$$

In Example 1 of Section 1.7 we stated that if 10 million dollars is deposited for 1 year at 6 per cent, then the amount in the account cannot exceed \$10,618,365.47, no matter how many times during the year the interest is compounded. If interest is compounded continuously at 6 per cent, the amount in the account after 1 year is

$$\begin{aligned} A = Pe^{rt} &= (10{,}000{,}000)e^{(.06)(1)} \\ &\approx (10{,}000{,}000)(1.061836547) \\ &= \$10{,}618{,}365.47. \end{aligned}$$

(From tables, $e^{.06}$ is slightly less than 1.061836547.) The reader may wish to compare the answer above with Table 1.10, in which the results of compounding daily, quarterly, and so on, are shown. ∎

APPLICATIONS

Example 3

Suppose that an amount y of some quantity has value 0 at time $t = 0$ and that, as time passes, y gradually increases but remains less than A. In this example we shall mention some situations in which this type of bounded growth might occur. An equation that has provided good experimental results for this type of growth is

$$y = A(1 - e^{-kt}). \qquad (6.4)$$

In (6.4) A and k are positive constants and time $t \geq 0$. Notice that if $t = 0$, then

$$y = A(1 - e^{0}) = 0.$$

The expression e^{-kt} is 1 when $t = 0$, and e^{-kt} decreases but remains positive as t increases; therefore, the expression $(1 - e^{-kt})$ is always less than 1, and hence y is always less than A. However, $y \to A$ as $t \to \infty$ because

$$\lim_{t\to\infty} A(1 - e^{-kt}) = \lim_{t\to\infty}\left(A - \frac{A}{e^{kt}}\right) = A - 0 = A.$$

In Figure 6.3 equation (6.4) is graphed for the case when $A = 10$ and $k = \frac{1}{2}$. If k were smaller, the curve would approach $y = A$ less quickly.

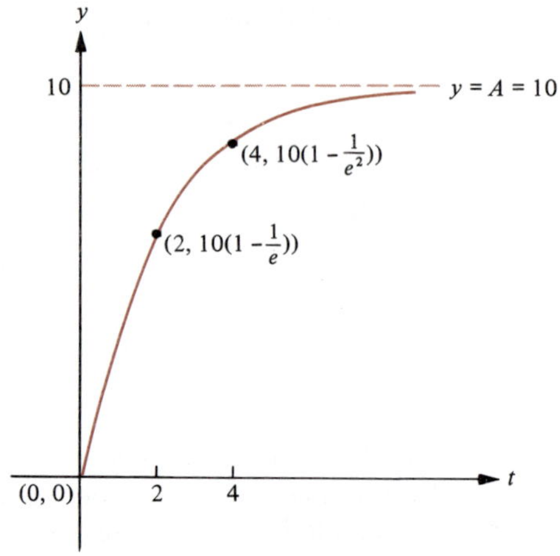

Figure 6.3. $y = 10(1 - e^{-(1/2)t})$.

A curve of the type in Figure 6.3 is sometimes called a **learning curve.** It shows how learning can take place rapidly at first and then level out as it nears a maximum possible level A. Perhaps y represents the number of items produced by a worker t days after starting to learn a new job. On pp. 29–32 of his *Essentials of Behavior* Hull [13] discusses his formula for habit strength H as a function of the number N of reinforcements from the beginning of learning. The formula is of the form

$$H = 1 - 10^{-lN},$$

where l is a positive constant, or

$$H = 1 - e^{-kN},$$

where $10^{-l} = e^{-k}$. The formula, therefore, is of the same type as (6.4).

Equation (6.4) might be a useful model in any situation in which a growing quantity has an upper bound A. Over extended periods of time the magnitude of a quantity is usually limited by various factors. For example, the size of a population of objects is limited by the space in which the population must live; the size of a certain individual animal is limited by hereditary factors. The growth of plants and the production of crops certainly can be enhanced by fertilizer, but there is an upper bound to this growth and production. Equation (6.4) has proved useful in studying such situations. A derivation of (6.4) will be given in Example 8 of Section 6.4. ∎

EXERCISES

Suggested minimum assignment: Exercises 1, 2, 7, 11, and 14.

1. Graph $f(x) = 3^x$. State the domain and range of f. Is f an increasing function or is f a decreasing function? State the value of $\lim_{x \to -\infty} 3^x$.
2. Graph $f(x) = (\frac{1}{3})^x = 3^{-x}$. State the domain and range of f. Is f an increasing function or is f a decreasing function? State the value of $\lim_{x \to \infty} 3^{-x}$.
3. (*a*) Graph the exponential function with base 4.
   (*b*) Graph the exponential function with base $\frac{1}{4}$.
4. Sketch $y = 2^{x+1}$. What is the relationship of this graph to the graph of $y = 2^x$?
5. Solve for x, if possible.
   (*a*) $3^x = 81$. (*b*) $3^x = \frac{1}{81}$. (*c*) $3^x = -81$.
6. What is meant by $2^{\sqrt{3}}$?
7. Simplify. (*a*) $4^{5/2}$. (*b*) $(4^9)(4^{-7})$.
   (*c*) $(3^{10})^{1/5}$. (*d*) $\dfrac{5^6}{5^3}$.
8. True or false: $2^{3/2} > e$. Give a reason for your answer.
9. Evaluate $(1 + x)^{1/x}$ to three decimal places for $x = \frac{1}{5}$.
10. What is the correct value of e to five decimal places?
11. Evaluate $\lim_{x \to 0} (1 + x)^{2/x}$.
12. Evaluate $\lim_{x \to \infty} 2\left(1 + \dfrac{1}{x}\right)^x$.
13. (*a*) Suppose that $P = \$100$ is invested in a savings account at an annual interest rate of $r = .04$. If interest is compounded four times a year, how much money is in the account after 1 year?
    (*b*) Rework part (*a*) if interest is compounded continuously. (Use $e^{.04} = 1.0408$.)
14. Find the amount to which \$800 will accumulate if deposited for 2 years at 5 per cent compounded continuously. (Use $e^{.1} = 1.105171$.)
15. Sketch (6.4) for the case when $A = 15$ and $k = \frac{1}{3}$.
16. If digital computer facilities are available, verify that

$$(1.001)^{1000} \approx 2.71692.$$

6.2 Logarithmic Functions

We shall find that there is a close relation between exponential functions and **logarithmic functions.** Before logarithmic functions can be understood, however, we shall have to introduce the idea of the **logarithm of a positive number a to the base b.**

▶ Definition 3

Let a be a positive real number and let b be any positive real number except 1. Then $\log_b a$ denotes **the logarithm of a to the base b** and

$$\log_b a = c \qquad \text{means that} \qquad b^c = a.$$

Example 1

$$\log_2 8 = 3, \text{ because } 2^3 = 8.$$

$$\log_{10}\left(\tfrac{1}{100}\right) = -2, \text{ because } (10)^{-2} = \tfrac{1}{100}. \blacksquare$$

In Definition 3, $\log_b a$ has a unique value for given numbers a and b because we know from Section 6.1 that there is exactly one number c such that $b^c = a$.

When the base is the number e, we shall adopt the following standard notation:

$$\log_e a = \ln a.$$

For example, when $\ln 2$ is written, the base e is understood. Logarithms with base e are called **natural logarithms.** Logarithms with base 10 are called **common logarithms.**

Example 2

$$\ln\left(\frac{1}{e}\right) = -1,$$
$$\ln 1 = 0,$$
$$\ln e = 1,$$
$$\ln e^2 = 2. \blacksquare$$

▶ Definition 4

Let b be any positive real number except 1. The function g, whose domain is the set of all positive real numbers, determined by

$$g(x) = \log_b x$$

is called the **logarithmic function with base b.**

Example 3

Graph the logarithmic function g determined by

$$y = g(x) = \log_2 x.$$

One procedure is to rewrite $y = \log_2 x$ as

$$x = 2^y,$$

and then to assign values to y and compute the corresponding values of x as shown in Table 6.3. The graph of g is shown in Figure 6.4. ▮

Table 6.3

x	$\frac{1}{8}$	$\frac{1}{4}$	$\frac{1}{2}$	1	2	4	8
$y = \log_2 x$	-3	-2	-1	0	1	2	3

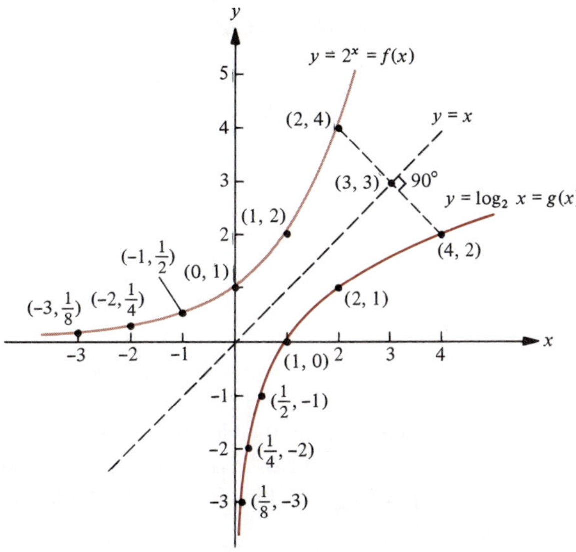

Figure 6.4. *Inverse functions.*

The domain of g in Example 3 is the set of all positive real numbers, and the range of g is the set of all real numbers. As Figure 6.4 suggests, a logarithmic function is continuous over the domain of positive real numbers.

There is a definite relationship between the functions f and g, where

$$f(x) = 2^x \qquad \text{and} \qquad g(x) = \log_2 x.$$

Observe that the point (2, 4) satisfies the equation $y = f(x)$, whereas the point (4, 2) satisfies the equation $y = g(x)$. Examination of Figure 6.4 reveals that this sort of inversion of the two coordinates of a point always occurs. It is customary to call f and g **inverse functions**—we say that g is the inverse of f and that f is the inverse of g. As Figure 6.4 indicates, the line $y = x$ is the perpendicular bisector of a segment between (x, y) and (y, x) whenever $x \neq y$; when inverse functions are plotted together, the combination is a configuration symmetric about the line $y = x$.

In Theorem 1 some of the basic properties of logarithms are listed.

▶ Theorem 1

Assume that a and b are positive numbers other than 1, M and N are positive numbers, and c is any real number. Then the following properties of logarithms are valid:

(*i*) $\log_b 1 = 0$.

(*ii*) $\log_b b = 1$.

(*iii*) $\log_b a = \dfrac{1}{\log_a b}$.

(*iv*) $\log_b b^c = c$.

(*v*) $b^{\log_b M} = M$.

(*vi*) $\log_b MN = \log_b M + \log_b N$.

(*vii*) $\log_b \dfrac{M}{N} = \log_b M - \log_b N$.

(*viii*) $\log_b M^c = c \log_b M$.

(*ix*) If $\log_b M = \log_b N$, then $M = N$.

(*x*) $\log_a M = \dfrac{\log_b M}{\log_b a}$.

The proofs of these properties depend primarily upon Definition 3. As an example, part (*vii*) is proved next. In words, part (*vii*) asserts that the logarithm of a quotient equals the logarithm of the numerator minus the logarithm of the denominator.

Proof of part (*vii*): Let $x = \log_b M$. Then $b^x = M$ by Definition 3. Let $y = \log_b N$. Then $b^y = N$ by Definition 3.

$$\frac{M}{N} = \frac{b^x}{b^y} = b^{x-y}.$$

By Definition 3, $\dfrac{M}{N} = b^{x-y}$ means that

$$\log_b \frac{M}{N} = x - y,$$

which upon substitution for x and y becomes

$$\log_b \frac{M}{N} = \log_b M - \log_b N. \blacktriangleleft$$

Example 4

(*a*) $\log_3 21 = \log_3 3 + \log_3 7 = 1 + \log_3 7$.

(*b*) $\ln \dfrac{7}{5} = \ln 7 - \ln 5$.

(*c*) $\log_{10} 3^4 = 4 \log_{10} 3$.

(*d*) $\ln 4 = \dfrac{\log_{10} 4}{\log_{10} e}$.

(*e*) $\ln e^x = x$.

(*f*) $e^{\ln x} = x$. ∎

Example 5

Solve for x:

$$\log_2 x = \log_2 7 + 2\log_2 3 - \log_2 5,$$

$$\log_2 x = \log_2 \frac{(7)(3)^2}{5} = \log_2 \frac{63}{5},$$

$$x = \frac{63}{5}. \ ∎$$

Example 6

Given that $\ln 7 = 1.946$ and $\ln 11 = 2.398$. Find $\ln(1/\sqrt{77})$.

$$\begin{aligned}\ln \frac{1}{\sqrt{77}} &= \ln 1 - \ln (77)^{1/2} \\ &= 0 - \frac{1}{2}\ln 77 \\ &= -\frac{1}{2}\ln [(7)(11)] \\ &= -\frac{1}{2}(\ln 7 + \ln 11) \\ &= -\frac{1}{2}(1.946 + 2.398) \\ &= -2.172. \ ∎\end{aligned}$$

APPLICATIONS

Example 7

In Example 2 of Section 6.1 we found that if a principal P were compounded continuously at an annual interest rate r, then the amount A after t years is given by the formula

$$A = Pe^{rt}. \tag{6.5}$$

Sometimes in a problem the unknown quantity in (6.5) is t; since natural logarithms have been introduced in this section, we are now able to solve (6.5) for t as follows:

$$e^{rt} = \frac{A}{P},$$

$$rt = \ln \frac{A}{P},$$

$$t = \frac{1}{r}\ln \frac{A}{P}. \tag{6.6}$$

Suppose, for example, that one wondered how many years it would take for a principal $P = \$700$ to accumulate to an amount $A = \$1100$ at 5 per cent compounded continuously. By (6.6)

$$t = \frac{1}{.05} \ln \frac{1100}{700}$$
$$= 20 \ln \frac{11}{7}$$
$$= 20(\ln 11 - \ln 7).$$

By use of the values of $\ln 7$ and $\ln 11$ from Example 6 of this section,

$$t = 20(2.398 - 1.946) = 9.04 \text{ years.} \blacksquare$$

Example 8

Suppose that in an experiment the data shown in Table 6.4, which relate the variables x and Y, are observed:

Table 6.4

x	1	2	4	5	7	8
Y	12	20	41	60	134	202

Apparently, Y increases rapidly as x increases. It is conjectured that x and Y are approximately related by an equation of the form

$$Y = cb^x,$$

where c is positive and the base b is greater than 1. How can one determine whether this conjecture is true?

By Theorem 1(*vi*) and (*viii*), if $Y = cb^x$, then

$$\ln Y = \ln c + x \ln b.$$

If we let $y = \ln Y$, it follows that if x is plotted against y, the graph is a line with y-intercept $\ln c$ and slope $\ln b$. Conversely, if

$$y = \ln c + x \ln b,$$

then

$$Y = e^y = e^{\ln c + x \ln b}$$
$$= e^{\ln c} e^{\ln b^x}$$
$$= cb^x.$$

Therefore the graph, if x is plotted against y, will appear to be a line if and only if the variables x and $Y = e^y$ are related by an equation of the form $Y = cb^x$.

For the data given at the beginning of this example, the values of $y = \ln Y$ can be found from a set of tables and we obtain the data given in Table 6.5. Then Figure 6.5 is constructed; the points certainly appear to lie close to the line shown in the figure. Hence we conclude that the

Table 6.5

x	1	2	4	5	7	8
Y	12	20	41	60	134	202
$y = \ln Y$	2.5	3.0	3.7	4.1	4.9	5.3

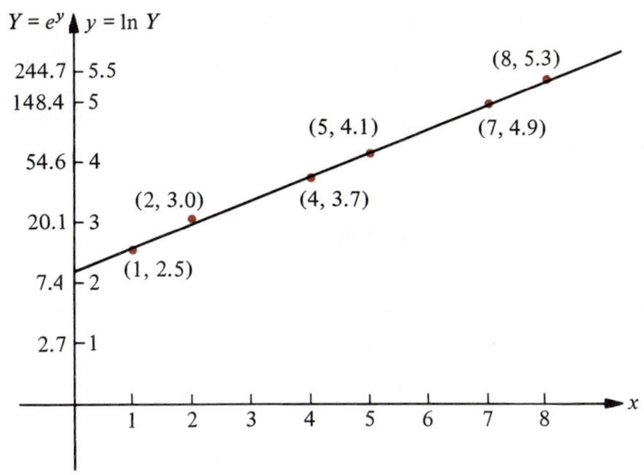

Figure 6.5

relationship between x and Y on the interval $[1, 8]$ can be approximated by an equation of the form $Y = cb^x$. By estimating the slope $\ln b$ and the y-intercept $\ln c$ from the graph, one could obtain an estimate of b and c; more accuracy could be obtained by using a technique that we shall learn in Example 5 of Section 7.3.

In the discussion above we talked about plotting x against $\ln Y$. We could have used $\log_b Y$ instead of $\ln Y$, where b is any legitimate base. Special graph paper is available for plotting values of one variable against values of the logarithm of another variable—it is called **semilogarithmic paper.**

Applications to the biological sciences of the ideas in this example may be found on pp. 152–158 of Smith's *Biomathematics,* Vol. 1 [21]. ▮

EXERCISES

Suggested minimum assignment: Exercises 1, 2, 3, 7(*a*) (*c*), and 9(*c*) (*g*) (*k*) (*m*).

1. Find each of the following:
   (*a*) $\log_{10} 1000$. (*b*) $\ln e^3$. (*c*) $\log_2 \frac{1}{4}$.
2. Solve for x, if possible:
   (*a*) $\log_2 16 = x$. (*b*) $\log_x 16 = 2$.
   (*c*) $\log_5 x = -1$. (*d*) $\log_5 (-1) = x$.
3. Let $f(x) = e^x$ and $g(x) = \ln x$. Graph f and g on the same set of axes.

Are f and g inverse functions? State the domain and range of f. State the domain and range of g.

4. Graph the exponential function with base $\frac{3}{2}$ and the logarithmic function with base $\frac{3}{2}$ on the same set of axes. What is the relationship of these functions?
5. Sketch $y = \log_{1/2} x$.
6. Sketch $y = 2 \log_2 (x - 1)$.
7. Solve for x:

   (*a*) $\ln x = \ln 7 + \ln 8$. (*b*) $\ln x = \ln 7 - \ln 8$.
   (*c*) $\ln x = 2 \ln 7$. (*d*) $x = e^{\ln 5}$.

   (*e*) $\ln x = \dfrac{1}{\log_{10} e}$.
8. Solve for x:

   (*a*) $\log_3 x = (\log_3 1 + \log_3 3)(\log_3 9)$.
   (*b*) $\log_5 x = \log_5 3 - 2 \log_5 2 + 3 \log_5 5$.
9. Given that $\ln 7 = 1.946$ and $\ln 11 = 2.398$. Find each of the following:

   (*a*) $\ln 77$. (*b*) $\ln \frac{11}{7}$. (*c*) $\ln \frac{49}{11}$.
   (*d*) $\ln 7^2$. (*e*) $(\ln 7)^2$. (*f*) $\ln 7 - \ln 11$.
   (*g*) $\ln \sqrt{7}$. (*h*) $\ln \frac{1}{11}$. (*i*) $\dfrac{1}{\log_7 e}$.
   (*j*) $\log_{11} e$. (*k*) $e^{1.946}$. (*l*) $e^{2 \ln 7}$.
   (*m*) $\log_7 11$.
10. If we use the idea of an integral with a variable upper limit from optional Section 5.6, it is possible to define $\ln x$ by

$$\ln x = \int_1^x \frac{1}{t}\, dt, \qquad x > 0.$$

   (*a*) Use this definition to show that $\ln 1 = 0$.
   (*b*) Find $D_x (\ln x)$. *Hint:* Use Theorem 5 of Chapter 5.
   (*c*) Why is the logarithmic function with base e continuous throughout its domain?
   (*d*) Why is the logarithmic function with base e increasing throughout its domain?
11. Prove Theorem 1(*vi*).
12. Sketch the function f defined by $y = f(x) = 2x + 1$. Also sketch the inverse function g of function f on the same graph and find $g(x)$.
13. (*a*) Why is part (*iv*) of Theorem 1 a special case of part (*viii*)?
   (*b*) Why is part (*iii*) of Theorem 1 a special case of part (*x*)?
14. Solve (6.2) on p. 230 for t.
15. The equation of a demand curve is $y = f(x) = 20 - 3x$, where $0 \leq x \leq 6$. Graph f and the inverse function g of f on the same axes. [*Hint:* First show that g is given by $y = g(x) = \frac{20}{3} - \frac{1}{3}x$, where $2 \leq x \leq 20$.] The combination is a configuration symmetric about what line?

16. Suppose for $x = 2$, $x = 4$, and $x = 5$ that Y has the values 1, 10, and 30, respectively. Plot x against $y = \ln Y$ (use $\ln 10 = 2.3$ and $\ln 30 = 3.4$). What conclusion can be drawn? (See Example 8.)

6.3 Differentiation Formulas

In this section formulas are obtained for differentiation of the logarithmic and exponential functions. These formulas, together with the fundamental knowledge about derivatives learned in Chapters 2 and 3, provide the means to solve many new applied problems. To derive the formula in Theorem 2 for differentiating a *logarithmic function,* we start with the definition of the derivative of a function (Definition 2 of Section 2.2) and proceed with the standard limit process. The formula for differentiating an *exponential function* then will be stated in Theorem 3; the derivation of the formula involves a shortcut process that makes use of Theorem 2.

▶ Theorem 2

If $y = \log_b u$ and $u = f(x) > 0$, where f is differentiable, then

$$D_x y = (\log_b e)\frac{1}{u}D_x u.$$

Proof: First we will find a formula for $D_x(\log_b x)$, and then the chain rule will be applied.

STATEMENT	REASON
1. Let $g(x) = \log_b x$, and let $c > 0$ be an arbitrary element of the domain of g. Then $g'(c)$ $= \lim_{\Delta x \to 0} \frac{\log_b (c + \Delta x) - \log_b c}{\Delta x}$	1. Definition 2, Section 2.2.
2. $= \lim_{\Delta x \to 0} \frac{1}{\Delta x} \log_b \left(\frac{c + \Delta x}{c}\right)$	2. Theorem 1(*vii*), Section 6.2.
3. $= \lim_{\Delta x \to 0} \frac{c}{c\,\Delta x} \log_b \left(1 + \frac{\Delta x}{c}\right)$	3. Change of form.
4. $= \lim_{\Delta x \to 0} \frac{1}{c} \log_b \left(1 + \frac{\Delta x}{c}\right)^{c/\Delta x}$	4. Theorem 1(*viii*), Section 6.2.
5. $= \lim_{z \to 0} \frac{1}{c} \log_b (1 + z)^{1/z}$	5. Let $z = \frac{\Delta x}{c}$.
6. $= \frac{1}{c}\left[\lim_{z \to 0} \log_b (1 + z)^{1/z}\right]$	6. Theorem 2(*v*), Section 1.8.
7. $= \frac{1}{c} \log_b \left[\lim_{z \to 0} (1 + z)^{1/z}\right]$	7. Follows from continuity of logarithmic function.

8. $= \frac{1}{c}\log_b e.$	8. Definition 2, Section 6.1.
9. $D_x(\log_b x) = (\log_b e)\frac{1}{x}.$	9. Statements 1–8 and because c is an arbitrary element of the domain of g.
10. $D_x y = D_u y\, D_x u.$	10. Theorem 2(*iii*), Section 2.4.
11. $D_x y = (\log_b e)\frac{1}{u}D_x u.$	11. Statements 9 and 10. ◀

Example 1

Find $D_x y$ if $y = \log_2 (x^2 + 1)$.

Apply Theorem 2 with $b = 2$ and $u = x^2 + 1$ to obtain

$$D_x y = (\log_2 e)\frac{1}{x^2 + 1}(2x)$$

$$= 2(\log_2 e)\frac{x}{x^2 + 1}. \quad ▮$$

The most important special case of Theorem 2 occurs when $b = e$. Fortunately, the formula is simpler in this case because $\log_b e$ equals 1 when $b = e$. Hence we have the following corollary of Theorem 2:

$$D_x \ln u = \frac{1}{u}D_x u. \tag{6.7}$$

Moreover, if $u = x$, we have

$$D_x \ln x = \frac{1}{x}.$$

Example 2

Find the slope of $f(x) = \ln (2x + 1)$ at $x = 3$.

By (6.7)

$$f'(x) = \frac{1}{2x + 1}D_x(2x + 1) = \frac{2}{2x + 1}.$$

The desired slope, given by $f'(3)$, is found to be

$$f'(3) = \tfrac{2}{7}. \quad ▮$$

Example 3

Find y' if $y = (\ln x)/x$.

By the quotient rule

$$y' = \frac{(x)(1/x) - (\ln x)(1)}{x^2} = \frac{1 - \ln x}{x^2}. \quad ▮$$

The reader is familiar with the formula $D_x x^n = nx^{n-1}$, where n is a real number [see Theorem 1(*ii*) of Section 2.3]. This formula is proved below for $x > 0$ by a process called **logarithmic differentiation.**

$$\begin{aligned}
y &= x^n,\\
\ln y &= n \ln x,\\
D_x(\ln y) &= D_x(n \ln x),\\
\frac{1}{y} D_x y &= n\left(\frac{1}{x}\right),\\
D_x y &= \frac{ny}{x} = \frac{nx^n}{x} = nx^{n-1}.
\end{aligned}$$

The steps in the process of logarithmic differentiation illustrated above are

1. Take logarithm (preferably with base e) of both sides of the equation.
2. Differentiate both sides (note use of chain rule on left side).
3. Solve for $D_x y$ and simplify.

Logarithmic differentiation is a process to consider whenever an exponent occurs in the given function. Hence it is useful in the proof of Theorem 3 when we differentiate an exponential function. Some other situations in which the process is used will be given in the exercises.

Theorem 3

If $y = b^u$ and $u = f(x)$, where f is differentiable, then

$$D_x y = (\ln b) b^u D_x u.$$

Outline of Proof: Use logarithmic differentiation to show that $D_x b^x = (\ln b) b^x$, and then apply the chain rule. ◀

The use of base e in Theorem 3 has the advantage (as it did in Theorem 2) of simplifying the formula. Very important special cases of Theorem 3 are

$$D_x e^u = e^u D_x u \tag{6.8}$$

and

$$D_x e^x = e^x. \tag{6.9}$$

Example 4

Find $f'(x)$ if $f(x) = e^{3x} + 1/5^x$.

$$\begin{aligned}
f(x) &= e^{3x} + 5^{-x},\\
f'(x) &= e^{3x} D_x(3x) + (\ln 5)5^{-x} D_x(-x)\\
&= 3e^{3x} - (\ln 5)5^{-x}. \ \blacksquare
\end{aligned}$$

Example 5

In a suitable medium a full-grown bacterial cell divides by fission into two equal cells. Suppose that the number y of cells present at time t is given by

$$y = a(2)^{t/k}.$$

In this formula a and k are constants—a is the number of cells present initially and k is the average time required for a cell to double. Find the rate of change dy/dt.

By Theorem 3,

$$\frac{dy}{dt} = a(\ln 2)(2)^{t/k} D_t\left(\frac{t}{k}\right)$$
$$= \left(\frac{a}{k}\right)(\ln 2)(2)^{t/k}. \blacksquare$$

Example 6

Find all maximum and minimum points of f if $y = f(x) = xe^{-x}$.

The domain of f is the set of all real numbers. Even though f is not a rational function, we shall assume (as is the case) that the methods of Chapter 3 are applicable. By (6.8) and the product rule,

$$f'(x) = x(-e^{-x}) + e^{-x}(1) = e^{-x}(1 - x).$$

Since e^{-x} is never zero, the only critical number is $x = 1$. Therefore, $(1, 1/e)$ is the only possible maximum or minimum point. See Table 6.6 and Figure 6.6. $\blacksquare$

Table 6.6

Interval	Calculation of sign of $f'(x)$	Result	Conclusion
$(-\infty, 1)$	$(+)(+)$	$+$; f increasing	$(1, 1/e)$ is maximum point
$(1, \infty)$	$(+)(-)$	$-$; f decreasing	

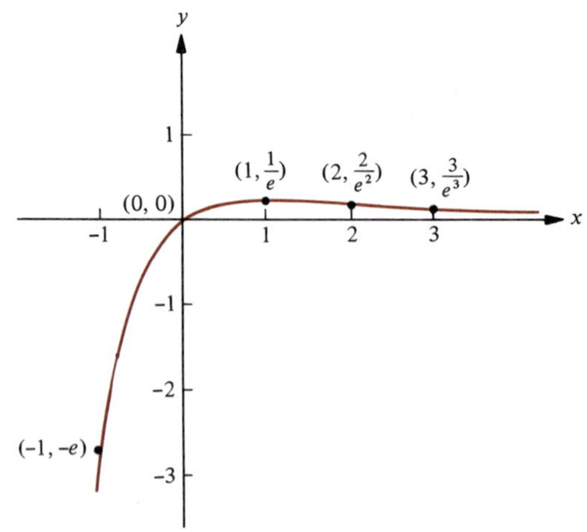

Figure 6.6. $y = xe^{-x}$.

APPLICATIONS

Example 7

In many applications data are studied over a period of time, and an attempt is made to discover an equation that approximately agrees with the data. In other words, an attempt is made to find a mathematical model that agrees with the observations. Such a model can then be used to study the situation to an extent that is almost impossible from the data alone. The type of equation that is hypothesized, particularly in the biological sciences, often involves logarithmic or exponential functions.

As an example suppose that a biologist is studying the growth of infants. He has found that the following formula predicts rather accurately the relationship between the age t of a child in months and the weight x of the child in ounces:

$$t = 2 + 2 \ln \frac{x}{2a - x}.$$

The number a is a constant and stands for the weight at age 2 months. The formula is claimed to be valid only between the ages of 2 months and 6 months.

(*a*) Solve for x and verify that $x = a$ when $t = 2$.

$$\begin{aligned} \ln \frac{x}{2a - x} &= \frac{t}{2} - 1, \\ \frac{x}{2a - x} &= e^{(t/2)-1}, \\ x + xe^{(t/2)-1} &= 2ae^{(t/2)-1}, \\ x &= \frac{2ae^{(t/2)-1}}{1 + e^{(t/2)-1}}. \end{aligned}$$

When $t = 2$

$$x = \frac{2ae^0}{1 + e^0} = \frac{2a}{2} = a.$$

(*b*) A particular child weighs 200 ounces at age 2 months. What is its weight expected to be at age 4 months?

To answer this question, it is necessary to find x, given that $a = 200$ and $t = 4$.

$$\begin{aligned} x &= \frac{2(200)e^{2-1}}{1 + e^{2-1}} \\ &= \frac{(400)(2.71828)}{3.71828} \\ &= 292 \text{ ounces.} \end{aligned}$$

(*c*) Find the rate of growth dx/dt when $t = 4$. Assume again that $a =$ 200.

Starting with

$$x = \frac{400e^{(t/2)-1}}{1 + e^{(t/2)-1}},$$

we find dx/dt by using the quotient rule and also (6.8) of this section.

$$\frac{dx}{dt} = 400\left[\frac{(1 + e^{(t/2)-1})e^{(t/2)-1}(\frac{1}{2}) - e^{(t/2)-1}e^{(t/2)-1}(\frac{1}{2})}{(1 + e^{(t/2)-1})^2}\right].$$

Rather than attempt to simplify, it is probably more efficient to substitute $t = 4$ immediately.

$$\begin{aligned}\left.\frac{dx}{dt}\right|_{t=4} &= 400\frac{(1 + e)(e)(\frac{1}{2}) - (e)(e)(\frac{1}{2})}{(1 + e)^2}\\ &= 400\frac{\frac{1}{2}e}{(1 + e)^2}\\ &= 200\frac{2.71828}{(3.71828)^2}\\ &= 39 \text{ ounces per month.} \blacksquare\end{aligned}$$

EXERCISES

Suggested minimum assignment: Exercises 1(*a*) (*d*) (*g*) (*k*) (*l*), 2, 4(*a*), 5(*a*), and 6(*c*).

1. Find $D_x y$ in each of the following:
   (*a*) $y = \log_3 (2x + 1)$. (*b*) $y = \log_{10} (x^3 + 2)$.
   (*c*) $y = \ln (2x + 5)$. (*d*) $y = \ln \sqrt{x}$.
   (*e*) $y = (1 + \ln x)^8$. (*f*) $y = x^4 \ln x$.
   (*g*) $y = 3^{2x}$. (*h*) $y = x2^x$.
   (*i*) $y = \dfrac{e^x - e^{-x}}{2}$. (*j*) $y = e^{\sqrt{2x+1}}$.
   (*k*) $y = \dfrac{e^{3x}}{9 + e^{3x}}$. (*l*) $y = \sqrt{1 + e^{2x}}$.
2. Prove Theorem 3.
3. Use logarithmic differentiation to show that if $u > 0$ is a differentiable function of x and n is any real number, then

$$D_x u^n = nu^{n-1} D_x u.$$

4. Use logarithmic differentiation to find $D_x y$ in each of the following. [Assume that $x > 0$ in parts (*a*) and (*b*) and that $(1 - 3x) > 0$ in part (*c*).]
   (*a*) $y = x^x$. (*b*) $y = (x)^{e^x}$. (*c*) $y = \dfrac{\sqrt{x^2 + 1}}{(1 - 3x)^7}$.
5. Find all maximum and minimum points of each of the given functions.

Assume that the methods of Chapter 3 are applicable.
(*a*) $y = f(x) = x^2 \ln x$. (*b*) $y = f(x) = x \ln x$.
(*c*) $y = f(x) = e^{-x^2}$. (*d*) $y = f(x) = 2xe^{-x^2}$.

6. Find the slope of each of the following at the given point.

(*a*) $y = \log_2 x$, (8, 3). (*b*) $y = e^{2x} + \dfrac{1}{e^{2x}}$, (0, 2).

(*c*) $y = e^{2/x}$, (2, *e*).

7. (*a*) In Example 7 find x if $a = 200$ and $t = 5$. (Use $e^{1.5} = 4.4817$.)
(*b*) In Example 7 it was found that $\left.\dfrac{dx}{dt}\right|_{t=4} = 39$ ounces per month, but from Example 7 and part (*a*) of this exercise it can be concluded that the child gained only 35 ounces during its fifth month. How can you account for the difference?

8. In Example 6 (see Figure 6.6), find all points of inflection. Assume that the methods of Section 3.5 are applicable.

9. Consider the equation $2^x + x - 4 = 0$, which has a root between $x = 1$ and $x = 2$. Use Newton's method (see Example 8 of Section 2.3) to find a second approximation x_2 of the root if the first approximation is $x_1 = 1.5$. (Use $\sqrt{2} = 1.414$ and $\ln 2 = .693$.)

6.4 Integration Formulas

The main integration formula studied so far is the power rule for integration in Section 4.2. This rule is essentially the power rule for differentiation (Section 2.3) in reverse. In this section we shall take the formulas in Section 6.3 for differentiation of the exponential and logarithmic functions and reverse them in the proper manner to obtain two new integration formulas. With these new formulas the basic ideas of Chapters 4 and 5 can be applied to many more problems.

▶ Theorem 4

Let $u = f(x)$, where f is differentiable. Then

$$\int b^{f(x)} f'(x)\, dx = \int b^u\, du = \frac{b^u}{\ln b} + C.$$

Proof: It is understood above that the base b is a positive real number other than 1. In the following calculation Theorem 4 is verified; notice the use of Theorem 3 of Section 6.3.

$$D_x\left(\frac{b^u}{\ln b} + C\right) = D_x\left(\frac{b^{f(x)}}{\ln b} + C\right)$$
$$= \frac{1}{\ln b}(\ln b) b^{f(x)} f'(x)$$
$$= b^{f(x)} f'(x). \blacktriangleleft$$

Important special cases of Theorem 4 are

$$\int e^u\,du = e^u + C \tag{6.10}$$

and

$$\int e^x\,dx = e^x + C. \tag{6.11}$$

The difficulties in handling the du correctly in Theorem 4 are similar to those encountered in Section 4.2—it is suggested that the reader review the examples in Section 4.2 before proceeding.

Example 1

$$\begin{aligned}\int 3^{2x}\,dx &= \frac{1}{2}\int 3^{2x}(2\,dx)\\ &= \frac{1}{2}\int 3^u\,du = \frac{1}{2}\frac{3^u}{\ln 3} + C\\ &= \frac{3^{2x}}{2\ln 3} + C. \quad \blacksquare\end{aligned}$$

Example 2

$$\begin{aligned}\int \frac{e^{1/x}}{x^2}\,dx &= -\int e^{1/x}\left(-\frac{1}{x^2}\,dx\right)\\ &= -\int e^u\,du = -e^u + C\\ &= -e^{1/x} + C. \quad \blacksquare\end{aligned}$$

The answer to Example 2 can be checked by differentiating the answer to obtain the original integrand:

$$D_x(-e^{1/x} + C) = -e^{1/x}\left(-\frac{1}{x^2}\right) = \frac{e^{1/x}}{x^2}.$$

Example 3

$$\begin{aligned}\int [e^x + x(1 + e^{x^2})]\,dx &= \int e^x\,dx + \int x\,dx + \int xe^{x^2}\,dx\\ &= e^x + \frac{x^2}{2} + \frac{1}{2}\int e^{x^2}(2x\,dx)\\ &= e^x + \frac{x^2}{2} + \frac{1}{2}\int e^u\,du\\ &= e^x + \frac{x^2}{2} + \frac{1}{2}e^u + C\\ &= e^x + \frac{x^2}{2} + \frac{1}{2}e^{x^2} + C. \quad \blacksquare\end{aligned}$$

Example 4

Find the area bounded by $y = e^{-x}$, $y = x + 1$, and $x = 1$. (See Figure 6.7.)

$$\begin{aligned}
A &= \int_0^1 [(x+1) - e^{-x}]\,dx \\
&= \int_0^1 x\,dx + \int_0^1 1\,dx + \int_0^1 e^{-x}(-dx) \\
&= \left(\frac{x^2}{2} + x + e^{-x}\right)\Big|_0^1 \\
&= \left(\frac{1}{2} + 1 + e^{-1}\right) - (0 + 0 + 1) \\
&= \frac{1}{2} + \frac{1}{e}. \blacksquare
\end{aligned}$$

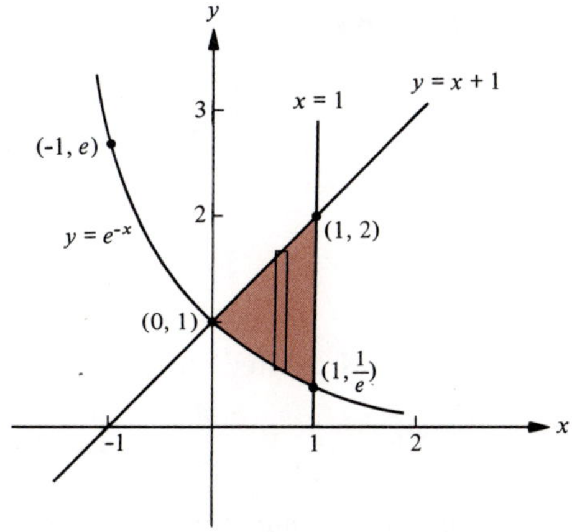

Figure 6.7

In the last section we learned that $D_x \ln x = 1/x$; this formula is valid only when $x > 0$ because $\ln x$ is not defined if $x \leq 0$. To prepare for the next theorem, it is necessary to investigate the possibility of finding $D_x \ln |x|$. Since $\ln |x|$ has meaning except when $x = 0$, we split the problem into two cases:

1. $x > 0$.

$$D_x \ln |x| = D_x \ln x = \frac{1}{x}.$$

2. $x < 0$.

$$D_x \ln |x| = D_x \ln(-x) = \frac{1}{-x} D_x(-x) = \frac{1}{-x}(-1) = \frac{1}{x}.$$

Since the result is the same in either case, it can be concluded that

$$D_x \ln |x| = \frac{1}{x}, \qquad x \neq 0.$$

As an immediate consequence the following formula can be written:

$$\int \frac{1}{x}\,dx = \ln|x| + C, \qquad x \neq 0.$$

Theorem 5 is a generalization of the above result; it can be verified in a manner similar to that used in the verification of Theorem 4 of this section.

Theorem 5

Let $u = f(x)$, where f is differentiable. Then

$$\int \frac{du}{u} = \ln|u| + C, \qquad u \neq 0.$$

The proof is left as Exercise 6. ◀

Example 5

$$\begin{aligned}\int \frac{x\,dx}{16 - x^2} &= -\frac{1}{2}\int \frac{-2x\,dx}{16 - x^2}\\ &= -\frac{1}{2}\int \frac{du}{u} = -\frac{1}{2}\ln|u| + C\\ &= -\frac{1}{2}\ln|16 - x^2| + C. \blacksquare\end{aligned}$$

It was pointed out in Section 4.2 that the power rule for integration,

$$\int u^n\,du = \frac{u^{n+1}}{n+1} + C,$$

is valid for any real number n except $n = -1$. Theorem 5 takes care of the exceptional case when $n = -1$.

At this stage the three basic tools at our disposal for doing integration problems are Theorem 4, Theorem 5, and the power rule for integration. Example 6 is designed to demonstrate the use of all three of these formulas.

Example 6

$$\begin{aligned}&\int e^x\left[1 + \frac{1}{e^x + 1} + (e^x + 1)^2\right]dx\\ &= \int e^x\,dx + \int \frac{e^x\,dx}{e^x + 1} + \int (e^x + 1)^2 e^x\,dx\\ &= \int e^x\,dx + \int \frac{du}{u} + \int u^2\,du\\ &= e^x + \ln|u| + \frac{u^3}{3} + C\\ &= e^x + \ln|e^x + 1| + \frac{(e^x + 1)^3}{3} + C.\end{aligned}$$

In this answer it is acceptable to write $\ln(e^x + 1)$ instead of $\ln|e^x + 1|$ because the quantity $e^x + 1$ is always positive. ∎

Example 7

Find the area bounded by $y = 4/(2x + 1)$, $y = 0$, $x = -2$, and $x = -1$. (See Figure 6.8.)

$$\begin{aligned} A &= \int_{-2}^{-1} \left(0 - \frac{4}{2x+1}\right) dx \\ &= -2 \int_{-2}^{-1} \frac{2\,dx}{2x+1} \\ &= -2 \ln|2x+1| \Big|_{-2}^{-1} \\ &= -2(\ln 1 - \ln 3) \\ &= 2 \ln 3. \end{aligned}$$ ∎

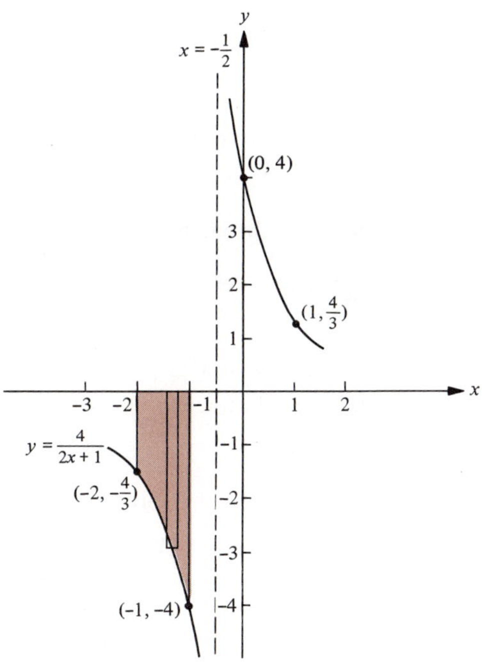

Figure 6.8

In using Theorem 5, it is important to avoid numbers that make the denominator zero. In Example 7 the denominator $2x + 1$ equals zero when $x = -\frac{1}{2}$ (this corresponds to the discontinuity in Figure 6.8 at $x = -\frac{1}{2}$). Observe that the interval $[-2, -1]$ of integration does not include $-\frac{1}{2}$.

APPLICATIONS

Example 8

Before reading this example, the reader should review the equation

$$y = A(1 - e^{-kt})$$

for bounded growth that was presented in Example 3 of Section 6.1. A reasonable assumption to make when y cannot exceed the value A is that the rate of growth dy/dt is proportional to the difference $A - y$. In this example we shall show that the solution of the differential equation

$$\frac{dy}{dt} = k(A - y),$$

subject to the condition that $y = 0$ when $t = 0$, is

$$y = A(1 - e^{-kt}).$$

Recall from Section 4.3 that our procedure for solving a differential equation is to separate the variables and then integrate.

$$\frac{dy}{A - y} = k\,dt,$$

$$-\int \frac{-dy}{A - y} = \int k\,dt,$$

$$-\ln|A - y| = kt + C.$$

Since $y = 0$ when $t = 0$,

$$-\ln|A| = 0 + C$$

and

$$C = -\ln|A|.$$

Because A and $A - y$ are positive, the absolute value signs may be omitted and the solution may be written

$$-\ln(A - y) = kt - \ln A.$$

This result can be solved for y as follows:

$$\begin{aligned}\ln(A - y) &= \ln A - kt,\\ A - y &= e^{\ln A - kt},\\ A - y &= e^{\ln A}e^{-kt},\\ A - y &= Ae^{-kt},\\ y &= A - Ae^{-kt} = A(1 - e^{-kt}). \quad \blacksquare\end{aligned}$$

Example 9

Often a single mathematical model will apply to a variety of applied problems. In fact, to the observant researcher, the application of a model in one discipline may suggest the application of the same model in other disciplines. The differential equation

$$\frac{dy}{dt} = k(A - y)$$

of Example 8 may be thought of as a special case of a differential equation of the form

$$\frac{dy}{dt} = ay + b. \tag{6.12}$$

In this example we shall solve (6.12) and mention some other applications in which this type of equation might arise.

Rewrite (6.12) as

$$\frac{dy}{y + (b/a)} = a\,dt.$$

This separates the variables, so we proceed to integrate each side of the equation and then to solve for y.

$$\ln\left|y + \frac{b}{a}\right| = at + C,$$

$$\left|y + \frac{b}{a}\right| = e^{at+C},$$

$$y = \pm e^{at+C} - \frac{b}{a},$$

$$y = ce^{at} - \frac{b}{a}, \tag{6.13}$$

where $c = \pm e^C$. (The reader may wish to verify that the result agrees with the solution in Example 8 with $a = -k$, $b = kA$, and $c = -A$.)

Suppose that the size y of a population of objects will be affected by the birth rate, death rate, immigration (entrance) rate, and emmigration (exit) rate. Under the assumptions that the birth rate r_B and the death rate r_D are proportional to the size of the existing population, and that the immigration and emmigration rates (r_i and r_e, respectively) are constant, we have

$$\begin{aligned}\frac{dy}{dt} &= r_B - r_D + r_i - r_e\\ &= k_1 y - k_2 y + r_i - r_e\\ &= (k_1 - k_2)y + (r_i - r_e)\\ &= ay + b,\end{aligned}$$

where $a = k_1 - k_2$ and $b = r_i - r_e$. We have found that the solution of this differential equation is

$$y = ce^{at} - \frac{b}{a},$$

and thus the size of the population at time t is given by

$$y = ce^{(k_1-k_2)t} - \frac{r_i - r_e}{k_1 - k_2}.$$

On pp. 293–298 of Batschelet [3] the applications mentioned above, which lead to equations of the form of (6.12), are discussed along with applications concerning nerve excitation, diffusion, and cooling. ∎

Example 10

Extensive use of exponential functions, logarithmic functions, and definite integrals is made in *The Growth and Structure of Human Populations, A Mathematical Investigation,* by Coale [8]. He defines a stable population as a population that is attained if fertility and mortality schedules remain unchanged over a long period of time, and he obtains formulas for the calculation of many different quantities associated with a stable population. For example, let x be the variable representing age, and let the value h of x be the highest age in the stable population; also, suppose that the rate r of increase in the stable population is known and the proportion $p(x)$ surviving to age x is known. On p. 17 Coale shows that the birth rate b is given by

$$b = \frac{1}{\int_0^h e^{-rx} p(x)\, dx}. \quad ∎$$

EXERCISES

Suggested minimum assignment: Exercises 1(*a*) (*c*) (*e*), 2(*c*) (*d*), 3(*b*), and 7(*b*) (*e*).

1. Perform the indicated integration.

(*a*) $\int 7^{4x}\, dx.$ (*b*) $\int 2^{1+x^2} x\, dx.$

(*c*) $\int \frac{e^x - e^{-x}}{2}\, dx.$ (*d*) $\int \frac{dx}{2x + 1}.$

(*e*) $\int \frac{x^2\, dx}{x^3 + 2}.$ (*f*) $\int \frac{\ln x}{x}\, dx.$

2. Perform the indicated integration.

(*a*) $\int \frac{dx}{2^x}.$ (*b*) $\int 3xe^{x^2}\, dx.$

(*c*) $\int \frac{e^{2x}}{e^{2x} + 1}\, dx.$ (*d*) $\int \frac{e^{2x}}{(e^{2x} + 1)^2}\, dx.$

(*e*) $\int \frac{2t + 3}{t^2 + 3t + 1}\, dt.$ (*f*) $\int \frac{1/x}{\ln x}\, dx.$

3. Evaluate the following definite integrals.

(a) $\int_0^3 e^{x/3}\,dx.$ (b) $\int_0^{e^2-1} \frac{dt}{t+1}.$ (c) $\int_0^{\ln 2} (e^x - 1)^{10} e^x\,dx.$

4. Sketch the graph of $y = \ln |x|$.
5. By long division verify that $\frac{x}{x+1} = 1 - \frac{1}{x+1}$. Use this identity to find $\int \frac{x}{x+1}\,dx$.
6. Prove Theorem 5.
7. Draw a graph and find the area bounded by the given curves.
   (a) $y = e^x$, $y = 0$, $x = 0$, $x = 2$.
   (b) $y = \ln x$, $x = 0$, $y = 0$, $y = 1$.
   (c) $y = 3^x$, $x = 0$, $y = 3x$.
   (d) $y = \frac{1}{2x}$, $y = 0$, $x = 1$, $x = 2$.
   (e) $y = \frac{1}{1-x}$, $x = -4$, $x = 0$, $y = 0$.
8. Suppose that y represents the ratio of voters who would vote for candidate Q if an election were held today, and that x represents the amount of money that Q spends on television advertisements. Assume that dy/dx is directly proportional to the amount of money spent and to the ratio of those who would not vote for Q today. In other words, assume that

$$\frac{dy}{dx} = kx(1 - y),$$

where k is the constant of proportionality. Solve this equation for y as a function of x, using methods similar to those demonstrated in Examples 8 and 9. Assume that $y = .3$ when $x = 0$ in order to determine the constant of integration.

6.5 Exponential Growth and Decay

At the outset of this chapter we stated that in many applications the assumption is made that the rate of change dy/dt of the amount y of a quantity with respect to time t is proportional to the amount y that is present. In other words, we assumed the differential equation

$$\frac{dy}{dt} = ky, \tag{6.14}$$

where k is the constant of proportionality. We also stated that the solution of (6.14) is

$$y = Ae^{kt}, \tag{6.15}$$

where $A > 0$ is the value of y at a given starting time $t = 0$. If $k > 0$, (6.15) is called the **exponential growth law**; if $k < 0$, (6.15) is called the **exponential decay law.**

Example 1

In this example we obtain (6.15) from (6.14) using methods of Section 4.3. Recall from Section 4.3 that (6.14) is called a first-order differential equation and that this type of equation can often be solved by separating the variables and integrating.

$$\frac{dy}{y} = k\,dt,$$

$$\int \frac{dy}{y} = k \int dt.$$

Applying Theorem 5 of Section 6.4, we obtain

$$\ln|y| = kt + C. \tag{6.16}$$

Next find C. Since $y = A$ when $t = 0$,

$$\ln|A| = C,$$

and hence (6.16) becomes

$$\ln|y| = kt + \ln|A|.$$

Because the amount y of the quantity present is assumed to always be positive, the absolute value signs may be deleted to yield

$$\ln y = kt + \ln A.$$

Next we solve for y as follows:

$$\ln y - \ln A = kt,$$

$$\ln \frac{y}{A} = kt,$$

$$\frac{y}{A} = e^{kt},$$

$$y = Ae^{kt}. \blacksquare$$

W. F. Libby [15] received a Nobel prize for his efforts in showing that by determination of the radioactivity of carbon it is possible to establish accurately the age of organic matter. When a plant or animal dies, the intake of radioactive carbon ceases and the radioactive decay of carbon-14 takes over. Suppose that carbon-14 is the only radioactive substance in a given specimen. It is known that the rate at which carbon-14 disintegrates is proportional to the amount that remains, and hence (6.14) applies. It is also known that it takes about 5760 years for one-half of a quantity of carbon-14 to decay—in other words, the **half-life** of carbon-14 is approximately 5760 years.

Example 2

Suppose that a piece of wood found in an archaeological dig is compared with a piece of new wood and it is determined that the specimen contains

85 per cent of its original amount A of carbon-14. Estimate the age of the specimen. (See Table 6.7.)

Table 6.7

y	t
A	0
$\frac{1}{2}A$	5760
$.85A$	?

The amount y of carbon-14 remaining after t years by (6.15) is

$$y = Ae^{kt}.$$

The constant of proportionality k can be determined by using the fact that the half-life of carbon-14 is 5760 years.

$$\begin{aligned}\tfrac{1}{2}A &= Ae^{k(5760)},\\ \tfrac{1}{2} &= e^{5760k},\\ 5760k &= \ln \tfrac{1}{2},\\ k &= \tfrac{1}{5760}(\ln 1 - \ln 2) = \tfrac{1}{5760}(-\ln 2).\end{aligned}$$

Therefore, the equation

$$y = Ae^{-t \ln 2/5760}$$

is valid for any problem of this type in which the radioactive substance is carbon-14. Since 85 per cent of A remains in the given specimen,

$$\begin{aligned}.85A &= Ae^{-t \ln 2/5760},\\ \ln .85 &= \frac{-t \ln 2}{5760},\\ t &= \frac{-5760 \ln .85}{\ln 2}\\ &= \frac{-5760(-.16252)}{.69315}\\ &= 1351 \text{ years.} \ \blacksquare\end{aligned}$$

Example 3 involves exponential growth rather than exponential decay. We shall not assume (6.15), but rather we shall reemphasize the entire process by starting with (6.14).

Example 3

Suppose that the number of bacteria in a culture increases at a rate proportional to the number present. If the initial count of bacteria is 500 and after 2 hours the count is 800, how many are expected to be present after 3 hours? Also find the number of hours elapsed until the count reaches 1500.

We assume the equation

$$\frac{dy}{dt} = ky,$$

where y is the number of bacteria present after t hours and k is the constant of proportionality. The known and required data are summarized in Table 6.8.

Table 6.8

y	t
500	0
800	2
?	3
1500	?

The procedure in this type of problem is to find a solution of the given differential equation that relates y and t and that agrees with Table 6.8. Then if either y or t is given, the other one can be found.

$$\frac{dy}{y} = k\,dt,$$

$$\int \frac{dy}{y} = k \int dt,$$

$$\ln|y| = kt + C. \tag{6.17}$$

Since $y = 500$ when $t = 0$,

$$\ln 500 = k(0) + C$$

and

$$C = \ln 500.$$

When we substitute $C = \ln 500$ in (6.17) and note that $|y| = y$ because $y > 0$, we obtain

$$\ln y = kt + \ln 500. \tag{6.18}$$

The constant k can be determined by substituting $y = 800$ and $t = 2$:

$$\ln 800 = 2k + \ln 500,$$
$$2k = \ln 800 - \ln 500 = \ln \tfrac{800}{500},$$
$$k = \tfrac{1}{2} \ln \tfrac{8}{5}.$$

Equation (6.18) becomes

$$\ln y = (\tfrac{1}{2} \ln \tfrac{8}{5})t + \ln 500. \tag{6.19}$$

Equation (6.19) can be solved for y as follows:

$$y = e^{[(1/2)\ln(8/5)]t + \ln 500}$$
$$= e^{[(1/2)\ln(8/5)]t} e^{\ln 500}$$
$$= 500\, e^{[(1/2)\ln(8/5)]t}, \tag{6.20}$$

which is of the form of (6.15).

If $t = 3$, y can be found from (6.20),

$$\begin{aligned} y &= 500\, e^{(3/2)\ln(8/5)} = 500\, e^{\ln(8/5)^{3/2}} \\ &= 500(\tfrac{8}{5})^{3/2} = 500(\tfrac{8}{5})\sqrt{8/5} \\ &= 800(1.2649) = 1012 \text{ bacteria.} \end{aligned}$$

If $y = 1500$, t can be found from (6.20),

$$\begin{aligned} 1500 &= 500\, e^{[(1/2)\ln(8/5)]t}, \\ e^{\ln(8/5)^{t/2}} &= 3, \\ \left(\frac{8}{5}\right)^{t/2} &= 3, \\ \frac{t}{2}\ln\frac{8}{5} &= \ln 3, \end{aligned}$$

$$t = \frac{2\ln 3}{\ln 1.6} = \frac{2(1.0986)}{.4700} = 4.675 \text{ hours.} \blacksquare$$

APPLICATIONS

The mathematical techniques necessary for solving problems involving (6.14) or (6.15) have been illustrated in the preceding examples. In Example 4 we mention some other settings in which the type of analysis learned in this section might be applicable.

Example 4

(*a*) Straight line depreciation was mentioned briefly at the outset of Section 1.5. Sometimes an object, however, such as a new car, seems to depreciate more rapidly at first than later on. In this case it might be logical to assume that the rate of change of value of the object is proportional to its value. In other words, an equation of the type (6.14) might be assumed under some circumstances when computing depreciation.

(*b*) The equation $A = Pe^{rt}$ was used when dealing with problems in which interest was compounded continuously (see Example 2 of Section 6.1 and Example 7 of Section 6.2). This equation has the same form as (6.15). In fact, (6.15) is sometimes called the **compound interest law.**

(*c*) In studying population growth, the assumption that the rate of change of population is proportional to the population is often valid over limited periods of time. In other words, equation (6.14) might be assumed in this type of problem for some interval of time. Example 3 can be considered to be in this category. Some discussion of the underlying assumptions made when using (6.14) may be found on pp. 7–8 of *An Introduction to Mathematical Ecology* by Pielou [18].

(*d*) On pp. 285–292 Batschelet [3] discusses several applications of (6.14), including the following one adapted from p. 305 of Ackerman [1]. Let x be the number of ionizing particles that bombard a given unit of tissue during radiation treatment. Let A be the original number of undamaged molecules of a chemical compound present in the given unit of tissue, and let y be the number of undamaged molecules after exposure to radiation. Note that if $x = 0$, then $y = A$, and as x increases, y decreases. It may be assumed that

$$\frac{dy}{dx} = ky,$$

where k is a negative constant of proportionality; this equation is of the form (6.14), although the independent variable is not time, as it has been in our other applications.

(*e*) In a chemical reaction (6.14) may be applicable in determining the amount of a substance present after a certain amount of time. Suppose that 2 pounds of a substance are present initially in a chemical reaction, but after 30 minutes only 1.2 pounds remain. If (6.14) is assumed, it can be shown (Exercise 10) that after 3 hours only slightly less than .1 pound of the substance is present. ▮

Example 5

In an article [7] entitled "Turnover and Tenure in the Canadian House of Commons 1867–1968," Casstevens and Denham develop a measure c of the turnover rate of the deputies. The measure is obtained by starting with the differential equation

$$\frac{dM}{dt} = -cM, \tag{6.21}$$

where M represents the number of deputies still holding office after t units of time have elapsed from some initial point in time; the constant c is the constant of proportionality and represents the turnover rate per Canadian deputy. The method requires the reasonable assumption (6.21) that the rate of change of M with respect to t will be directly proportional to M. The solution of the differential equation (6.21) is

$$M = M_0 e^{-ct}, \tag{6.22}$$

where M_0 is the number of deputies in the house at the beginning of a particular session (265 in 1953).

Taking the natural logarithm of both sides of (6.22), one gets a linear equation in t and $\ln M$ (see Example 8 of Section 6.2). Historical data such as those given in Table 6.9 (where t measures the number of general elections) can be used in the linear equation to obtain the slope $(-c)$ of the line of best fit for the 1953 house. (Use the method of least squares, to be explained in Example 5 of Section 7.3.) In this manner one can develop a turnover rate c for each house following a general election. A table of such rates for the Canadian House of Commons shows that the rates increased slightly over the years 1867–1958. The half-life and expectation can be calculated from the turnover rates.

Table 6.9. 1953 HOUSE OF COMMONS.

t	0	1	2	3	4	5	6
M	265	146	73	48	32	21	9

Several interesting possibilities arise from this discussion: (1) Turnover rates c_i for different legislative bodies (or different subsets of the same body) can be calculated and compared [reasons for any differences (or similarities) can be investigated]; (2) institutions other than legislatures, such as universities, can be compared with respect to their faculty turnover rates c_i; (3) the effect of a proposed change on the turnover rate for a certain institution can be ascertained in advance by comparing the turnover rates c_i of those similar institutions that have already effected the change with those that have not effected the change; and (4) in those instances in which the turnover rates have very little change, a degree of predictive value can be obtained. ∎

EXERCISES

Suggested minimum assignment: Exercises 1, 3, and 7.

1. Solve (6.14) for y in terms of t if it is known that $y = 900$ when $t = 0$ and that $y = 1500$ when $t = 5$.
2. Solve (6.14) for y in terms of t if it is known that $y = 20$ when $t = 0$ and that $y = 5$ when $t = 10$.
3. The rate of decay of a radioactive element is proportional to the amount present. Suppose that a sample weighs 200 milligrams. At the end of 2 years 150 milligrams remain. How much will remain after 3 years? (Use $\sqrt{3} = 1.73205$.)
4. Rework Example 2 if the specimen contains 91 per cent (instead of 85 per cent) of its original amount of carbon-14. (Use $\ln .91 = -.09431$.)
5. In this exercise use the values of $\sqrt{\frac{8}{5}}$ and $\ln 1.6$ given in Example 3.
   (*a*) In Example 3 find y when $t = 1$.
   (*b*) In Example 3 find t if $y = 1000$. (Use $\ln 2 = .69315$.)
6. In 18 years a quantity of a radioactive element decays to 65 per cent of the original amount present. Find the half-life of the element. (Use $\ln 2 = .69315$ and $\ln .65 = -.43078$.)
7. A new car is purchased for \$4000. During the next 10 years the rate of change in the value of the car is expected to be proportional to its value, and after 10 years the car is expected to be worth \$100 scrap value. What will the value of the car be 5 years from now? (Use $\sqrt{10} = 3.162$.)
8. The population of a certain state has increased from 3.2 million to 4.8 million during the last 20 years. Assume the exponential growth law; what is the population expected to be 5 years from now? (Use $\sqrt[4]{1.5} = 1.107$.)

9. A radioactive element present in freshly cut hay exists to an extent estimated to be five times the safe level for feeding to dairy cows. The element is iodine-131, which has a half-life of 8 days. Assume the exponential decay law; how many days must pass before it is safe to feed the hay to the cows? (Use $\ln 2 = .69315$ and $\ln 5 = 1.60944$.)
10. Verify the statement made in Example 4(*e*).

6.6 Integration Using Table of Integration Formulas

In practice, integration problems may be encountered that require formulas that have not been studied in this text. To help remedy this situation, a list of 25 selected integration formulas is given on p. 307; only the first seven of these formulas should look familiar. These 25 formulas should suffice for many applications as well as for the examples and exercises of this section. Much longer tables of integration formulas are given in handbooks of mathematical tables.* The use of a table of integration formulas is not as easy as one might expect. Some basic knowledge of integration is needed in order to select the right formula and to use it correctly.

Example 1

$$\int \frac{dx}{9 - 4x^2}.$$

A study of the table of integration formulas on p. 307 should reveal that formula 9 is the one to use. First we rewrite

$$\int \frac{dx}{9 - 4x^2} = \frac{1}{2} \int \frac{2\,dx}{(3)^2 - (2x)^2}.$$

Then formula 9 can be applied with $a = 3$, $u = 2x$, and $du = 2\,dx$ to obtain

$$\int \frac{dx}{9 - 4x^2} = \frac{1}{2}\left(\frac{1}{2(3)} \ln \left|\frac{3 + 2x}{3 - 2x}\right|\right) + C$$
$$= \frac{1}{12} \ln \left|\frac{3 + 2x}{3 - 2x}\right| + C.$$

An alternative procedure in this example is to write

$$\int \frac{dx}{9 - 4x^2} = \frac{1}{4} \int \frac{dx}{(\frac{3}{2})^2 - x^2}$$

and to apply formula 9 with $a = \frac{3}{2}$, $u = x$, and $du = dx$. (In Exercise 1 the reader is asked to carry out the problem by this method.) ▮

Example 2

$$\int \frac{dx}{8 + 4x - 4x^2}.$$

* A total of 596 formulas are given in *Standard Mathematical Tables,* 19th ed., Chemical Rubber Company, Cleveland, Ohio, 1971, pp. 395–444.

It may appear that this problem does not fit any of the 25 formulas in the table on p. 307 and that a more comprehensive table of integration formulas is needed. However, by completing the square the problem may be rewritten as

$$\int \frac{dx}{9 - (4x^2 - 4x + 1)} = \int \frac{dx}{(3)^2 - (2x - 1)^2},$$

and formula 9 can be used again exactly as in Example 1 except that $u = 2x - 1$ instead of $u = 2x$. The answer is

$$\int \frac{dx}{8 + 4x - 4x^2} = \frac{1}{2}\left[\frac{1}{2(3)} \ln \left|\frac{3 + (2x - 1)}{3 - (2x - 1)}\right|\right] + C$$

$$= \frac{1}{12} \ln \left|\frac{1 + x}{1 - x}\right| + C. \blacksquare$$

Example 3

$$\int x^2 e^{5x}\, dx.$$

Prepare to use formula 18 with $u = 5x$, $du = 5\, dx$, and $n = 2$ by observing that

$$\int x^2 e^{5x}\, dx = \tfrac{1}{125} \int (5x)^2 e^{5x} (5\, dx).$$

Hence, by formula 18,

$$\int x^2 e^{5x}\, dx = \tfrac{1}{125}\left[(5x)^2 e^{5x} - 2 \int (5x)^1 e^{5x} 5\, dx\right].$$

The problem is not completed, but formula 18, known as a **reduction formula,** has reduced the power of x from the second power to the first power. Another application of the formula will reduce the exponent from 1 to 0. Note that $n = 1$ this time, whereas $u = 5x$ and $du = 5\, dx$ are the same.

$$\int x^2 e^{5x}\, dx = \tfrac{1}{125}(5x)^2 e^{5x} - \tfrac{2}{125}\left[5x e^{5x} - \int e^{5x} 5\, dx\right].$$

The only remaining integration can be done by formula 6, which was studied in Section 6.4.

$$\int x^2 e^{5x}\, dx = \tfrac{1}{5}x^2 e^{5x} - \tfrac{2}{25} x e^{5x} + \tfrac{2}{125} e^{5x} + C. \blacksquare$$

APPLICATIONS

Example 4

The purpose of this example is to illustrate the use of the table of integration formulas in developing a mathematical model for the study of epidemics. A pertinent reference is p. 20 of Bailey [2].

Let x represent the number of animals susceptible to a certain disease at time t, and let the total population be $n + 1$. In the interval of time from t to $t + \Delta t$, the change in the number of susceptibles is

$$x(t + \Delta t) - x(t),$$

and the average rate of change is

$$\frac{x(t + \Delta t) - x(t)}{\Delta t}.$$

As $\Delta t \to 0$ we obtain the instantaneous rate of change

$$\lim_{\Delta t \to 0} \frac{x(t + \Delta t) - x(t)}{\Delta t} = \frac{dx}{dt}.$$

If one assumes that this rate is proportional to both the number of susceptibles and the number of nonsusceptibles (infectors), we obtain the differential equation

$$\frac{dx}{dt} = -\beta x(n + 1 - x),$$

where $-\beta$ is the constant of proportionality. (The minus sign is used because, in an epidemic, the number of susceptibles is decreasing and it is desired to have $\beta > 0$.) Rearranging the latter equation to separate the variables, we get

$$\frac{dx}{x(n + 1 - x)} = -\beta\, dt$$

or

$$-\beta t = \int \frac{dx}{x(n + 1 - x)}.$$

The integration can be done by formula 11 on p. 307 with $u = x$, $du = dx$, $b = -1$, and $a = n + 1$. This yields

$$-\beta t = -\frac{1}{n + 1} \ln \left| \frac{n + 1 - x}{x} \right| + C.$$

We assume that at $t = 0$ there are n susceptibles and one infectious animal. Since x is positive and not more than n, the absolute value sign can be dropped; also we can substitute $x = n$ and $t = 0$ in order to find C.

$$0 = -\frac{1}{n + 1} \ln \left(\frac{n + 1 - n}{n} \right) + C,$$

$$C = \frac{1}{n + 1} \ln \frac{1}{n} = \frac{-1}{n + 1} \ln n.$$

Hence

$$-\beta t = -\frac{1}{n + 1} \ln \left(\frac{n + 1 - x}{x} \right) - \frac{1}{n + 1} \ln n$$

and

$$\beta t = \frac{1}{n + 1} \left[\ln \left(\frac{n + 1 - x}{x} \right) + \ln n \right].$$

We proceed to solve for x in terms of t as follows:

$$\ln\left[\frac{n(n+1)-nx}{x}\right] = (n+1)\beta t,$$

$$\frac{n(n+1)-nx}{x} = e^{(n+1)\beta t},$$

$$nx + xe^{(n+1)\beta t} = n(n+1),$$

$$x(n + e^{(n+1)\beta t}) = n(n+1),$$

$$x = \frac{n(n+1)}{n + e^{(n+1)\beta t}}.$$

From the last result it can be shown that as $t \to \infty$, $x \to 0$. This means that, under our assumptions, the disease will eventually spread through the entire population; the rapidity of spreading, or diffusion, through the population is determined by β. ▮

Example 5

Variations of the model presented in Example 4 have been adapted in other ways. On pp. 147–174 Richardson [19] uses a variation to study the diffusion of "war fever" among a population. On pp. 392–401 Dodd [9] uses a variation to study the diffusion of rumors or messages among a population. ▮

EXERCISES

Suggested minimum assignment: Exercises 1, 4, 5, 8, and 11.

1. Find

$$\frac{1}{4}\int \frac{dx}{(\frac{3}{2})^2 - x^2}$$

by using formula 9 of the table of integration formulas with $a = \frac{3}{2}$, $u = x$, and $du = dx$. (See Example 1.)

In Exercises 2–9 perform the indicated integration, using the table of integration formulas on p. 307. State which formula or formulas are used.

2. $\int \sqrt{9 + 4x^2}\, dx$.

3. $\int \sqrt{4x^2 - 4x + 10}\, dx$. [*Hint:* $4x^2 - 4x + 10 = (2x-1)^2 + 9$; see Exercise 2.]

4. $\int \frac{dx}{16x^2 - 9}$.

5. $\int \frac{x\, dx}{(3 + 2x)^2}$.

6. $\int \frac{2x\,dx}{5 + 4x}$.

7. $\int xe^{2x}\,dx$.

8. $\int x^3e^x\,dx$. Also check the answer by differentiation.

9. $\int x^2 \ln 2x\,dx$.

10. Use the result of Example 3 to evaluate $\int_0^1 x^2e^{5x}\,dx$.

11. Evaluate $\int_{1/2}^{e^2/2} \ln 2x\,dx$.

12. An interesting discussion of population models is given in *The Encyclopedia of the Biological Sciences,* 2nd ed., Van Nostrand Reinhold Company, New York, 1970, pp. 755–757. One model is the equation

$$\frac{dN}{dt} = rN\left(1 - \frac{N}{K}\right).$$

In this equation N is the number of individuals in the population, t is time, r is the intrinsic rate of natural increase, and K is the environmental capacity in terms of the number of animals. Show that

$$\frac{N}{1 - N/K} = Ce^{rt}.$$

6.7 Integrals with Infinite Limits

In our previous discussion of a definite integral, $\int_a^b f(x)\,dx$, the numbers a and b were assumed to be real numbers. In this section we shall give meaning to certain integrals of the types

$$\int_a^\infty f(x)\,dx, \qquad \int_{-\infty}^b f(x)\,dx, \qquad \int_{-\infty}^\infty f(x)\,dx,$$

for which the upper limit, the lower limit, or both the upper and lower limits are infinite. These types of integrals are known as **improper integrals;** there are also other types of improper integrals that will not be discussed in this book.

Example 1

Consider the improper integral

$$\int_1^\infty \frac{4}{x^2}\,dx,$$

with an infinite upper limit.

First, we imagine definite integrals having upper limits that get larger and larger, such as $\int_1^{100} 4/x^2\,dx$, $\int_1^{1000} 4/x^2\,dx$, and, in general, $\int_1^t 4/x^2\,dx$, where t increases without bound. (The particular letter t used here has no special significance—a different letter could be used.) If $\int_1^\infty 4/x^2\,dx$ is to be assigned any meaning, it seems logical that it should be defined by

$$\int_1^\infty \frac{4}{x^2}\,dx = \lim_{t\to\infty} \int_1^t \frac{4}{x^2}\,dx,$$

provided that the limit exists.

We calculate as follows:

$$\begin{aligned}\lim_{t\to\infty} \int_1^t \frac{4}{x^2}\,dx &= \lim_{t\to\infty} \int_1^t 4x^{-2}\,dx \\ &= \lim_{t\to\infty} \frac{4x^{-1}}{-1}\bigg|_1^t \\ &= \lim_{t\to\infty} \left(-\frac{4}{t} + 4\right) \\ &= 0 + 4 \\ &= 4.\end{aligned}$$

Therefore, we shall agree to write

$$\int_1^\infty \frac{4}{x^2}\,dx = 4.$$

Geometrically, the answer may be interpreted as the area to the right of $x = 1$ between the curve $y = 4/x^2$ and the x-axis. See Figure 6.9. As $t \to \infty$ the area given by $\int_1^t 4/x^2\,dx$ keeps increasing and approaches the limit of 4. ▮

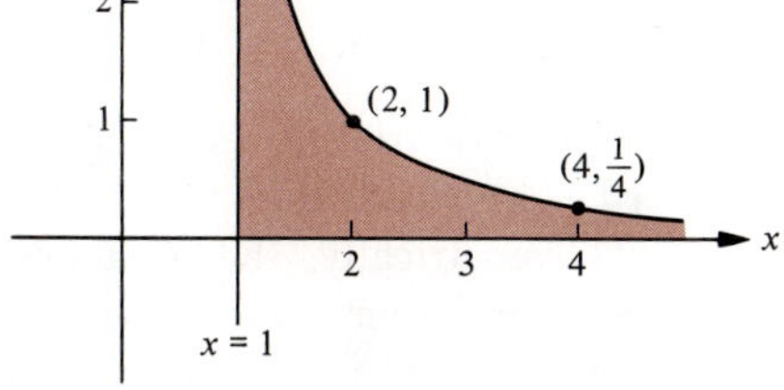

Figure 6.9

In Example 1 the function f given by $f(x) = 4/x^2$ is continuous on $[1, \infty)$. By this statement we mean that f is continuous on any interval $[1, t]$, where $t > 1$. By Theorem 1 of Section 5.2, $\int_1^t 4/x^2\,dx$ exists for any real number $t > 1$. As we study improper integrals of the type $\int_a^\infty f(x)\,dx$, we shall insist that f be continuous on $[a, \infty)$. Therefore, in this book we shall not consider improper integrals such as $\int_0^\infty 4/x^2\,dx$ or $\int_{-1}^\infty 4/x^2\,dx$, because the rational function f given by $f(x) = 4/x^2$ is discontinuous at $x = 0$ (there is a vertical asymptote at $x = 0$); a discussion of such improper integrals may be found in more advanced texts.

Definition 5

If f is continuous on $[a, \infty)$, then

$$\int_a^\infty f(x)\,dx = \lim_{t\to\infty} \int_a^t f(x)\,dx,$$

provided that the latter limit exists. If $\lim\limits_{t\to\infty} \int_a^t f(x)\,dx$ exists, $\int_a^\infty f(x)\,dx$ is said to **converge;** otherwise it is said to **diverge.**

The improper integral $\int_1^\infty 4/x^2\,dx$ of Example 1 converges. Next we give an example of an improper integral that diverges.

Example 2

Evaluate $\int_0^\infty 4/(x + 1)\,dx$, if possible.

The function f given by $f(x) = 4/(x + 1)$ is continuous on $[0, \infty)$. Therefore,

$$\int_0^\infty \frac{4}{x+1}\,dx = \lim_{t\to\infty} \int_0^t \frac{4}{x+1}\,dx,$$

provided that this latter limit exists.

$$\begin{aligned} \lim_{t\to\infty} \int_0^t \frac{4}{x+1}\,dx &= \lim_{t\to\infty} 4 \ln |x + 1| \Big|_0^t \\ &= \lim_{t\to\infty} (4 \ln |t + 1| - 4 \ln 1) \\ &= 4 \lim_{t\to\infty} \ln |t + 1|. \end{aligned}$$

However, $\lim\limits_{t\to\infty} \ln |t + 1|$ does not exist because $\ln |t + 1|$ can be made as large as desired by selecting a large enough value of t. We conclude that $\int_0^\infty 4/(x + 1)\,dx$ does not exist and say that it diverges.

Geometrically, the area under $y = 4/(x + 1)$ to the right of $x = 0$ becomes larger without bound as x increases (see Figure 6.10). One might say that there is an infinite amount of area under the curve in Figure 6.10 but only a finite amount of area under the curve in Figure 6.9; this interest-

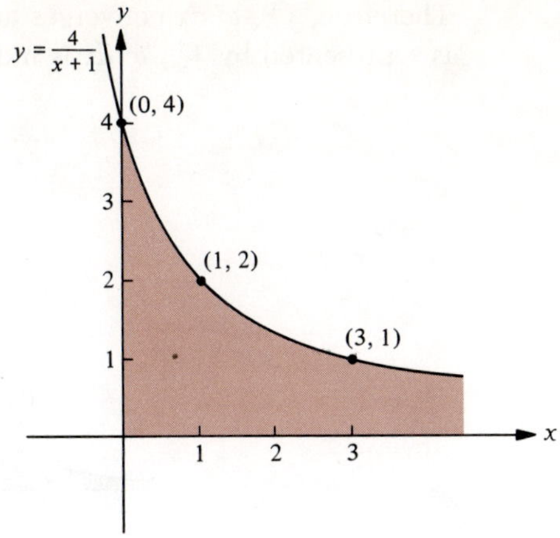

Figure 6.10

ing fact is difficult to believe at first, and it certainly is not obvious from the figures. ∎

The next definition is similar to Definition 5.

▶ Definition 6

If f is continuous on $(-\infty, b]$, then

$$\int_{-\infty}^{b} f(x)\,dx = \lim_{v\to-\infty} \int_{v}^{b} f(x)\,dx,$$

provided that the latter limit exists. If $\lim_{v\to-\infty} \int_v^b f(x)\,dx$ exists, $\int_{-\infty}^{b} f(x)\,dx$ is said to **converge;** otherwise it is said to **diverge.**

Example 3

Determine whether $\int_{-\infty}^{1} e^x\,dx$ converges or diverges; if it converges, find its value.

$$\int_{-\infty}^{1} e^x\,dx = \lim_{v\to-\infty} \int_{v}^{1} e^x\,dx,$$

provided that this latter limit exists. Recall that an exponential function is continuous for all real numbers, and hence Definition 6 applies for this example.

$$\begin{aligned}
\lim_{v\to-\infty} \int_{v}^{1} e^x\,dx &= \lim_{v\to-\infty} e^x \Big|_{v}^{1} \\
&= \lim_{v\to-\infty} (e - e^v) \\
&= e - 0 \\
&= e.
\end{aligned}$$

Therefore, $\int_{-\infty}^{1} e^x\,dx$ converges and its value is e. The region whose area is represented by $\int_{-\infty}^{1} e^x\,dx$ is indicated in Figure 6.11. ▮

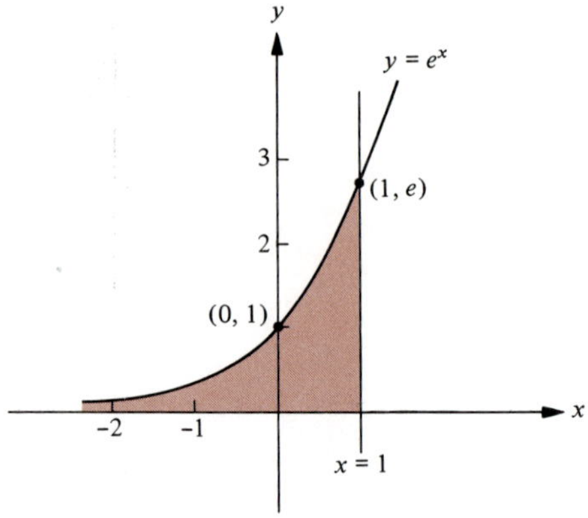

Figure 6.11

In Definition 7 we consider the type of improper integral for which both the upper and lower limits are infinite.

▶ **Definition 7**

Suppose that f is continuous for all real numbers. Then

$$\int_{-\infty}^{\infty} f(x)\,dx = \int_{-\infty}^{0} f(x)\,dx + \int_{0}^{\infty} f(x)\,dx, \tag{6.23}$$

provided that the latter two improper integrals each converge, in which case $\int_{-\infty}^{\infty} f(x)\,dx$ is said to **converge.** If either (or both) of the improper integrals on the right of (6.23) diverges, then $\int_{-\infty}^{\infty} f(x)\,dx$ is said to **diverge.**

In (6.23) of Definition 7 any real number c can be used instead of 0. One should attempt to evaluate *separately* each of the improper integrals on the right side of (6.23), using Definitions 6 and 5, respectively; $\int_{-\infty}^{\infty} f(x)\,dx$ converges if *both* of the improper integrals on the right converge, and in this case equals the sum of their values.

Example 4

Consider $\int_{-\infty}^{\infty} e^{-x^2/2}\,dx$.

The function f given by $f(x) = e^{-x^2/2}$ is continuous for all real numbers. Therefore,

$$\int_{-\infty}^{\infty} e^{-x^2/2}\,dx = \int_{-\infty}^{0} e^{-x^2/2}\,dx + \int_{0}^{\infty} e^{-x^2/2}\,dx, \tag{6.24}$$

provided that the latter two improper integrals converge. By Definitions 6 and 5 the improper integrals on the right side of (6.24) are equal to

$$\lim_{v\to-\infty}\int_v^0 e^{-x^2/2}\,dx$$

and

$$\lim_{t\to\infty}\int_0^t e^{-x^2/2}\,dx,$$

respectively, provided these limits exist. At this stage we can go no further with (6.24) because

$$\int e^{-x^2/2}\,dx$$

cannot be found by any of the integration formulas we have studied. By methods beyond the scope of this book it can be shown that both of the improper integrals on the right side of (6.24) converge and that each of them has the value $\frac{1}{2}\sqrt{2\pi}$. Therefore, $\int_{-\infty}^{\infty} e^{-x^2/2}\,dx$ converges and

$$\int_{-\infty}^{\infty} e^{-x^2/2}\,dx = \tfrac{1}{2}\sqrt{2\pi} + \tfrac{1}{2}\sqrt{2\pi} = \sqrt{2\pi}.$$

The answer can be interpreted geometrically as the area of the region beneath the curve $y = e^{-x^2/2}$ and above the x-axis. See Figure 6.12. ▮

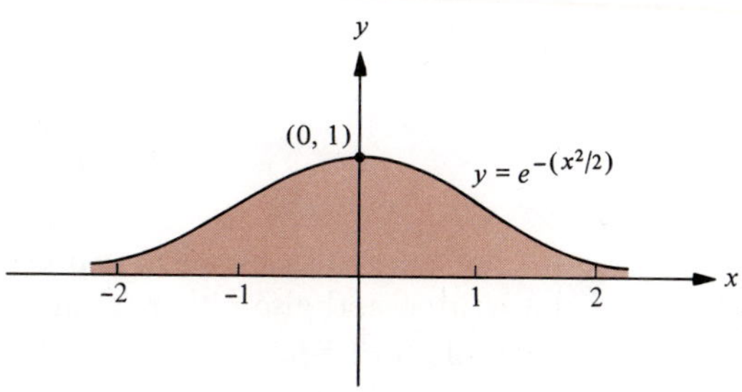

Figure 6.12

In statistics the result of Example 4 turns out to be very important; it is related to a study of probability. We shall pursue this idea further in Example 5.

APPLICATIONS

*Example 5 Normal Distribution

Suppose that the set of outcomes for an experiment is the sample space U, and suppose that a continuous measurement X with domain U and range S has been defined. Assume that S is the set of all real numbers, so that the given continuous measurement is *not* bounded. In Example 4 of Section 5.2 a probability density function p for a *bounded* continuous measurement was defined; if the bounded requirement is removed, then a probability density function p must be a continuous function such that

1. $p(x) > 0$ for any real number x.

2. $\int_{-\infty}^{\infty} p(x)\,dx = 1$.

There is no change in the four properties required for the corresponding cumulative probability distribution function F when the continuous measurement is not bounded (see Example 8 of Section 1.8). Also, for the unbounded case, the relationships between the two functions F and p are given by the two equations

$$F(x) = \int_{-\infty}^{x} p(s)\,ds$$

and

$$F'(x) = p(x),$$

where x is *any* real number (compare with Example 3 of Section 5.6). The probability of an event E, where $E = \{s : s \leq c\}$ is given by

$$P(E) = F(c) = \int_{-\infty}^{c} p(x)\,dx.$$

If $E_1 = \{s : c \leq s < d\}$, then

$$P(E_1) = F(d) - F(c) = \int_{c}^{d} p(x)\,dx.$$

If $E_2 = \{s : s > c\}$, then

$$P(E_2) = \int_{c}^{\infty} p(x)\,dx = 1 - F(c).$$

(In the inequalities used in expressing E, E_1, and E_2, either $<$ or $\leq$ can be written and also either $>$ or $\geq$.) The mean of the measurement (see Example 5 of Section 5.2) is found by

$$\mu = \int_{-\infty}^{\infty} xp(x)\,dx.$$

The variance σ^2 and the standard deviation σ (see Example 6 of Section 5.2) are determined from

$$\sigma^2 = \int_{-\infty}^{\infty} (x - \mu)^2 p(x)\,dx.$$

In statistics many different probability density functions for unbounded continuous measurements are studied. The most important of these, because it represents quite accurately the distribution of a large number of outcomes for many practical experiments, is called a **normal probability density function** and is given by

$$p(x) = \frac{1}{\sqrt{2\pi}\,\sigma} e^{-(x-\mu)^2/2\sigma^2}, \tag{6.25}$$

where μ is the mean of the measurement and $\sigma > 0$ is the standard deviation of the measurement. In this case the measurement is said to have the **normal distribution,** which is sometimes also called the **Gaussian distribution.** If one measures the height in inches of a large number of adult males, a certain "normal" amount of deviation of the results from the mean is expected; the normal probability density function, with appropriate values of μ and σ, usually will reflect the situation quite accurately. On p. 95 of his *Statistics in Psychology and Education* Garrett [10] lists several instances for which there is evidence of the accuracy of the normal distribution; his five major headings are biological statistics, anthropometrical data, social and economic data, psychological measurements, and errors of observation.

We shall focus our attention on the normal probability density function with $\mu = 0$ and $\sigma = 1$; ordinarily, an understanding of this case will enable one to handle all the other cases also. If $\mu = 0$ and $\sigma = 1$, (6.25) reduces to

$$p(x) = \frac{1}{\sqrt{2\pi}} e^{-x^2/2};$$

in this case p is called the **standardized normal probability density function.** The function p is continuous for all real numbers, and $p(x) > 0$ for any real number x. From Example 4

$$\int_{-\infty}^{\infty} e^{-x^2/2}\, dx = \sqrt{2\pi},$$

and hence

$$\int_{-\infty}^{\infty} \frac{1}{\sqrt{2\pi}} e^{-x^2/2}\, dx = \int_{-\infty}^{\infty} p(x)\, dx = 1.$$

Therefore, p does meet the requirements of a probability density function; its graph is shown in Figure 6.13. The corresponding cumulative probability distribution function F is given by

$$F(x) = \int_{-\infty}^{x} p(s)\, ds = \int_{-\infty}^{x} \frac{1}{\sqrt{2\pi}} e^{-s^2/2}\, ds.$$

The graph of F is shown in Figure 6.14. Tables of function values $p(x)$ and

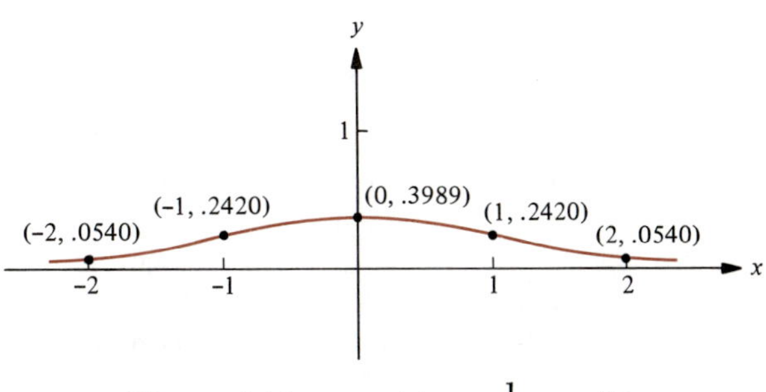

Figure 6.13. $y = p(x) = \dfrac{1}{\sqrt{2\pi}} e^{-x^2/2}$.

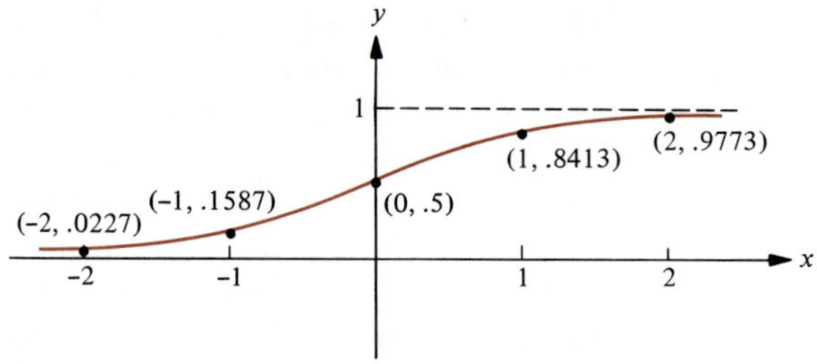

Figure 6.14. $y = F(x) = \int_{-\infty}^{x} \frac{1}{\sqrt{2\pi}} e^{-s^2/2}\, ds.$

$F(x)$ have been carefully computed and may be found in many statistics books; a few of these values are given in Table 6.10.

Suppose that a measurement has the **standard normal distribution**—in other words, $\mu = 0$ and $\sigma = 1$. What is the probability that an outcome is between -1 and 1? The answer is

$$\int_{-1}^{1} p(x)\, dx = F(1) - F(-1).$$

Table 6.10 does not include negative values of x (such as -1). Because of symmetry (see Figures 6.13 and 6.14) we have

$$p(-x) = p(x)$$

and

$$F(-x) = 1 - F(x).$$

Therefore,

$$\begin{aligned} F(1) - F(-1) &= .8413 - (1 - .8413) \\ &= .6826. \end{aligned}$$

Table 6.10. VALUES OF $p(x)$ AND $F(x)$ FOR THE STANDARD NORMAL DISTRIBUTION.

x	$p(x)$	$F(x)$
0	.3989	.5000
.25	.3867	.5987
.50	.3521	.6915
.75	.3011	.7734
1.00	.2420	.8413
1.25	.1827	.8944
1.50	.1295	.9332
1.75	.0863	.9599
2.00	.0540	.9773
2.50	.0175	.9938
3.00	.0044	.9987
3.50	.0009	.9998

The result above means that slightly over 68 per cent of the outcomes are within one standard deviation of the mean (this is also true for normal distributions with different values of μ and σ). Similarly, it can be shown for a normal distribution that about 95.46 per cent of the outcomes are within two standard deviations of the mean.

Suppose that a company produces a certain type of light bulb. The life in hours of bulbs of this type is assumed to have a normal distribution with $\mu = 600$ and $\sigma = 30$. What is the probability that a given bulb will last longer than 645 hours? In other words, what is the probability that the outcome will be more than $\frac{45}{30}$ or $\frac{3}{2}$ standard deviations above the mean? If one converts to the standard normal distribution, what is the probability of the event $E = \{s : s > 1.5\}$?

$$\begin{aligned} P(E) = \int_{1.5}^{\infty} p(x)\,dx &= 1 - F(1.5) \\ &= 1 - .9332 \\ &= .0668. \end{aligned}$$

In other words, only 6.68 per cent of the bulbs are expected to last longer than 645 hours. ▮

EXERCISES

Suggested minimum assignment: Exercises 1, 3, 5, 10, 11, and 13.

In Exercises 1–6 determine whether the given improper integral converges or diverges. If it converges, find its value and draw a sketch showing the region whose area is given by the improper integral.

1. $\displaystyle\int_0^{\infty} e^{-x}\,dx.$
2. $\displaystyle\int_1^{\infty} \frac{dx}{x}.$
3. $\displaystyle\int_{-\infty}^{0} \frac{dx}{(x-2)^2}.$
4. $\displaystyle\int_{-\infty}^{3} 2^x\,dx.$
5. $\displaystyle\int_{-\infty}^{\infty} \frac{x\,dx}{x^2+1}.$
6. $\displaystyle\int_{-\infty}^{\infty} e^x\,dx.$

In Exercises 7–14 determine whether the given improper integral converges or diverges. If it converges, find its value.

7. $\displaystyle\int_3^{\infty} \frac{dx}{x^3}.$
8. $\displaystyle\int_1^{\infty} \frac{4\,dx}{x^{3/2}}.$
9. $\displaystyle\int_1^{\infty} \frac{4\,dx}{x^{2/3}}.$
10. $\displaystyle\int_{-\infty}^{-1} \frac{dx}{2x-1}.$
11. $\displaystyle\int_0^{\infty} xe^{-x^2}\,dx.$
12. $\displaystyle\int_{-\infty}^{\infty} \frac{x\,dx}{\sqrt{16+x^2}}.$

13. $\displaystyle\int_{-\infty}^{\infty} \frac{x\,dx}{(x^2+4)^2}$.

14. $\displaystyle\int_{-\infty}^{-5} \frac{2x^3\,dx}{x^4+1}$.

15. Evaluate $\displaystyle\int_{-\infty}^{\infty} x(1/\sqrt{2\pi})e^{-x^2/2}\,dx$. With respect to Example 5, what is represented by the given improper integral?

16. In Figure 6.14 what is the slope of the curve at (0, .5)?

17. Suppose that the heights in inches of adult women are normally distributed with $\mu = 64$ and $\sigma = 3$.
   (*a*) What is the probability that a woman selected at random is less than $62\frac{1}{2}$ inches tall?
   (*b*) What percentage of women are 5 feet 10 inches or taller?
   (*c*) What percentage of women are between 5 feet 1 inch and 5 feet 7 inches?

NEW VOCABULARY

exponential function with base *b* 6.1
compounded continuously 6.1
logarithm of a positive number to the base *b* 6.2
natural logarithms 6.2
common logarithms 6.2
logarithmic function with base *b* 6.2
inverse functions 6.2
logarithmic differentiation 6.3
exponential growth law 6.5
exponential decay law 6.5
half-life 6.5
reduction formula 6.6
improper integral 6.7
improper integral converges 6.7
improper integral diverges 6.7

7 Functions of More Than One Variable

In previous chapters we have given many illustrations of quantities that were dependent upon only one variable. In many practical situations, however, a quantity often depends upon several variables. For example, the cost of an item may be thought of as depending upon labor, materials, overhead expenses, and transportation expenses. The sales of an item may be dependent upon the price of the item and the amount spent on advertising. Productivity is often thought of as dependent upon the capital invested as well as the labor force. The demand for an item is related to its price and also to the income of the consumer; the demand might also depend upon the price of a similar competing item. Such situations are necessarily more complicated than those in which there is only one independent variable. Our purpose in this chapter, therefore, is to present a brief introduction to the geometry and calculus of functions of more than one variable and to present techniques for the solution of certain optimization problems that cannot be solved by the methods of Chapter 3. Such optimization problems will include those in which we seek to maximize or minimize certain functions that are subject to specified restrictions; the solution of these problems will be accomplished by the Lagrange multiplier method, the discovery of which is credited to Joseph Louis Lagrange (1736–1813).

Prerequisites: Section 6.3 is a prerequisite to 7.1–7.3. Sections 3.4 and 7.3 are needed in 7.4.
Suggested Sections for Moderate Emphasis: 7.1–7.3
Suggested Sections for Minimum Emphasis: 7.1, 7.2

7.1 Introduction

We have frequently expressed a function f of one variable by an equation

$$y = f(x).$$

The graph was drawn in two dimensions and was called a curve. The variable x was called the independent variable, and y was known as the dependent variable. In this chapter a function f of two variables can be expressed by an equation

$$z = f(x, y).$$

The graph of such a function will be a **surface** lying in three-dimensional space. If $z = f(x, y)$ expresses the dependence of z upon x and y, then z is called the **dependent variable,** whereas both x and y are **independent variables.**

Example 1

Let $z = f(x, y) = 9 - x^2 - y^2$. Then

$$f(1, 2) = 9 - 1 - 4 = 4$$

and

$$f(0, 0) = 9 - 0 - 0 = 9. \blacksquare$$

The proper functional notation is shown in Example 1. Given values of the independent variables, one can determine (as above) a *unique* value of the dependent variable; in this basic respect the idea of a function of more than one independent variable is similar to that of a function of one independent variable. The graph of the function in Example 1 will be given in Figure 7.3.

In order to discuss functions of more than one variable, it is helpful to learn the fundamentals of graphing in three dimensions and to see a few examples. In Figure 7.1 the x-axis, y-axis, and z-axis of a three-dimensional coordinate system are shown. Each axis is perpendicular to each other axis, although this may not appear to be the case, since we are attempting to draw a three-dimensional figure on a two-dimensional piece of paper. The x-axis can be thought of as being perpendicular to the plane of the paper, and the units on the x-axis are marked about 70 per cent as far apart as on the other axes in order to give the best perspective. A few points (x, y, z) are also plotted in Figure 7.1; notice that the x-coordinate is given first, then the y-coordinate, and finally the z-coordinate.

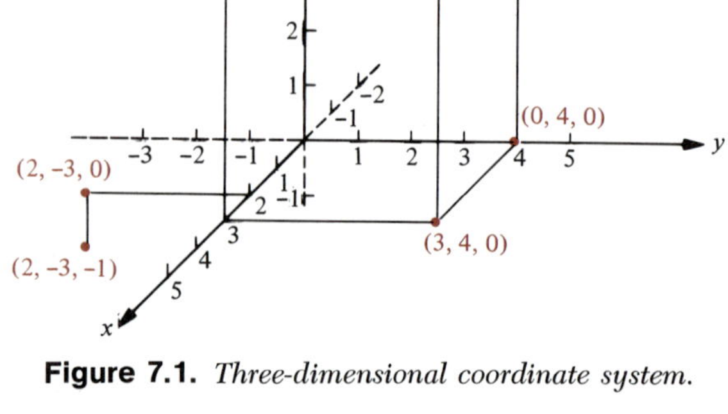

Figure 7.1. *Three-dimensional coordinate system.*

The plane determined by the x and y axes is called the **xy-plane;** every point in the xy-plane has a z-coordinate of zero, so an equation of the xy-plane is

$$z = 0.$$

In Figure 7.2 the xy-plane, the xz-plane, the yz-plane, and some planes parallel to these **coordinate planes** are shown along with their equations.

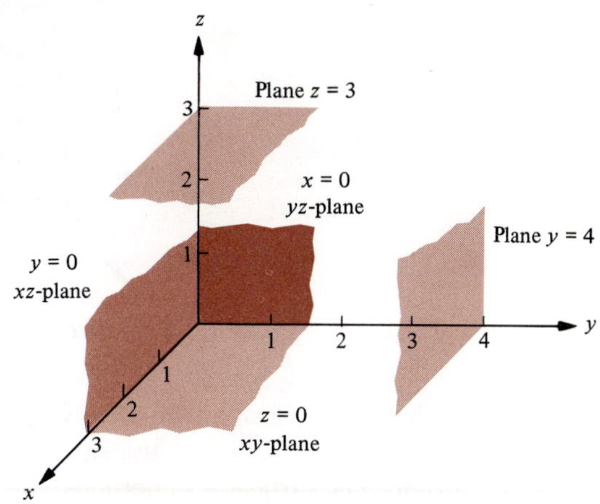

Figure 7.2

Example 2

Sketch $z = 9 - x^2 - y^2$.

To obtain the curve of intersection of the surface $z = 9 - x^2 - y^2$ in the xy-plane, we solve simultaneously the equations

$$\begin{cases} z = 9 - x^2 - y^2, \\ z = 0. \end{cases}$$

The curve of intersection is the circle

$$x^2 + y^2 = 9,$$

with center at the origin and radius 3; this circle is shown in Figure 7.3. Similarly, the equation of the surface is solved simultaneously with $x = 0$, and then with $y = 0$, to obtain the parabolas shown in the yz-plane and xz-plane, respectively. In general, surfaces can be drawn by sketching the curves of intersection with the coordinate planes or with planes parallel to the coordinate planes. These curves of intersection give a clear outline of the surface that opens downward and has (0, 0, 9) as a point on the surface that is higher than all surrounding points; we say that (0, 0, 9) is a (*relative*) *maximum point.* ▮

Example 3

In Figure 7.4 a sketch of the surface

$$z = y^2 - x^2$$

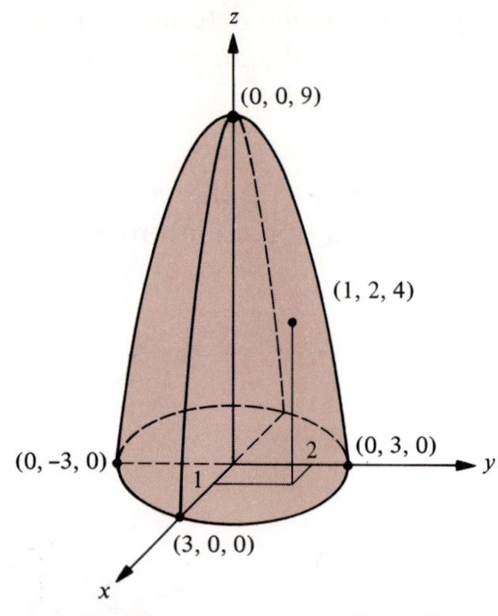

Figure 7.3. $z = 9 - x^2 - y^2$.

is shown. The graph is relatively difficult to draw, but the interesting thing is the general shape of the surface, particularly the appearance near the origin. The surface is sometimes called a **saddle surface,** and the origin may be called a **saddle point.** The origin is the minimum point on the parabola $z = y^2$ shown in the yz-plane, but it is the maximum point on the parabola $z = -x^2$ shown in the xz-plane. There are points near the origin that are higher and points that are lower; therefore, the origin is neither a (relative) maximum point nor a (relative) minimum point. ▮

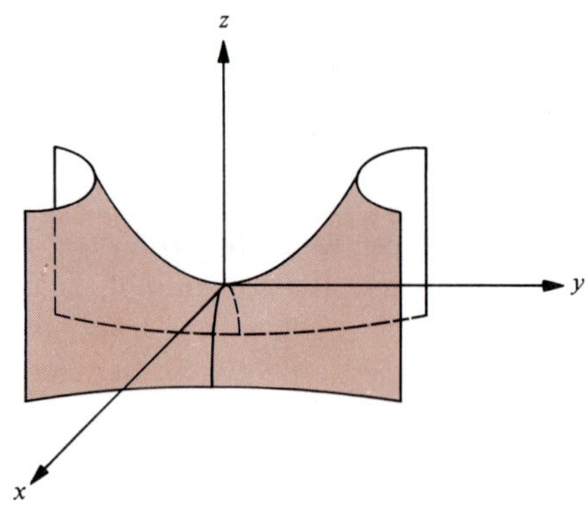

Figure 7.4. $z = y^2 - x^2$.

We shall not attempt a lengthy discussion of limits and continuity for functions of more than one variable; we shall make a few comments of an intuitive nature, however, concerning the example

$$z = f(x, y) = 9 - x^2 - y^2.$$

For any point in the xy-plane near (1, 2), the z-coordinate of the surface is close to 4 (see Figure 7.3); in a manner that is analogous with our presentation of limits of functions of a single variable, this concept can be expressed by saying that the *limit* of f, as (x, y) approaches (1, 2), is 4. That is,

$$\lim_{(x, y)\to(1,2)} (9 - x^2 - y^2) = 4.$$

This statement would be correct even if there were a hole in the surface at (1, 2, 4), in which case we would say that the surface had a discontinuity at this point. In the given problem, however, the limit of f as $(x, y) \to (1, 2)$ is equal to the value of f at (1, 2); that is,

$$\lim_{(x, y)\to(1,2)} f(x, y) = f(1, 2) = 4,$$

and thus f is said to be *continuous* at (1, 2). In fact, f is continuous at each point of the xy-plane.

Example 4

As an example of a function of *three* independent variables, suppose that the profit P made by a builder of homes for building one house is thought of as depending upon the revenue R received for the house, the labor cost L charged to the house, and the cost M of materials used in building the house. Since profit equals revenue minus cost, we have

$$P = f(R, L, M) = R - L - M.$$

Because there are three independent variables, a sketch will not be given. It is still possible to define the concepts of limit, continuity, and partial derivative (see Section 7.2) for the function even though a geometric interpretation is not readily available. ▮

EXERCISES

Suggested minimum assignment: Exercises 1(*a*), 3(*a*), 4, 6, 8, 9, and 11(*b*).

1. Suppose that $f(x, y) = 2x^2 + y^2 + 2xy - 3$.
   (*a*) Find $f(2, 1)$. (*b*) Find $f(-1, 1)$.
2. Suppose that $f(x, y) = x + 2^{xy}$.
   (*a*) Find $f(3, 1)$. (*b*) Find $f(0, 2)$.
3. Suppose that $f(w, x, y) = w^2 + wxy + \log_2 x$.
   (*a*) Find $f(1, 2, 3)$. (*b*) Find $f(2, 1, 3)$.
4. Plot all the given points (x, y, z) on the same drawing and label them clearly:

   $$(3, 0, 0), (3, 6, 0), (3, 6, 5), (1, 10, 3).$$

5. Repeat Exercise 4 for the following points:

   $$(5, 1, 0), (1, 4, -1), (0, 0, 6).$$

6. (*a*) What is an equation of the xy-plane?
   (*b*) What is an equation of the xz-plane?
   (*c*) What is an equation of the yz-plane?
   (*d*) What is an equation of the plane containing the points $(0, 0, 3)$, $(2, 5, 3)$, and $(1, -4, 3)$?
7. Sketch the plane $x + y + z = 4$ by drawing the lines of intersection of this plane with each of the coordinate planes.
8. (*a*) Sketch the surface $z = x^2 + y^2$ by drawing the curves of intersection of this surface with each of the planes $x = 0$, $y = 0$, and $z = 4$.
   (*b*) What point on the surface is a relative minimum point?
   (*c*) Does the surface have any points of discontinuity?
9. Suppose that the cost C of an item depends upon w, x, and y according to the formula

$$C = f(w, x, y) = w^2 + x + y.$$

   Name the dependent and independent variables.
10. Your choice of a new car is a function of many variables. Name some of these variables.
11. In a three-dimensional drawing it can be shown that the distance d between points (x_1, y_1, z_1) and (x_2, y_2, z_2) is given by

$$d = \sqrt{(x_2 - x_1)^2 + (y_2 - y_1)^2 + (z_2 - z_1)^2}.$$

   Find the distance between each of the following pairs of points.
   (*a*) $(4, 7, 1)$ and $(0, 5, 4)$.
   (*b*) $(-2, 1, 5)$ and $(4, 3, 2)$.

7.2 Partial Derivatives

For functions of more than one variable, the important concept of a partial derivative plays a role analogous to that of the ordinary derivative discussed in previous chapters. We now define a partial derivative of a function of two variables, and later in this section we shall give a geometric interpretation.

▶ Definition 1

Let $z = f(x, y)$. The functions determined by the following limits, if the limits exist, are called the **partial derivative of f with respect to x** and the **partial derivative of f with respect to y,** respectively:

$$(i)\quad \frac{\partial f}{\partial x} = \lim_{\Delta x \to 0} \frac{f(x + \Delta x, y) - f(x, y)}{\Delta x}.$$

$$(ii)\quad \frac{\partial f}{\partial y} = \lim_{\Delta y \to 0} \frac{f(x, y + \Delta y) - f(x, y)}{\Delta y}.$$

The key thing to observe in (*i*) is that the definition is similar to the definition of the ordinary derivative in Section 2.2, except that we have a second independent variable y that is kept fixed. Thus $\partial f/\partial x$ can be found by using ordinary differentiation rules provided that y is treated as a constant. Similarly, $\partial f/\partial y$ can be calculated using previous techniques by remembering to treat x as a constant.

Example 1

Find $\partial f/\partial x$ and $\partial f/\partial y$ if

$$z = f(x, y) = x^4y^3 + 3x.$$

$$\frac{\partial f}{\partial x} = y^3(4x^3) + 3 = 4x^3y^3 + 3.$$

$$\frac{\partial f}{\partial y} = x^4(3y^2) + 0 = 3x^4y^2. \blacksquare$$

Some comments about notation need to be made. In a problem like Example 1 it is customary to use the notation $\partial z/\partial x$ or $f_x(x, y)$ as well as $\partial f/\partial x$. Hence in Example 1

$$\frac{\partial z}{\partial x} = \frac{\partial f}{\partial x} = f_x(x, y) = 4x^3y^3 + 3.$$

Also,

$$\frac{\partial z}{\partial y} = \frac{\partial f}{\partial y} = f_y(x, y) = 3x^4y^2.$$

If these partial derivatives are to be evaluated for $x = -1$ and $y = 2$, then the notation that we shall use is

$$\left.\frac{\partial z}{\partial x}\right|_{(-1,2)} = \left.\frac{\partial f}{\partial x}\right|_{(-1,2)} = f_x(-1, 2) = 4(-1)(8) + 3 = -29.$$

Also,

$$\left.\frac{\partial z}{\partial y}\right|_{(-1,2)} = \left.\frac{\partial f}{\partial y}\right|_{(-1,2)} = f_y(-1, 2) = 3(1)(4) = 12.$$

The reader should think of a symbol such as $\partial f/\partial x$ as one quantity—no separate meanings will be assigned to ∂f and ∂x.

For a function of more than two variables it is possible to calculate a partial derivative with respect to each of the independent variables. In differentiating with respect to any one variable, all other independent variables should be treated as constants.

Example 2

A **utility function** of several commodities gives a quantitative measure of the desirability for a particular consumer to possess various quantities of these products. Suppose that a utility function of three commodities is given by

$$u = f(w, x, y) = 5w + 6x + 8y - \frac{w^2}{4} - \frac{x^2}{2} - y^2 + xy + 2wy.$$

Find $\partial u/\partial w$, $\partial u/\partial x$, and $\partial u/\partial y$, which are called the **marginal utilities** of the three products.

Holding x and y fixed and differentiating with respect to w, we calculate

$$\frac{\partial u}{\partial w} = 5 - \frac{1}{2}w + 2y.$$

Similarly,

$$\frac{\partial u}{\partial x} = 6 - x + y$$

and

$$\frac{\partial u}{\partial y} = 8 - 2y + x + 2w. \blacksquare$$

Let us return to the case of a function of two variables given by

$$z = f(x, y)$$

and give a geometric interpretation of partial derivatives. In particular, we shall use as an illustration

$$z = f(x, y) = 9 - x^2 - y^2$$

from Examples 1 and 2 of Section 7.1; attention will be focused on the point (1, 2, 4) of this surface.

$$f_x(x, y) = -2x,$$
$$f_x(1, 2) = -2.$$

In Figure 7.5 a portion of the surface $z = 9 - x^2 - y^2$ is shown. With y held constant ($y = 2$), we can interpret $f_x(1, 2)$ in a manner similar to the interpretation given for ordinary derivatives. That is, $f_x(1, 2) = -2$ represents the rate of change of z with respect to x at (1, 2, 4) with y held

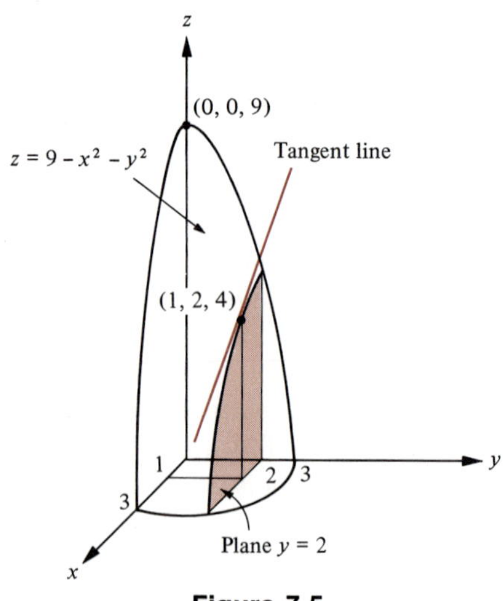

Figure 7.5

constant. In the plane $y = 2$, in other words, the slope of the tangent line (see Figure 7.5) to the curve of intersection of $z = 9 - x^2 - y^2$ and $y = 2$ at (1, 2, 4) is -2. Similarly, for a geometric interpretation of $\partial f/\partial y$, consider

$$f_y(x, y) = -2y$$

and

$$f_y(1, 2) = -4.$$

In the plane $x = 1$, the slope of the tangent line (see Figure 7.6) to the curve of intersection of $z = 9 - x^2 - y^2$ and $x = 1$ at (1, 2, 4) is -4.

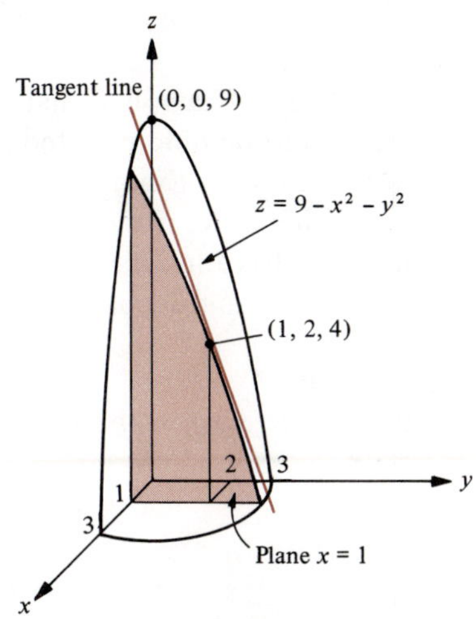

Figure 7.6

For a function of more than two variables no similar geometric interpretation is available. A partial derivative with respect to one of the variables, evaluated at a point, may be thought of as a rate of change with respect to that variable with all other independent variables being held constant.

A partial derivative of a function of several variables is also a function of the same variables and may be differentiated again to obtain a second partial derivative. This type of calculation will be important in the next section; we illustrate the technique and notation in Examples 3 and 4.

Example 3

Given $z = f(x, y) = ye^{2x} + x^2$.

$$\frac{\partial z}{\partial x} = \frac{\partial f}{\partial x} = f_x(x, y) = 2ye^{2x} + 2x.$$

$$\frac{\partial^2 z}{\partial x^2} = \frac{\partial^2 f}{\partial x^2} = f_{xx}(x, y) = 4ye^{2x} + 2.$$

$$\left.\frac{\partial^2 z}{\partial x^2}\right|_{(0,1)} = \left.\frac{\partial^2 f}{\partial x^2}\right|_{(0,1)} = f_{xx}(0, 1) = 4(1)e^0 + 2 = 6.$$

$$\frac{\partial}{\partial y}\left(\frac{\partial z}{\partial x}\right) = \frac{\partial^2 z}{\partial y\,\partial x} = \frac{\partial^2 f}{\partial y\,\partial x} = f_{xy}(x, y) = 2e^{2x}.$$

$$\frac{\partial z}{\partial y} = \frac{\partial f}{\partial y} = f_y(x, y) = e^{2x}.$$

$$\frac{\partial}{\partial x}\left(\frac{\partial z}{\partial y}\right) = \frac{\partial^2 z}{\partial x\,\partial y} = \frac{\partial^2 f}{\partial x\,\partial y} = f_{yx}(x, y) = 2e^{2x}.$$

$$\frac{\partial^2 z}{\partial y^2} = \frac{\partial^2 f}{\partial y^2} = f_{yy}(x, y) = 0. \blacksquare$$

Notice in Example 3 that $\partial^2 f/\partial y\,\partial x$ and $f_{xy}(x, y)$ each technically mean that f is to be differentiated partially *first with respect to x and then with respect to y*. In Example 3 the answers for $f_{xy}(x, y)$ and $f_{yx}(x, y)$ were the same; therefore, if the differentiation had been done in the wrong order, it would have made no difference in this particular problem. In general, whenever f and its partial derivatives are continuous, $f_{xy}(x, y) = f_{yx}(x, y)$.

Example 4

If $z = f(w, x, y) = (wx/y) + w \ln w$, find $\partial z/\partial w$ and $\partial^2 z/\partial y\,\partial w$.

$$\begin{aligned}\frac{\partial z}{\partial w} &= \frac{x}{y} + w\left(\frac{1}{w}\right) + \ln w \\ &= \frac{x}{y} + 1 + \ln w.\end{aligned}$$

Notice that x and y were treated as constants, and the product rule was used to differentiate $w \ln w$.

$$\begin{aligned}\frac{\partial^2 z}{\partial y\,\partial w} &= \frac{\partial}{\partial y}\left(\frac{x}{y} + 1 + \ln w\right) \\ &= x\left(-\frac{1}{y^2}\right) + 0 + 0 \\ &= -\frac{x}{y^2}. \blacksquare\end{aligned}$$

APPLICATIONS

Example 5

The productivity or output of any company may usually be thought of as a function of several inputs. As a simple example, suppose that the number P of ironing boards produced by a firm in one day depends upon the number x of workers in the labor force on that day and the number y of machines of a certain type that are used. Suppose that from many days of experience it has been found that

$$P = f(x, y) = x^2 + 5xy + 2y^2$$

gives an accurate estimate of P when x and y are known. Assume that $5 \leq x \leq 50$ and $2 \leq y \leq 15$. If 20 workers are present on a given day and 5 machines are operating, then the number of ironing boards that are expected to be completed is given by

$$f(20, 5) = (20)^2 + 5(20)(5) + 2(5)^2 = 950.$$

We define $\partial P/\partial x$ to be the **marginal productivity of labor** and $\partial P/\partial y$ to be the **marginal productivity of machines.** Since $P = x^2 + 5xy + 2y^2$,

$$\frac{\partial P}{\partial x} = 2x + 5y$$

and

$$\frac{\partial P}{\partial y} = 5x + 4y.$$

The interpretation below of these partial derivatives in this example is analogous to the interpretation of marginal cost in Example 2 of Section 2.6 (that example should be reviewed before proceeding). If $x = 20$ and $y = 5$, then $\partial P/\partial x|_{(20,5)}$ gives the approximate number of additional ironing boards that can be produced in 1 day if one worker is added to the labor force, and the number of machines is kept fixed at 5; the computation yields

$$\left.\frac{\partial P}{\partial x}\right|_{(20,5)} = 2(20) + 5(5) = 65.$$

On the other hand, if x is kept fixed at 20 and the number of machines is increased from 5 to 6, then the approximate number of extra ironing boards produced in 1 day is

$$\left.\frac{\partial P}{\partial y}\right|_{(20,5)} = 5(20) + 4(5) = 120. \blacksquare$$

EXERCISES

Suggested minimum assignment: Exercises 1, 7, 9, 12, 13, and 15(*b*).

1. If $z = f(x, y) = x^3 + 3xy^2 + y^3$, find $\dfrac{\partial f}{\partial x}$ and $\dfrac{\partial f}{\partial y}$.
2. If $z = f(w, x, y)$, define $\dfrac{\partial f}{\partial w}$ in a manner similar to Definition 1.
3. If $z = f(x, y) = x^2 + e^{xy}$, find $\dfrac{\partial z}{\partial x}$ and $\dfrac{\partial z}{\partial y}$.
4. If $z = f(x, y) = \dfrac{x}{y} + \dfrac{y}{x}$, find $f_x(x, y)$ and $f_y(x, y)$.
5. If $z = f(x, y) = 2x^3y^2 - 7xy + 3$, find $f_x(2, 1)$ and $f_y(2, 1)$.
6. If $z = f(w, x, y) = wxy + \ln wxy$, find $\dfrac{\partial f}{\partial w}$.

7. If $z = f(w, x, y) = e^{xyw^2} + \dfrac{x}{w + y}$, find $\dfrac{\partial z}{\partial x}$ and $\dfrac{\partial z}{\partial y}$.

8. Let a utility function be given by

$$u = f(w, x, y) = 5wxy - w^2 - x^2 - y^2.$$

Find the marginal utility $\dfrac{\partial u}{\partial x}$.

9. If $z = f(x, y) = 9 - x^2 - y^2$, give a geometric interpretation of $f_y(2, 1)$. Draw a sketch.

10. In the plane $y = 3$, find the slope of the tangent line to the curve of intersection of $z = x^2 + y^2$ and the plane $y = 3$ at $(1, 3, 10)$.

11. Let $z = x^3y^3 + 2x^2y$. Show that $\dfrac{\partial^2 z}{\partial y\, \partial x} = \dfrac{\partial^2 z}{\partial x\, \partial y}$.

12. Let $z = f(x, y) = ye^x + 2xe^y$. Show that $f_{xy}(x, y) = f_{yx}(x, y)$.

13. Let $z = xy - y \ln x$. Find $\left.\dfrac{\partial^2 z}{\partial x^2}\right|_{(1,4)}$ and $\left.\dfrac{\partial^2 z}{\partial y^2}\right|_{(1,4)}$.

14. Let $z = f(x, y) = xe^x + x(1 + 2y)^4$. Find $f_{xx}(x, y)$ and $f_{yy}(x, y)$.

15. If $z = f(w, x, y) = (w + 2x + 3y)^8$, calculate each of the following:

(a) $\dfrac{\partial^2 f}{\partial x^2}$. (b) $\dfrac{\partial^2 z}{\partial y\, \partial w}$. (c) $\dfrac{\partial^2 z}{\partial w\, \partial x}$.

16. In Example 5 find $\left.\dfrac{\partial P}{\partial x}\right|_{(10,3)}$ and interpret the answer.

17. Suppose that the output of a company is given in terms of three inputs by

$$P = f(w, x, y) = 2w + 3x + 4y + wx.$$

Find the marginal productivity of each of the inputs.

7.3 Maximum and Minimum Points

In Chapter 3 we defined a relative maximum point and a relative minimum point for functions of one variable. For functions of two variables, rather than make formal definitions of relative maxima and relative minima, we shall use examples to illustrate the analogous concepts. As in Chapter 3 the word "relative" will often be omitted by agreement. In this section we seek only maxima and minima that occur at *interior* points of the domain. Referring back to Figure 7.3, we say that the function f defined by

$$z = f(x, y) = 9 - x^2 - y^2$$

has the maximum point $(0, 0, 9)$ and has no minimum points. The point $(0, 0, 9)$ is called a maximum point because all nearby points on the surface are *less* than 9 units above the xy-plane. If a surface has several "humps" in it, there might be many maximum and minimum points—for example, a putting green on a golf course often has many maximum and minimum points.

Observe that for $z = f(x, y) = 9 - x^2 - y^2$ it follows that

$$\left.\frac{\partial f}{\partial x}\right|_{(0,0)} = 0$$

and

$$\left.\frac{\partial f}{\partial y}\right|_{(0,0)} = 0.$$

These results are consistent with our geometric interpretation in Section 7.2 and with Figure 7.3; notice that the parabolas shown in Figure 7.3 are the curves of intersection of the surface with the planes $y = 0$ and $x = 0$, and in these planes the tangent lines to these curves of intersection have slope 0 at (0, 0, 9). According to the next definition, the point (0, 0, 9) is a **critical point** of f.

▶ Definition 2

A point on a surface $z = f(x, y)$ at which $\partial f/\partial x$ and $\partial f/\partial y$ are both zero is called a **critical point** of f.

In Definition 3 of Section 3.3 we required that the value of the derivative equal 0 at a critical point, whereas in Definition 2 we require that *both* first partial derivatives of f equal 0 at a critical point. In Chapter 3 most of the discussion was restricted to rational functions in order that the derivatives would exist and be continuous at each interior point of the domain. *In this chapter we shall assume that the necessary partial derivatives exist and are continuous* in order to avoid theoretical complications; with this understanding, Theorem 1, which is stated without proof, is valid.

▶ Theorem 1

Let $z = f(x, y)$. Each maximum or minimum point of f is a critical point of f. ◀

Theorem 1 is analogous to Theorem 2 of Chapter 3; as before, the converse of Theorem 1 is *not* true. In other words, a critical point may be neither a maximum nor minimum point—the critical points are just the only possible *candidates* for being maximum or minimum points. Example 3 of Section 7.1 furnishes an example of this last statement. The saddle point at the origin is a critical point because $\partial z/\partial x|_{(0,0)} = 0$ and $\partial z/\partial y|_{(0,0)} = 0$, but it is neither a maximum nor a minimum point as is suggested by the graph. Fortunately, there is a theorem from advanced calculus that usually enables one to determine whether a critical point is a maximum point, a minimum point, or neither. This theorem is stated below; it involves second partial derivatives and is analogous to the second derivative test stated in Theorem 6 of Section 3.5.

Theorem 2

Suppose that $z = f(x, y)$ and $(x_1, y_1, z_1) = (x_1, y_1, f(x_1, y_1))$ is a critical point of f. Let constants A, B, C, and D be defined by

$$A = f_{xx}(x_1, y_1),$$
$$B = f_{xy}(x_1, y_1),$$
$$C = f_{yy}(x_1, y_1),$$

and

$$D = AC - B^2.$$

(*i*) If $D > 0$ and $A < 0$, then (x_1, y_1, z_1) is a maximum point.
(*ii*) If $D > 0$ and $A > 0$, then (x_1, y_1, z_1) is a minimum point.
(*iii*) If $D < 0$, then (x_1, y_1, z_1) is neither a maximum nor minimum point. (It is a saddle point.)
(*iv*) If $D = 0$, no conclusion may be made using this theorem. ◀

Example 1

Let us return to $z = f(x, y) = 9 - x^2 - y^2$ and verify by Theorem 2 that $(0, 0, 9)$ is a maximum point.

$$f_x(x, y) = -2x,$$
$$f_y(x, y) = -2y.$$

Setting these partial derivatives equal to 0, we find that there is a critical point at $x = 0$, $y = 0$. The critical point is $(0, 0, 9)$.

$$f_{xx}(x, y) = -2,$$
$$f_{xy}(x, y) = 0,$$
$$f_{yy}(x, y) = -2,$$
$$A = f_{xx}(0, 0) = -2,$$
$$B = f_{xy}(0, 0) = 0,$$
$$C = f_{yy}(0, 0) = -2,$$
$$D = AC - B^2 = (-2)(-2) - (0)^2 = 4.$$

It follows by Theorem 2(*i*) that $(0, 0, 9)$ is a maximum point. ▮

Example 2

Let us return to the function in Example 3 of Section 7.1 given by

$$z = f(x, y) = y^2 - x^2$$

and verify by Theorem 2 that $(0, 0, 0)$ is neither a maximum nor a minimum point.

$$f_x(x, y) = -2x,$$
$$f_y(x, y) = 2y.$$

If we equate these partial derivatives to 0, it follows that $(0, 0, 0)$ is the only critical point. By using the pattern of Example 1, it can easily be shown that $A = -2$, $B = 0$, $C = 2$, and $D = -4$. Since $D < 0$, Theorem 2(*iii*) guarantees that (0, 0, 0) is neither a maximum nor a minimum point (see Figure 7.4). ▮

Example 3

Suppose that the cost z in dollars of producing x units of gadget 1 and y units of gadget 2 per day is given by

$$z = f(x, y) = x^3 - xy + \frac{y^2}{2} - 10y + 1000.$$

Find nonnegative values of x and y, if possible, so that the cost is minimized.

$$\begin{aligned} f_x(x, y) &= 3x^2 - y, \\ f_y(x, y) &= -x + y - 10. \end{aligned}$$

Setting these partial derivatives equal to 0 in order to find critical points, we are led to the system

$$\begin{cases} 3x^2 - y = 0, \\ y - x - 10 = 0. \end{cases}$$

Substituting from the first equation into the second, we obtain

$$\begin{aligned} 3x^2 - x - 10 &= 0, \\ (3x + 5)(x - 2) &= 0, \\ x = -\tfrac{5}{3}, \qquad x &= 2. \end{aligned}$$

Since x must be nonnegative, $x = -\frac{5}{3}$ may be discarded. If $x = 2$, then

$$y = 3x^2 = 12,$$

and there is a critical point at $x = 2$, $y = 12$.

$$\begin{aligned} f_{xx}(x, y) &= 6x, \\ f_{xy}(x, y) &= -1, \\ f_{yy}(x, y) &= 1, \\ A = f_{xx}(2, 12) &= 12, \\ B = f_{xy}(2, 12) &= -1, \\ C = f_{yy}(2, 12) &= 1, \\ D = AC - B^2 &= 11. \end{aligned}$$

By Theorem 2(*ii*) there is a minimum at $x = 2$, $y = 12$ since $D > 0$ and $A > 0$. The minimum cost is

$$f(2, 12) = 8 - (2)(12) + \tfrac{144}{2} - 120 + 1000 = \$936. ▮$$

For the reader who has studied the characteristic value problem in matrix algebra, a procedure is stated in Exercise 12 that can be used in place of Theorem 2 in Examples 1–3.

Detection of maximum and minimum points for functions of three or more variables is generally a more difficult problem. Definition 2 can be generalized; that is, all first partial derivatives can be equated to zero to obtain critical points as before. Moreover, the critical points obtained in this manner are the *only* candidates for maximum or minimum points—this is because Theorem 1 can be generalized to cover functions of three or more variables. Formal tests for maxima and minima comparable to Theorem 2, however, are too complicated for this introductory text. Fortunately, in many applied problems, a critical point can be determined to be a maximum point or a minimum point by other means. We shall examine one example in which the existence of a minimum point is almost obvious.

Example 4

Find all maximum and minimum points of f if

$$z = f(w, x, y) = (w - 1)^2 + (x - 2)^2 + (y - 3)^2 + 4.$$

We calculate

$$\frac{\partial f}{\partial w} = 2(w - 1), \qquad \frac{\partial f}{\partial x} = 2(x - 2), \qquad \frac{\partial f}{\partial y} = 2(y - 3).$$

Setting these partial derivatives equal to 0, we find that the only critical point is at $w = 1$, $x = 2$, $y = 3$. Since

$$f(1, 2, 3) = 0 + 0 + 0 + 4 = 4,$$

the critical point is

$$(w, x, y, z) = (1, 2, 3, 4).$$

Because of the squared terms $(w - 1)^2$, $(x - 2)^2$, and $(y - 3)^2$, which must be nonnegative, it is clear that the value of f is never less than 4 and hence $(1, 2, 3, 4)$ is a minimum point. ∎

APPLICATIONS

Example 5

Suppose that points have been plotted on a two-dimensional graph from some given data. The art of finding the equation of a curve which in some sense comes closest to passing through the plotted points is known as **curve fitting.** In this example we shall begin with some points that nearly lie on a straight line and try to find a line $y = mx + b$ that best fits the data according to the **method of least squares.** The line $y = mx + b$ is called a **line of regression.**

Let the given points be

$$(4, 5), \quad (2, 3), \quad (-1, 1), \quad \text{and} \quad (-4, -2).$$

In Figure 7.7 we require the sum of the squares of the vertical distances between the given points and $y = mx + b$ to be a minimum. For example, the point (4, 5) has a y-coordinate of 5 and at $x = 4$ the line $y = mx + b$ has a y-coordinate of $4m + b$. The square of the difference can be written either $(4m + b - 5)^2$ or $(5 - 4m - b)^2$. This type of expression is found for each point, and the sum is the following function of m and b which is to be minimized:

$$f(m, b) = (4m + b - 5)^2 + (2m + b - 3)^2 + (-m + b - 1)^2 + (-4m + b + 2)^2,$$

$$\frac{\partial f}{\partial m} = 2(4m + b - 5)(4) + 2(2m + b - 3)(2) + 2(-m + b - 1)(-1) + 2(-4m + b + 2)(-4),$$

$$\frac{\partial f}{\partial b} = 2(4m + b - 5) + 2(2m + b - 3) + 2(-m + b - 1) + 2(-4m + b + 2).$$

Setting the partial derivatives equal to 0 yields the system

$$\begin{cases} 74m + 2b = 66, \\ 2m + 8b = 14. \end{cases}$$

The solution of this system is

$$m = \tfrac{125}{147}, \qquad b = \tfrac{226}{147},$$

at which the only critical point occurs. In Exercise 10 the reader is asked to use Theorem 2 of this section to verify that the critical point is a *minimum* point. Hence the line of regression is

$$y = \tfrac{125}{147}x + \tfrac{226}{147}. \quad \blacksquare$$

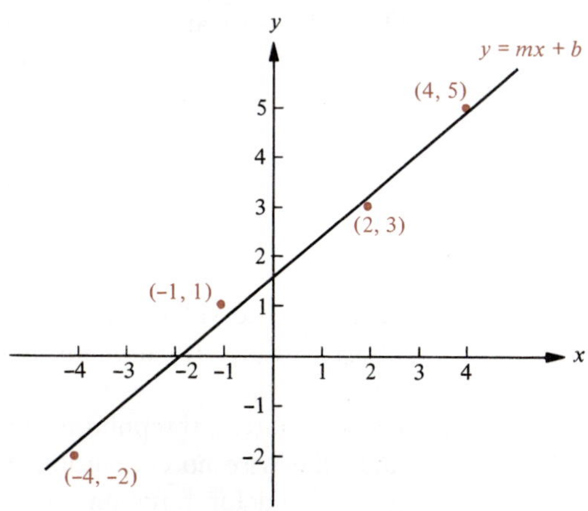

Figure 7.7. *Line of regression.*

EXERCISES

Suggested minimum assignment: Exercises 1, 4, 5, 8, and 9.

In Exercises 1–6 find all critical points and then use Theorem 2 to determine, if possible, whether each critical point is a maximum point, a minimum point, or neither.

1. $z = f(x, y) = 15x^2 - 40xy + 25y^2 + 20x + 500.$
2. $z = f(x, y) = (x - 1)^2 - 2(y - 2)^2 + 5.$
3. $z = f(x, y) = x^2 + 2y^2 - 2x - 8y + 14.$
4. $z = f(x, y) = 4x - 6y - x^2 - y^2 + 7.$
5. $z = f(x, y) = x^3 + x^2 + y^2 - xy + \frac{31}{16}.$
6. $z = f(x, y) = x^3 + y^3 + 9xy.$
7. A cost function is given by

$$C(x, y) = x^2 + xy + y^2 - 9x - 6y + 50.$$

Find x and y so as to minimize cost.

8. In one day a manufacturing process produces a number x of one item and a number y of a second item. The profit made from the sale of these items is given by

$$P(x, y) = xy + 6x + 11y - x^2 - 2y^2.$$

Find x and y so as to maximize profit.

9. Let $z = f(w, x, y) = 50 - (w + 2)^2 - (x - 4)^2 - (y - 6)^2$. Find all critical points after setting $\partial f/\partial w$, $\partial f/\partial x$, and $\partial f/\partial y$ equal to zero. By inspection tell whether each critical point is a maximum or minimum point.
10. In Example 5 verify by Theorem 2 that the point at which $m = \frac{125}{147}$ and $b = \frac{226}{147}$ is a minimum point.
11. Find the equation of the line of regression (see Example 5) for the points $(3, -5)$, $(0, -1)$, and $(-4, 3)$.
12. On p. 331 of Campbell [6] a procedure is discussed that can be used instead of Theorem 2 to determine, in certain cases, whether a critical point is a maximum point, a minimum point, or neither. Form the symmetric matrix

$$\begin{bmatrix} A & B \\ B & C \end{bmatrix},$$

where A, B, and C are the same as in Theorem 2. Then find the characteristic values of this matrix. If both characteristic values are positive, the critical point is a minimum point; if both values are negative, the point is a maximum point; if one value is positive and one is negative, the point is neither a maximum nor a minimum point; and otherwise no conclusion can be made. (This method can also be generalized for functions of more than two independent variables.)

In Example 3 test the critical point at $x = 2$, $y = 12$ by the matrix method.

7.4 Lagrange Multipliers

In many applied problems involving maximum and minimum points, there are one or more restrictions on the independent variables. Assume that these restrictions, or **constraints** as they are often called, can be given by means of equations. A sample constraint might be

$$w + x + y = 100{,}000.$$

This equation might express the fact that the sum of radio, television, and magazine advertising expenses for 1 month is \$100,000.

In this section we shall solve more maximum and minimum problems; the problems differ from those of Section 7.3, however, in that one or more constraints will appear. Suppose that a function of n variables ($n \geq 2$) expressed by

$$f(x_1, x_2, \ldots, x_n)$$

is to be maximized or minimized subject to just one constraint given by

$$g(x_1, x_2, \ldots, x_n) = 0.$$

(The case of more than one constraint will be considered later in this section.) The calculation of critical points for this type of problem will be accomplished by a method known as the **method of Lagrange multipliers.** The method is named after the French mathematician Joseph L. Lagrange (1736–1813). The method consists of the following steps:

1. Let a new function F be defined by

$$F(x_1, x_2, \ldots, x_n, \lambda) = f(x_1, x_2, \ldots, x_n) - \lambda g(x_1, x_2, \ldots, x_n),$$

where λ is called a **Lagrange multiplier.**

2. Form the system of $n + 1$ equations

$$\begin{cases} \dfrac{\partial F}{\partial x_1} = 0, \\ \dfrac{\partial F}{\partial x_2} = 0, \\ \quad \vdots \\ \dfrac{\partial F}{\partial x_n} = 0, \\ \dfrac{\partial F}{\partial \lambda} = 0. \end{cases}$$

3. Find all values of $x_1, x_2, \ldots, x_n, \lambda$ that satisfy the above system. Among the sets of values of $x_1, x_2, \ldots, x_n$ that satisfy the system can be found coordinates of all the maximum and minimum points if any such points exist. (The method of Lagrange multipliers does not distinguish maximum points from minimum points or from points that are neither.)

Example 1

We shall repeat Example 3 of Section 3.4, for which the critical points can be found by the method of Lagrange multipliers. In that problem a fenced area given by

$$f(x, y) = xy$$

was to be maximized subject to the restriction that the cost of the fence be \$1200; that is,

$$2x + 2x + 3y = 1200.$$

Let

$$F(x, y, \lambda) = xy - \lambda(4x + 3y - 1200).$$

(Notice that all terms of the constraint equation were transposed to one side to obtain the quantity multiplied by λ.) Next we form the system

$$\begin{cases} \dfrac{\partial F}{\partial x} = y - 4\lambda = 0, \\ \dfrac{\partial F}{\partial y} = x - 3\lambda = 0, \\ \dfrac{\partial F}{\partial \lambda} = -(4x + 3y - 1200) = 0. \end{cases}$$

From the first two equations

$$\lambda = \frac{y}{4} = \frac{x}{3},$$

and hence

$$y = \frac{4}{3}x.$$

This result can be substituted in the third equation to obtain

$$4x + 3\left(\frac{4}{3}x\right) - 1200 = 0,$$
$$8x = 1200,$$
$$x = 150.$$

Therefore,

$$\lambda = \frac{x}{3} = 50,$$

and

$$y = \frac{4}{3}x = 200.$$

The point $(x, y, f(x, y)) = (150, 200, 30{,}000)$ is a critical point. From Example 3 of Section 3.4 we know that it is a *maximum* point; this means that the height x of the surface $z = xy$ is greater when z is evaluated at the point $(x, y) = (150, 200)$ than at any other nearby point on the line $4x + 3y = 1200$. ∎

Observe in Example 1 that, in finding $\partial F/\partial x$, we hold λ as well as y fixed in doing the differentiation—F is a function of three variables, and λ is treated just like x and y in doing the routine partial differentiation necessary in forming the system of equations. The value 50 found for λ in solving the system was only a means to an end and was not a part of the final answer; however, there are situations in more advanced work in which the value of λ is significant.

In this brief discussion we do *not* present a method for testing a critical point to determine whether it is a maximum point, a minimum point, or neither. In many applied problems this is *not* a severe handicap, however, because the existence of a specific optimum can often be demonstrated from the nature of the problem; Example 1 is an illustration, since between the two extremes corresponding to very small x or very small y (in the domain), it follows that there must be a maximum area. It then can be reasoned that a specified optimum point, which is known to exist, must be one of the critical points, and it becomes a matter of identifying which one. In Example 1, (150, 200, 30,000) is the only candidate and hence the maximum area of 30,000 must occur when $x = 150$ and $y = 200$.

Example 2

Find positive values of w, x, and y that minimize

$$f(w, x, y) = x^3 - wx^2 + \frac{y^2}{2} - 10y + 1000$$

subject to the constraint

$$wx - y = 0.$$

$$F(w, x, y, \lambda) = x^3 - wx^2 + \frac{y^2}{2} - 10y + 1000 - \lambda(wx - y).$$

$$\begin{cases} \dfrac{\partial F}{\partial w} = -x^2 - \lambda x = 0, & (7.1) \\ \dfrac{\partial F}{\partial x} = 3x^2 - 2wx - \lambda w = 0, & (7.2) \\ \dfrac{\partial F}{\partial y} = y - 10 + \lambda = 0, & (7.3) \\ \dfrac{\partial F}{\partial \lambda} = -(wx - y) = 0. & (7.4) \end{cases}$$

From (7.1), $\lambda x = -x^2$, and since $x \neq 0$ it follows that $\lambda = -x$. Substituting $\lambda = -x$ in (7.2), we obtain

$$\begin{aligned} 3x^2 - 2wx + wx &= 0, \\ x(3x - w) &= 0, \end{aligned}$$

and since $x \neq 0$,

$$w = 3x.$$

Substituting $w = 3x$ in (7.4), we find that

$$y = wx = 3x^2.$$

In (7.3) replace y by $3x^2$ and λ by $-x$ to obtain

$$3x^2 - x - 10 = 0,$$

or

$$(3x + 5)(x - 2) = 0.$$

Since $x > 0$, it must follow that

$$x = 2.$$

Hence

$$\lambda = -x = -2,$$
$$w = 3x = 6,$$

and

$$y = 10 - \lambda = 12.$$

Therefore, if f has a minimum value, that value will occur when $w = 6$, $x = 2$, and $y = 12$. Example 3 of Section 7.3 is the same problem; in this section, however, w was kept and the Lagrange multiplier method used, whereas in Section 7.3 $w = y/x$ had already been substituted in $f(w, x, y)$. In Section 7.3 we found that *minimum* cost occurs at $w = 6$, $x = 2$, and $y = 12$. ▮

Suppose that $f(x_1, x_2, \ldots, x_n)$ is to be maximized or minimized subject to the m constraints

$$\begin{cases} g_1(x_1, x_2, \ldots, x_n) = 0, \\ g_2(x_1, x_2, \ldots, x_n) = 0, \\ \vdots \\ g_m(x_1, x_2, \ldots, x_n) = 0. \end{cases}$$

In this case the method of Lagrange multipliers is generalized in the following manner: Let

$$F(x_1, x_2, \ldots, x_n, \lambda_1, \lambda_2, \ldots, \lambda_m) = f(x_1, x_2, \ldots, x_n) - \lambda_1 g_1 - \lambda_2 g_2 - \cdots - \lambda_m g_m,$$

where g_i is an abbreviation for $g_i(x_1, x_2, \ldots, x_n)$. Then solve the system of $n + m$ equations

$$\begin{cases} \dfrac{\partial F}{\partial x_1} = 0, \\ \vdots \\ \dfrac{\partial F}{\partial x_n} = 0, \\ \dfrac{\partial F}{\partial \lambda_1} = 0, \\ \vdots \\ \dfrac{\partial F}{\partial \lambda_m} = 0. \end{cases}$$

Example 3

A cost function C of three nonnegative variables w, x, and y is given by

$$C(w, x, y) = wy - \tfrac{1}{2}x^2y + 2w + xy + 4y^2 - 9x + 35.$$

Minimize C subject to the two constraints

$$x^2 - 6y - 2w = 0$$

and

$$w + x + y = 10.$$

Let

$$F(w, x, y, \lambda_1, \lambda_2) = C(w, x, y) - \lambda_1(x^2 - 6y - 2w) - \lambda_2(w + x + y - 10).$$

Write the system of equations

$$\begin{cases} \dfrac{\partial F}{\partial w} = y + 2 + 2\lambda_1 - \lambda_2 = 0, \\ \dfrac{\partial F}{\partial x} = -xy + y - 9 - 2\lambda_1 x - \lambda_2 = 0, \\ \dfrac{\partial F}{\partial y} = w - \dfrac{1}{2}x^2 + x + 8y + 6\lambda_1 - \lambda_2 = 0, \\ \dfrac{\partial F}{\partial \lambda_1} = -(x^2 - 6y - 2w) = 0, \\ \dfrac{\partial F}{\partial \lambda_2} = -(w + x + y - 10) = 0. \end{cases}$$

It can be shown that the only solution of this system (for which w, x, and y are nonnegative) is given by $w = 5$, $x = 4$, $y = 1$, $\lambda_1 = -\frac{3}{2}$, and $\lambda_2 = 0$. The cost, when $w = 5$, $x = 4$, and $y = 1$, is

$$C(5, 4, 1) = 14.$$

Therefore, if there is a minimum value of C, then that value is 14. Values of w, x, and y that satisfy the constraints and that are close to (5, 4, 1) can be shown to yield a larger cost than 14, which suggests (but does not prove) that there is a *minimum* point at $w = 5$, $x = 4$, and $y = 1$. ▮

For the reader who has studied linear programming, the following comments may be beneficial. In linear programming the function to be optimized (maximized or minimized) has to be a linear function, whereas the function to be maximized or minimized in the method of Lagrange multipliers may be a linear function but very often is not. In linear programming the constraints must also be linear. When using Lagrange multipliers, the constraints do *not* have to be linear. Furthermore, in setting up a linear programming problem, one must arrange that the independent variables are nonnegative, whereas there is no such requirement in Lagrange multiplier problems. The methods of Lagrange multipliers and linear programming, along with the techniques of Chapter 3, provide useful tools for solving optimization problems.

APPLICATIONS

Example 4

Suppose that a utility function (see Example 2 of Section 7.2) of three products is given by

$$u(w, x, y) = 4wxy.$$

The costs of the three goods are \$1, \$2, and \$3, respectively. A consumer can spend \$54 and wishes to maximize the utility. The budget restriction is given by the equation

$$w + 2x + 3y = 54,$$

where w, x, and y must be nonnegative. Use the method of Lagrange multipliers to maximize the utility function subject to the budget constraint.

Let

$$F(w, x, y, \lambda) = 4wxy - \lambda(w + 2x + 3y - 54).$$

$$\begin{cases} \dfrac{\partial F}{\partial w} = 4xy - \lambda = 0, \\ \dfrac{\partial F}{\partial x} = 4wy - 2\lambda = 0, \\ \dfrac{\partial F}{\partial y} = 4wx - 3\lambda = 0, \\ \dfrac{\partial F}{\partial \lambda} = -(w + 2x + 3y - 54) = 0. \end{cases}$$

From the first three equations

$$\lambda = 4xy = 2wy = \tfrac{4}{3}wx.$$

Since $y \neq 0$ (if $y = 0$, the value of the utility function is 0, which clearly is not a maximum value of the utility function), we can divide $4xy = 2wy$ by y to obtain

$$w = 2x.$$

Also $2wy = \frac{4}{3}wx$ implies that

$$y = \tfrac{2}{3}x.$$

Substituting in the fourth equation of the system above, we obtain

$$\begin{aligned} -[2x + 2x + 3(\tfrac{2}{3}x) - 54] &= 0, \\ 6x &= 54, \\ x &= 9. \end{aligned}$$

Hence

$$\begin{aligned} y &= \tfrac{2}{3}x = 6, \\ \lambda &= 4xy = 216, \end{aligned}$$

and

$$w = 2x = 18.$$

If $w = 18$, $x = 9$, and $y = 6$ the budget constraint is satisfied, and if there is a maximum value of u, then such a value occurs at $(18, 9, 6)$. Nearby points, which satisfy the constraint, make the value of the utility function smaller than its value at $w = 18$, $x = 9$, and $y = 6$; this suggests (but does not prove) that there is a *maximum* point at $w = 18$, $x = 9$, and $y = 6$. ▮

Example 5

Several books give applications of the method of Lagrange multipliers to problems in business and economics. For the reader who wishes to study the method further or who wishes to see more applications in business and economics, we recommend the books by Kim [14], McAdams [16], and Theodore [22]. ▮

EXERCISES

Suggested minimum assignment: Exercises 1, 5, 9, 10, and 11.

In Exercises 1–10 use the method of Lagrange multipliers under the assumption that the desired optimum exists.

1. Minimize $f(x, y) = x^2 + y^2$ subject to $x + 2y = 10$.
2. Minimize $f(w, x, y) = 2w^2 + x^2 + y^2$ subject to $w + x + y = 50$.
3. Find three positive numbers whose sum is 36 and whose product is maximum.
4. Minimize $f(w, x, y) = w + x^2 + (y^2/2)$ subject to $w + 4x + y = 15$.
5. Find positive numbers w, x, and y so that $w + x + 2y$ is maximized subject to the restriction that $w^2 + x^2 + y^2 = 54$.
6. Find positive numbers w, x, and y so that $w^2 + x^3 + y^2$ is minimized subject to the constraint $w + 3x + y = 28$.
7. Maximize the utility function $f(x, y) = x\sqrt{y}$ subject to the budget constraint $2x + y = 3000$.
8. Rework Example 4 of Section 3.4.
9. Three sides of a rectangular plot of ground are to be fenced so that the total area of the plot is 5000 square feet. Find the length and width of the plot so that the number of feet of fencing is minimum.
10. The daily production P of an item is a function of three inputs as given by

$$P = wx + 6y - x^2 - y^2.$$

Find w, x, and y so as to maximize the daily production subject to the restriction

$$w + x + y = 10.$$

11. Rework Exercise 10 by replacing w with $10 - x - y$ (thus expressing P as a function of x and y) and then using methods of Section 7.3.
12. Maximize the volume lwh of a closed rectangular box subject to the

two constraints

$$w = \tfrac{1}{2}l$$

and

$$2wl + 2wh + 2lh = 300.$$

(The last equation requires that the total surface area of the box equal 300 square units.)

NEW VOCABULARY

surface 7.1
coordinate planes 7.1
saddle surface 7.1
saddle point 7.1
partial derivative 7.2
utility function 7.2
marginal utilities 7.2
critical point 7.3
constraint 7.4
method of Lagrange multipliers 7.4
Lagrange multiplier 7.4

References

[1] Ackerman, E., *Biophysical Science,* Prentice-Hall, Inc., Englewood Cliffs, N.J., 1962.

[2] Bailey, N. T. J., *The Mathematical Theory of Epidemics,* Charles Griffin & Company Ltd., London, 1957.

[3] Batschelet, Edward, *Introduction to Mathematics for Life Scientists,* Vol. 2, Biomathematics, Springer-Verlag New York Inc., New York, 1971.

[4] Bishir, John W., and Drewes, Donald W., *Mathematics in the Behavioral and Social Sciences,* Harcourt Brace Jovanovich, Inc., New York, 1970.

[5] Brems, Hans, *Quantitative Economic Theory,* John Wiley & Sons, Inc., New York, 1968.

[6] Campbell, Hugh G., *Linear Algebra with Applications: Including Linear Programming,* Appleton-Century-Crofts, New York, 1971.

[7] Casstevens, T. W., and Denham, W. A., III, "Turnover and Tenure in the Canadian House of Commons 1867–1968," *Canadian Journal of Political Science,* Vol. 3, No. 4, 1970.

[8] Coale, A. J., *The Growth and Structure of Human Populations, A Mathematical Investigation,* Princeton University Press, Princeton, N.J., 1972.

[9] Dodd, S. C., "Diffusion Is Predictable: Testing Probability Models for Laws of Interaction," *American Sociological Review,* Vol. 20, 1955.

[10] Garrett, Henry E., *Statistics in Psychology and Education,* 5th ed., David McKay Company, Inc., New York, 1958.

[11] Gray, Virginia, "Innovation in the States: A Diffusion Study," *American Political Science Review,* Vol. 67, No. 4, 1973.

[12] Hays, William L., *Quantification in Psychology,* Brooks/Cole Publishing Company, Monterey, Calif., 1967.

[13] Hull, Clark L., *Essentials of Behavior,* Yale University Press, New Haven, Conn., 1951.

[14] Kim, Thomas K., *Introductory Mathematics for Economic Analysis,* Scott, Foresman and Company, Glenview, Ill., 1971.

[15] Libby, W. F., *Radiocarbon Dating,* University of Chicago Press, Chicago, 1952.

[16] McAdams, A. K., *Mathematical Analysis for Management Decisions,* Macmillan Publishing Co., Inc., New York, 1970.

[17] Newman, James R., *The World of Mathematics,* Vol. 1, Simon and Schuster, New York, 1956.

[18] Pielou, E. C., *An Introduction to Mathematical Ecology,* John Wiley & Sons, Inc., New York, 1969.

[19] Richardson, L. F., "War Moods," *Psychometrika,* Vol. 13, 1948.

[20] Simon, William, *Mathematical Techniques for Physiology and Medicine,* Academic Press, Inc., New York, 1972.

[21] Smith, C. A. B., *Biomathematics,* Vol. 1, Hafner Press, New York, 1966.

[22] Theodore, Chris A., *Applied Mathematics: An Introduction,* rev. ed., Richard D. Irwin, Inc., Homewood, Ill., 1971.

[23] Thrall, R. M., Mortimer, J. A., Rebman, K. R., and Baum, R. F., *Some Mathematical Models in Biology,* Report No. 40241-R-7, rev. ed., University of Michigan Press, Ann Arbor, Mich., 1967.

Table of Integration Formulas

1. $\int du = u + C.$

2. $\int kf(x)\,dx = k\int f(x)\,dx.$

3. $\int [f(x) + g(x)]\,dx = \int f(x)\,dx + \int g(x)\,dx.$

4. $\int u^n\,du = \dfrac{u^{n+1}}{n+1} + C,\ n \neq -1.$

5. $\int b^u\,du = \dfrac{b^u}{\ln b} + C,\ b > 0,\ b \neq 1.$

6. $\int e^u\,du = e^u + C.$

7. $\int \dfrac{du}{u} = \ln|u| + C,\ u \neq 0.$

8. $\int u\,dv = uv - \int v\,du.$

9. $\int \dfrac{du}{a^2 - u^2} = \dfrac{1}{2a}\ln\left|\dfrac{a+u}{a-u}\right| + C.$

10. $\int \dfrac{du}{u^2 - a^2} = \dfrac{1}{2a}\ln\left|\dfrac{u-a}{u+a}\right| + C.$

11. $\int \dfrac{du}{u(a+bu)} = -\dfrac{1}{a}\ln\left|\dfrac{a+bu}{u}\right| + C.$

12. $\int \dfrac{du}{u^2(a+bu)} = -\dfrac{1}{au} + \dfrac{b}{a^2}\ln\left|\dfrac{a+bu}{u}\right| + C.$

13. $\int \dfrac{du}{u(a+bu)^2} = \dfrac{1}{a(a+bu)} - \dfrac{1}{a^2}\ln\left|\dfrac{a+bu}{u}\right| + C.$

14. $\int \dfrac{u\,du}{a+bu} = \dfrac{1}{b^2}(a + bu - a\ln|a+bu|) + C.$

15. $\int \dfrac{u^2\,du}{a+bu} = \dfrac{1}{b^3}[\tfrac{1}{2}(a+bu)^2 - 2a(a+bu) + a^2\ln|a+bu|] + C.$

16. $\int \dfrac{u\,du}{(a+bu)^2} = \dfrac{1}{b^2}\left(\dfrac{a}{a+bu} + \ln|a+bu|\right) + C.$

17. $\int \dfrac{u^2\,du}{(a+bu)^2} = \dfrac{1}{b^3}\left(a + bu - \dfrac{a^2}{a+bu} - 2a\ln|a+bu|\right) + C.$

18. $\int u^n e^u\,du = u^n e^u - n\int u^{n-1}e^u\,du,\ n > 0.$

19. $\int \frac{e^u}{u^n}\,du = -\frac{e^u}{(n-1)u^{n-1}} + \frac{1}{n-1}\int \frac{e^u\,du}{u^{n-1}},\ n > 0,\ n \neq 1.$

20. $\int \ln u\,du = u \ln u - u + C,\ u > 0.$

21. $\int u^n \ln u\,du = u^{n+1}\left[\frac{\ln u}{n+1} - \frac{1}{(n+1)^2}\right] + C,\ u > 0,\ n \neq -1.$

22. $\int u\sqrt{a+bu}\,du = \frac{6bu-4a}{15b^2}(a+bu)^{3/2} + C,\ a+bu > 0.$

23. $\int \frac{\sqrt{a^2-u^2}}{u}\,du = \sqrt{a^2-u^2} - a\ln\left|\frac{a+\sqrt{a^2-u^2}}{u}\right| + C,\ a^2 > u^2.$

24. $\int \sqrt{a^2+u^2}\,du = \frac{u}{2}\sqrt{a^2+u^2} + \frac{a^2}{2}\ln|u+\sqrt{a^2+u^2}| + C.$

25. $\int \sqrt{u^2-a^2}\,du = \frac{u}{2}\sqrt{u^2-a^2} - \frac{a^2}{2}\ln|u+\sqrt{u^2-a^2}| + C,\ u^2 > a^2.$

Answers to Odd-numbered Exercises

1.1 Page 3

1. $n = 100 + 50t$. **3.** $\Delta R = \frac{2}{50}$ is greater than $\Delta R = \frac{2}{100}$.

1.2 Page 8

1. π, 2π, and $\sqrt{2}$ are irrational; others rational. **3.** (a), (c), (d) are true; (b) is false.

5. **(a)** $(-1, 6)$, open;

(b) $[-2, 1)$, half-open;

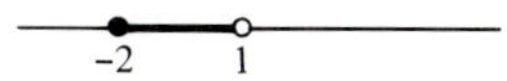

(c) $[-8, -6]$, closed;

(d) $(4, 7]$, half-open.

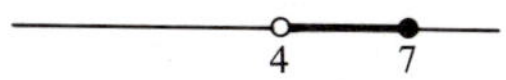

7. **(a)** $\{x: -3 < x < 8\}$; **(b)** $\{x: \pi \leq x \leq 5\}$; **(c)** $\{x: -19 \leq x \leq -5\}$; **(d)** $\{x: -3 < x \leq 3\}$. **9.** **(a)** 4; **(b)** 0; **(c)** 2; **(d)** $\frac{4}{5}$; **(e)** 7; **(f)** 8; **(g)** 2.
11. **(a)** 7; **(b)** 7; **(c)** 7; **(d)** 16.
13. Let x be the number of men who sign the petition, and let y be the number of women who sign the petition. Then $x + y \geq 100$ and also $y \geq 30$ in order for the petition to be considered.

1.3 Page 15

1. $(-3, 5)$, 2nd quadrant;
$(4, -1)$, 4th quadrant;
$(2, 7)$, 1st quadrant;
$(-5, -4)$, 3rd quadrant.

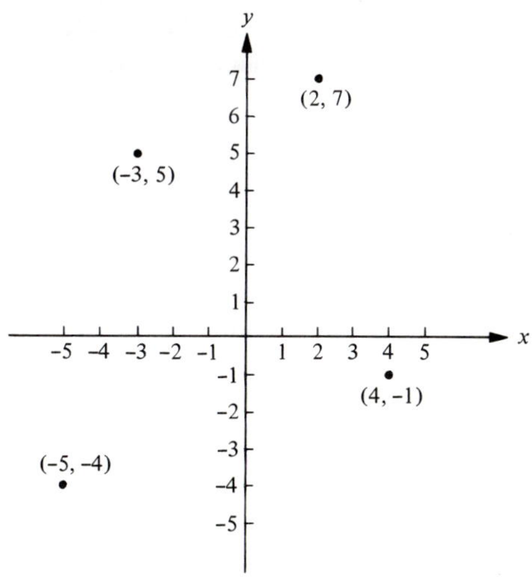

3. **(a)** 5; **(b)** $\sqrt{53}$; **(c)** 9. **5.** $\sqrt{113}$. **7.** **(a)** $\Delta x = 8$, $\Delta y = -7$; **(b)** $\Delta x = -8$, $\Delta y = 7$. **9.** Yes, because $1^4 + 2^3 - 8(1) - 1 = 0$.
11. $(x + 1)^2 + (y + 2)^2 = 9$. **13.** **(a)** Center $(7, -10)$, radius 2; **(b)** center $(0, 0)$, radius $\sqrt{29}$.
15. Yes, because the person is only 13 miles from the sender.

1.4 Page 24

1. $f(1) = 10$, $f(3) = 30$, $f(5) = 50$, $f(7) = 70$, $R = \{10, 30, 50, 70\}$.
3. $f(3) = 13$; $R = \{y : y \geq 4\}$.

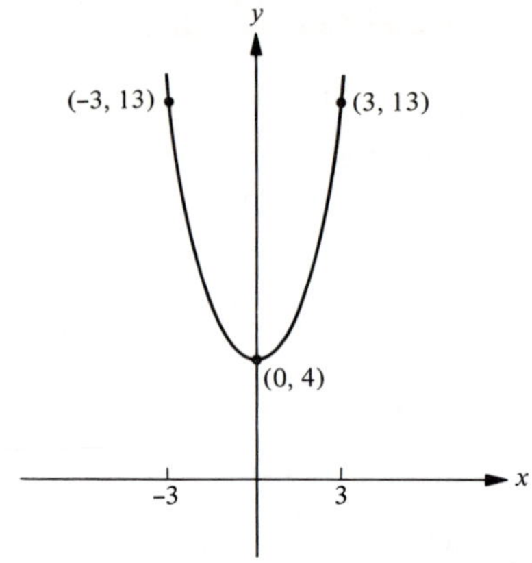

5. $D = \{1, 2, 3\}$; $R = \{4, 5, 6\}$. The number 2 has associated with it both 5 and 7—each element in the domain of a *function* has only *one* image.
7. y is the dependent variable and x is the independent variable; $g(5) = \frac{1}{5}$.
9. (**a**) $\{x : x \neq -2\}$; (**b**) set of all real numbers; (**c**) $\{x : x \neq \pm 1\}$; (**d**) $\{x : x \geq 0\}$. **11.** $f(-3) = -\frac{1}{3}$; $f(a + b) = \dfrac{1}{2a + 2b + 3}$. **13.** (**a**) 9; (**b**) 5; (**c**) 1; (**d**) $x^2 + 6x + 9$; (**e**) $6x + 9$; (**f**) $2x + h$.
15. $D = \{x : 0 \leq x \leq 10\}$; $C(5) = 44$. **17.** $P(x) = 2x - 40$; $P(50) = 60$; fixed cost 40; 20. **19.** (**a**) Yes, because the range of w is a subset of the set of real numbers. (**b**) No, because the range of X might not satisfy the two defining properties of a weighting function (weights must be nonnegative and total to 1).

1.5 Page 32

1. (**a**) 7; (**b**) -5; (**c**) 0; (**d**) no slope. **3.** $\frac{7}{11}$. **5.** (**a**) $y - 3 = 2(x - 5)$ or $y - 5 = 2(x - 6)$; (**b**) $y + 3 = \frac{5}{3}(x + 2)$ or $y - 2 = \frac{5}{3}(x - 1)$; (**c**) $y - 1 = -\frac{7}{5}(x - 3)$ or $y - 8 = -\frac{7}{5}(x + 2)$. **7.** (**a**) $y + 1 = \frac{2}{3}(x - 6)$; (**b**) $y - 2 = -5(x + 4)$.
9. (**a**) $y = 3x + 2$;

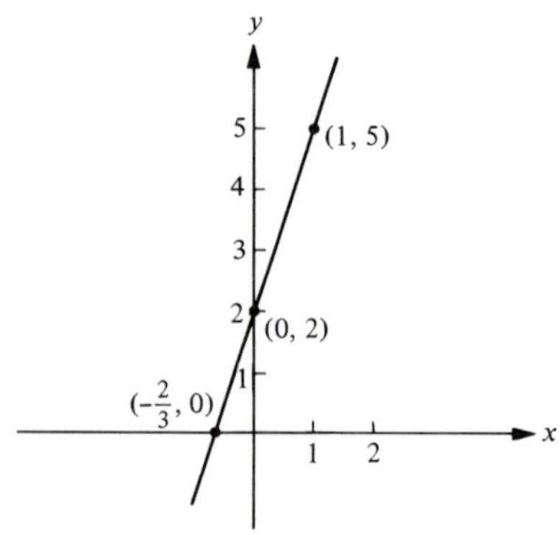

(b) $y = \frac{2}{3}x - 1$; **(c)** $y = -7x$.
11. **(a)** $y + 5 = \frac{3}{2}(x + 7)$ or $y - 1 = \frac{3}{2}(x + 3)$; **(b)** $y = \frac{3}{2}x + \frac{11}{2}$;
(c) $3x - 2y + 11 = 0$; **(d)** $\dfrac{x}{-\frac{11}{3}} + \dfrac{y}{\frac{11}{2}} = 1$. **13.** $5x + 2y - 24 = 0$.
15. $-\frac{3}{2}$. **17.** **(a)** 60 degrees Celsius; **(b)** 86 degrees Fahrenheit; **(c)** -40 degrees Fahrenheit.

1.6 Page 41

1. \$8405.00. **3.** **(a)** $10n$; **(b)** $\dfrac{1}{3n - 1}$. **5.** **(a)** Degree 5, set of all real numbers; **(b)** degree 8, set of all real numbers. **9.** \$8.50.
11.

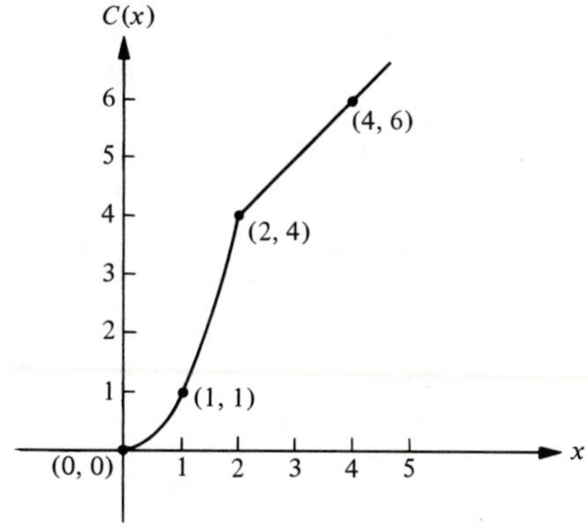

13. \$1849.11. **15.** $y = 1.3x = f(x)$, where x is the cost and y is the selling price.

1.7 Page 49
1. 3;

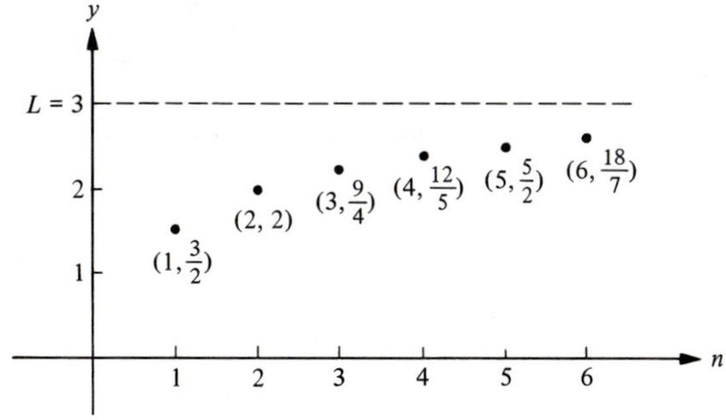

3. 1. **5.** **(a)** No limit; **(b)** limit is 1; **(c)** limit is 0; **(d)** limit is 0; **(e)** no limit.
7.
$$\begin{aligned}\lim_{x\to\infty} (f(x) - g(x)) &= \lim_{x\to\infty} (f(x) + (-1)g(x)) \\ &= \lim_{x\to\infty} f(x) + \lim_{x\to\infty} (-1)g(x) \quad \text{by Theorem 1}(i) \\ &= \lim_{x\to\infty} f(x) - \lim_{x\to\infty} g(x) \quad \text{by Theorem 1}(v).\end{aligned}$$

9. **(a)** 0; **(b)** 1; **(c)** 2; **(d)** 1; **(e)** 0. **11.** 7. **13.** **(a)** 12; **(b)** $\frac{1}{3}$.

1.8 Page 57

1. (**a**) 2; (**b**) 4. **3.** (**a**) 2; (**b**) 4; (**c**) 5; (**d**) does not exist; (**e**) 0; (**f**) 2; (**g**) 6.
5. (**a**) $\frac{19}{22}$; (**b**) 4. **7.** 8. **9.** (**a**) 2; (**b**) 5; (**c**) 0; (**d**) does not exist; (**e**) 2.

11.
$$
\begin{aligned}
F(x) &= 0 && \text{if } x < 2,\\
F(x) &= \tfrac{1}{36} && \text{if } 2 \le x < 3,\\
F(x) &= \tfrac{3}{36} && \text{if } 3 \le x < 4,\\
F(x) &= \tfrac{6}{36} && \text{if } 4 \le x < 5,\\
F(x) &= \tfrac{10}{36} && \text{if } 5 \le x < 6,\\
F(x) &= \tfrac{15}{36} && \text{if } 6 \le x < 7,\\
F(x) &= \tfrac{21}{36} && \text{if } 7 \le x < 8,\\
F(x) &= \tfrac{26}{36} && \text{if } 8 \le x < 9,\\
F(x) &= \tfrac{30}{36} && \text{if } 9 \le x < 10,\\
F(x) &= \tfrac{33}{36} && \text{if } 10 \le x < 11,\\
F(x) &= \tfrac{35}{36} && \text{if } 11 \le x < 12,\\
F(x) &= 1 && \text{if } x \ge 12.
\end{aligned}
$$

1.9 Page 64

1. No, because $\lim_{x\to 5} C(x)$ does not exist; hence $\lim_{x\to 5} C(x)$ does not equal $C(5)$.

3. Yes, because $\lim_{x\to 25} P(x) = P(25) = 2$. (Also P is a polynomial function.)

5. (**a**) No; (**b**) yes. **7.** (**a**) $\frac{1}{2}$; (**b**) because $\lim_{x\to 1} f(x) \neq f(1)$; (**c**) yes, $x = -1$.

9. (**a**) $x = 1$; (**b**) $x = 2$, $x = 3$; (**c**) $x = 0$, $x = -3$.

11.
$$
\begin{aligned}
\lim_{x\to c} (f(x) + g(x)) &= \lim_{x\to c} f(x) + \lim_{x\to c} g(x) && \text{by Theorem 2}(i)\\
&= f(c) + g(c) && \text{by Definition 9.}
\end{aligned}
$$

13. (**a**)

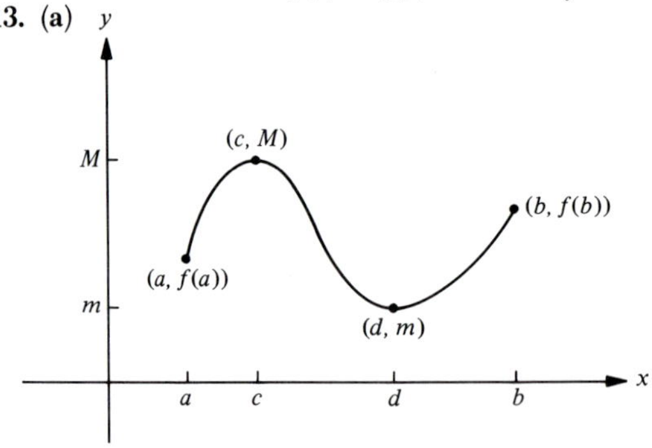

(**b**)

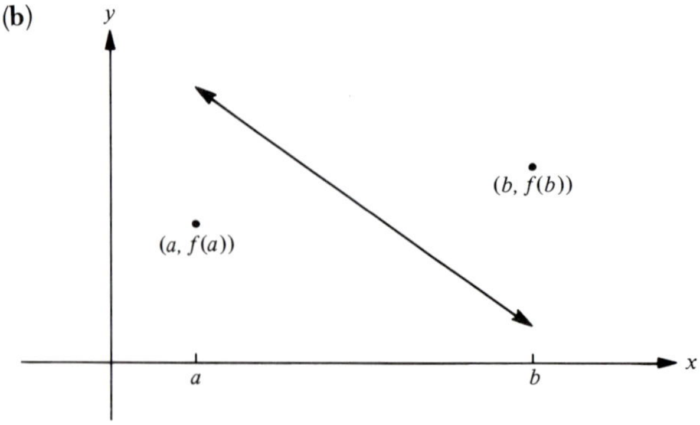

2.1 Page 75

1. **(a)** 1.25; **(b)** $-.19$.

3. Slope of secant line is $-\frac{5}{2}$.

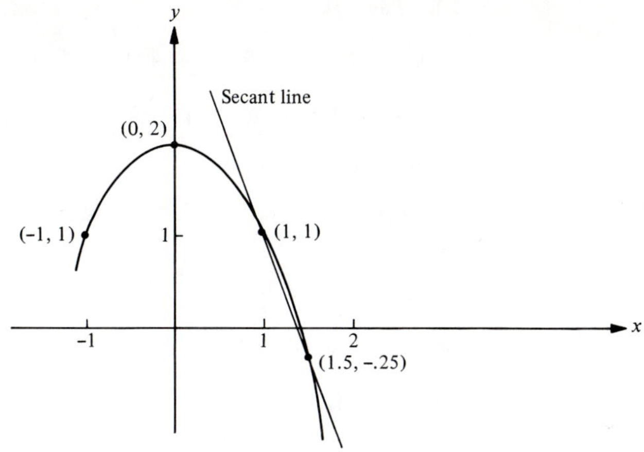

5. **(a)** $\Delta y = .728$; slope of secant line is 3.64; **(b)** $\Delta y = -1$; slope of secant line is 1; **(c)** 3.31. **7.** **(a)** .120601; **(b)** $-.119401$; **(c)** 12. **9.** -5; -7.

11. $f'(2) = \lim_{\Delta x \to 0} \frac{(2 + \Delta x)^3 - 8}{\Delta x} = \lim_{\Delta x \to 0} (12 + 6\,\Delta x + (\Delta x)^2) = 12.$

13. $3x + y + 4 = 0$.

15. There is no tangent line because $f'(1) = \lim_{\Delta x \to 0} \frac{\Delta y}{\Delta x}$ does not exist. Also, $\lim_{\Delta x \to 0} \frac{\Delta y}{\Delta x}$ does not exist because $\frac{\Delta y}{\Delta x} = 1$ whenever $\Delta x > 0$, but $\frac{\Delta y}{\Delta x} = -1$ whenever $\Delta x < 0$.

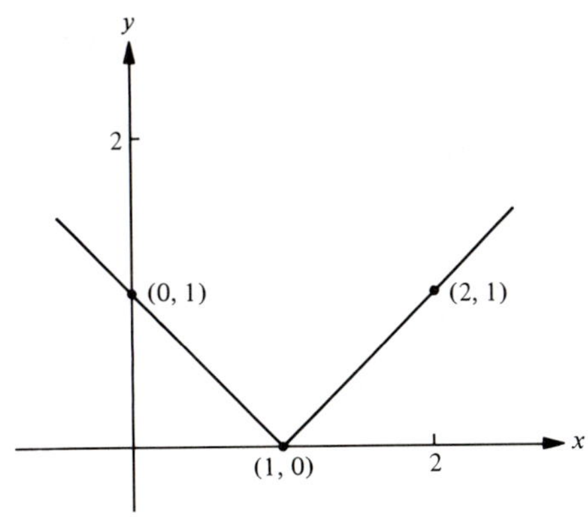

17. $x_1 = 5$.

2.2 Page 85

1. **(a)** -4; **(b)** -4. **3.** $f'(x) = 6x - 5$; domain of f' is the set of all real numbers. **5.** Yes; $f'(4) = -\frac{1}{9}$; no. **7.** $-1 - 2x$; -1. **9.** $\frac{7}{(2x + 1)^2}$.

11. -12. **13.** 2. **15.** **(a)** 3 feet per second; **(b)** 12 feet per second; **(c)** 7 feet per second. **17.** 2 seconds. **19.** .13.

2.3 Page 91

1. 0. **3.** $4x^3$. **5.** $\dfrac{1}{3x^{2/3}} + 1$. **7.** $15x^2 - 16x + 3$.

9. $5(x^2 + x)^4(2x + 1)$. **11.** 28. **13.** $-\frac{1}{9}$. **15.** $\frac{3}{8}$. **17.** $20 - \dfrac{x}{5}$.

19. $48 - \frac{3}{256}$. **21.** 168.

23. $$D_x(kf(x)) = \lim_{\Delta x \to 0} \frac{kf(x + \Delta x) - kf(x)}{\Delta x}$$

$$= k \lim_{\Delta x \to 0} \frac{f(x + \Delta x) - f(x)}{\Delta x} = kD_x f(x).$$

25. $x_2 = \frac{5}{4}$; $x_3 = \frac{239}{198} \approx 1.207$.

2.4 Page 97

1. $10x^4 + 6x$. **3.** $\dfrac{3x + 1}{\sqrt{2x + 1}}$. **5.** $\dfrac{2x}{(x^2 + 2)^2}$. **7.** $-6x^2 + 8x$.

9. -20. **11.** $-\frac{1}{2}$. **13.** $\dfrac{5x^2 + 10x - 2}{(x + 1)^2}$. **15.** **(a)** 129; **(b)** $\frac{5}{27}$.

17. $g(f(x)) = g(3x + 7) = (3x + 7)^2 + 1 = 9x^2 + 42x + 50$ and hence $D_x g(f(x)) = 18x + 42$; by the chain rule,

$$\frac{dy}{dx} = \frac{dy}{du}\frac{du}{dx} = (2u)(3) = 6u = 6(3x + 7) = 18x + 42.$$

19. Let n be a negative integer.

$D_x x^n = D_x \dfrac{1}{x^{-n}}$, where $-n$ is a positive integer.

$$D_x \frac{1}{x^{-n}} = \frac{(x^{-n})(0) - (1)(-n)x^{-n-1}}{(x^{-n})^2} = \frac{nx^{-n-1}}{x^{-2n}} = nx^{-n-1+2n} = nx^{n-1}.$$

21. $5t^4 + 4t - \dfrac{1}{t^2}$.

23. Yes; no.

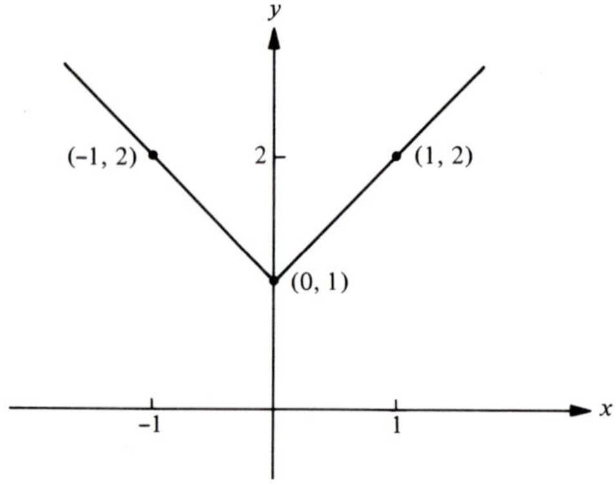

25. $\dfrac{x^2-1}{x^2}$; $E(x) = \dfrac{2x^2}{x^2+1}$.

2.5 Page 101

1. $28x^3$; $84x^2$; $168x$. **3.** $\frac{1}{2}x^{-1/2}$; $-\frac{1}{4}x^{-3/2}$; $\frac{3}{8}x^{-5/2}$; $-\frac{15}{16}x^{-7/2}$. **5.** $\dfrac{-3}{(x-4)^2}$; $\dfrac{6}{(x-4)^3}$. **7.** $210x^4$. **9.** $-\frac{49}{864}$. **11.** $-\dfrac{1}{2x^2}$. **13.** 24.

15. **(a)** Velocity is 1 meter per second; acceleration is 12 meters per second per second; **(b)** velocity is 192 meters per second; acceleration is 1792 meters per second per second. **17.** 14; speeding up.

2.6 Page 107

1. **(a)** $2x\,dx$; **(b)** $-\dfrac{3}{x^2}\,dx$; **(c)** dx; **(d)** $6x(1+x^2)^2\,dx$; **(e)** 0; **(f)** $(3x^2-8x)\,dx$.

3. 1.261; 1.2. **5.** $\dfrac{\pi}{3}(2.912)$; $.8\pi$. **7.** $20x^4 - 24x^2 + 14x - 8$; $80x^3 - 48x + 14$; $240x^2 - 48$. **9.** $C'(10) = 35$; $C(11) - C(10) = 34.75$; $P'(10) = 16$. **11.** $\dfrac{dy}{dx} = 27(3x+4)^8$; $\dfrac{d^2y}{dx^2} = 648(3x+4)^7$.

13. $9 + \frac{1}{18} \approx 9.056$.

15. Yes. Assume that x is the independent variable; then

$$dy = \left[\frac{d}{dx}(k)\right] dx = 0\,dx = 0.$$

2.7 Page 112

1. $-\dfrac{x}{y}$. **3.** $-\dfrac{y}{x}$. **5.** $\dfrac{y+8}{2y-x}$. **7.** $-\dfrac{4x^3+y^3}{3xy^2+5y^4}$. **9.** $-\dfrac{\sqrt{y}}{\sqrt{x}}$.

11. $-\dfrac{x}{y}$; $-\dfrac{4}{y^3}$. **13.** $-\dfrac{y}{x+2y}$; $\dfrac{24}{(x+2y)^3}$. **15.** $\dfrac{x}{y}$; $\dfrac{16}{y^3}$; $-\dfrac{48x}{y^5}$.

17. $\frac{2}{9}$. **19.** $-\dfrac{s+2t}{2s+t}$. **21.** $s = \frac{1}{8}$; $\frac{1}{64}$. **23.** $\dfrac{dh}{dr} = \dfrac{-2r-h}{r}$; $\dfrac{dr}{dh} = \dfrac{-r}{2r+h}$; $\dfrac{dr}{dh} = \dfrac{1}{dh/dr}$.

3.1 Page 120

1. \$245.10. **3.** **(a)** $[6, \infty)$; **(b)** $(-1, \infty)$; **(c)** $(-\infty, 20)$; **(d)** $(-\infty, 15]$.

5. **(a)**

Interval	Calculation of sign of $f(x)$	Result	Conclusion
$(0, 3)$	$(-)(-)$	$+$	$f(x)$ positive
$(3, 5)$	$(+)(-)$	$-$	$f(x)$ negative
$(5, \infty)$	$(+)(+)$	$+$	$f(x)$ positive

(b)

Interval	Calculation of sign of $f(x)$	Result	Conclusion
(0, 1)	$(-)(-)(-)$	$-$	$f(x)$ negative
(1, 2)	$(+)(-)(-)$	$+$	$f(x)$ positive
(2, 3)	$(+)(+)(-)$	$-$	$f(x)$ negative
(3, 5)	$(+)(+)(+)$	$+$	$f(x)$ positive

(c)

Interval	Calculation of sign of $f(x)$	Result	Conclusion
$(-\infty, -8)$	$\frac{(-)(+)}{(-)}$	$+$	$f(x)$ positive
$(-8, 1)$	$\frac{(-)(+)}{(+)}$	$-$	$f(x)$ negative
$(1, 4)$	$\frac{(-)(-)}{(+)}$	$+$	$f(x)$ positive
$(4, \infty)$	$\frac{(+)(-)}{(+)}$	$-$	$f(x)$ negative

7. **(a)** 0, 1; **(b)** $\frac{-2}{(x-1)^2}$; 1; **(c)** $\frac{4}{(x-1)^3}$; 1; **(d)** 1; **(e)** no. **9.** $f(p)$ is negative for $2 < p < 7$.

3.2 Page 126

1. Increasing: $(4, \infty)$; decreasing: $(0, 4)$. **3.** Increasing: $(-\infty, 4)$ and $(5, \infty)$; decreasing: $(4, 5)$. **5.** Decreasing: $(-\infty, 0)$ and $(0, \infty)$. **7.** Increasing: $(-2, 0)$ and $(3, \infty)$; decreasing: $(-\infty, -2)$ and $(0, 3)$. **9.** Increasing: $(-\infty, 0)$ and $(0, \infty)$. **11.** Increasing: $(-\infty, -2-\sqrt{3})$ and $(-2+\sqrt{3}, \infty)$; decreasing: $(-2-\sqrt{3}, -2+\sqrt{3})$. **13.** Increasing: $(0, 4)$ and $(4, \infty)$; decreasing: $(-\infty, 0)$. **15.** Increasing: $(-\infty, 0)$; decreasing: $(0, \infty)$.

17. Increasing: $(-\infty, -1)$, $(-\frac{1}{5}, 1)$, and $(1, \infty)$; decreasing: $(-1, -\frac{1}{5})$.
19. $y' = -\frac{100}{(x+1)^2}$, which is negative on $(0, 20)$.

3.3 Page 135

1. $-2, 2$. **3.** 0, 2.
5. $(-2, 16)$ maximum point; $(2, -16)$ minimum point. (See figure on p. 317.)

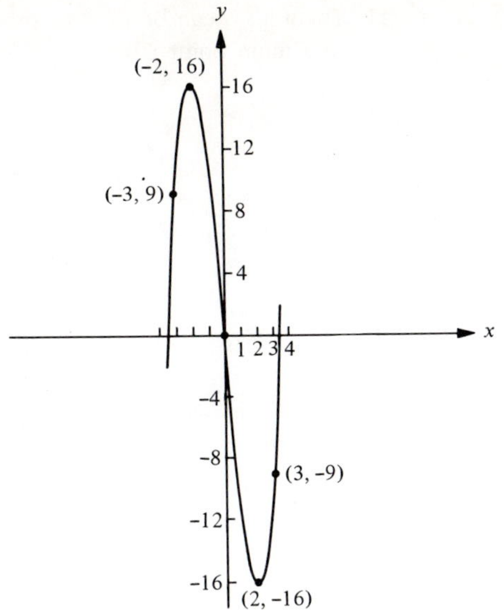

7. (0, 0) maximum point; (2, 4) minimum point. **9.** $(-1, 1)$ endpoint maximum; (0, 0) minimum point; (1, 1) endpoint maximum. **11.** $(0, -32)$ endpoint minimum; (3, 1) endpoint maximum. **13.** (0, 0) endpoint minimum; (2, 10) maximum point; (4, 8) endpoint minimum. **15.** $(-2, 30)$ maximum point; $(3, -\frac{65}{2})$ minimum point. **17.** $(5, -\frac{625}{3})$ minimum point. **19.** The function has no maximum or minimum points; change $>$ to $\geq$ and $<$ to $\leq$.

3.4 Page 142

1. 15. **3.** 95. **5.** $x = 5$. **7.** $x = 4$. **9.** $x = 225$ feet, $y = 300$ feet. **11.** $r = 3$ inches, $h = 6$ inches. **13.** $r = 2$ feet, $h = 2$ feet. **15.** 5 feet high, 10 feet wide. **17.** $\frac{c}{2}$. **19.** $x = 10$. **21.** **(a)** $s = 1$; **(b)** $x = 1$.

3.5 Page 150

1. $f''(x) = (x + 1)(x)(x - 1)$;

Interval	Calculation of sign of $f''(x)$	Result	Conclusion
$(-\infty, -1)$	$(-)(-)(-)$	$-$	f concave downward
$(-1, 0)$	$(+)(-)(-)$	$+$	f concave upward
$(0, 1)$	$(+)(+)(-)$	$-$	f concave downward
$(1, \infty)$	$(+)(+)(+)$	$+$	f concave upward

3. Concave upward: $(-3, 0)$ and $(3, \infty)$; concave downward: $(-\infty, -3)$ and $(0, 3)$. **5.** $(-2, 2)$. **7.** $(-2, -80)$ and $(2, -80)$. **9.** Concave upward on $(0, \infty)$; concave downward on $(-\infty, 0)$; $x = 0$ is not in the domain of the rational function f.

11. Theorem 6 can be used to show that $(0, 0)$ is a minimum point and $(\frac{2}{3}, \frac{4}{27})$ is a maximum point.

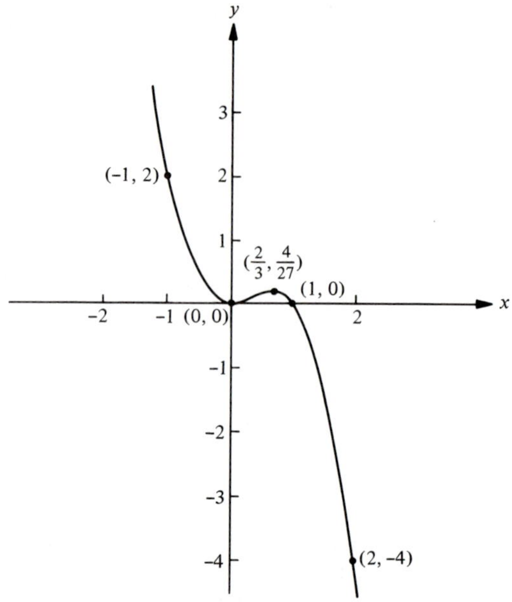

13. (**a**) $x = 3$; (**b**) $x = 12$. **15.** 12; $P''(12) = -20 < 0$; $R'(12) = C'(12) = 134$. **17.** $x = \dfrac{a}{2}$.

19. The function f'' is continuous at x_1, since f is a rational function. Since $f''(x_1) < 0$, there is an open interval containing x_1 throughout which $f''(x) < 0$ by Exercise 14 of Section 1.9. This interval can be chosen so that x_1 is the only critical number in the interval. By Theorem 1, f' is decreasing on this open interval. Since $f'(x_1) = 0$, $f'(x)$ changes from positive to zero to negative on this interval. By Theorem 3(*i*), $(x_1, f(x_1))$ is a maximum point.
21. (**a**) $x = 1$; (**b**) $x = 2$.

3.6 Page 159

1. (**a**) y-axis; (**b**) x-axis, y-axis, and origin; (**c**) x-axis, y-axis, and origin; (**d**) y-axis; (**e**) origin; (**f**) x-axis, y-axis, and origin; (**g**) y-axis; (**h**) origin; (**i**) x-axis; (**j**) none. **3.** (**a**) $y = \frac{2}{3}$; (**b**) $y = \frac{1}{5}$; (**c**) $y = 0$; (**d**) none.

5.

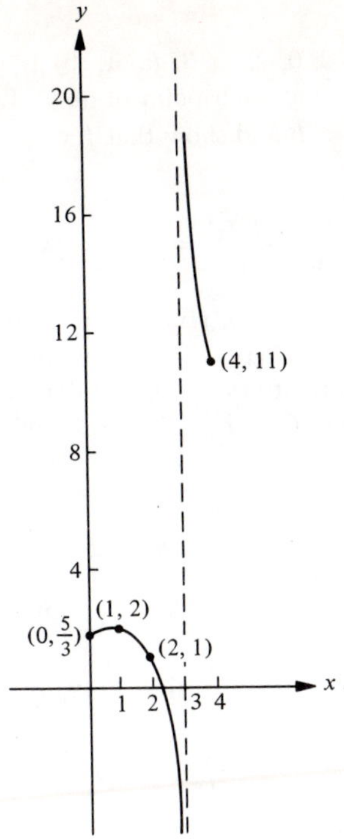
y
20
16
12
8
4
(4, 11)
(1, 2)
$(0, \frac{5}{3})$
(2, 1)
x
1 2 3 4

7.

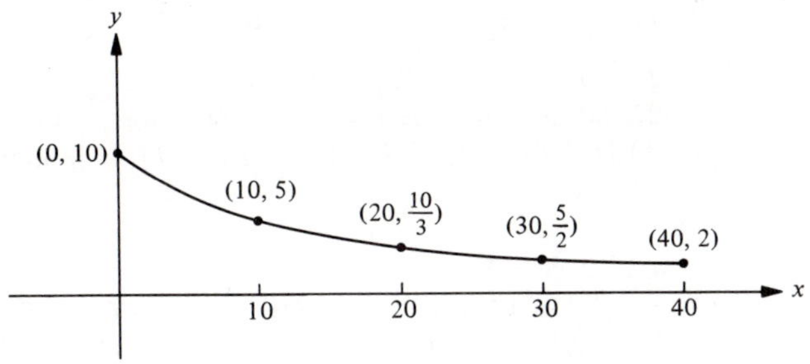
y
(0, 10)
(10, 5)
$(20, \frac{10}{3})$
$(30, \frac{5}{2})$
(40, 2)
x
10 20 30 40

9.

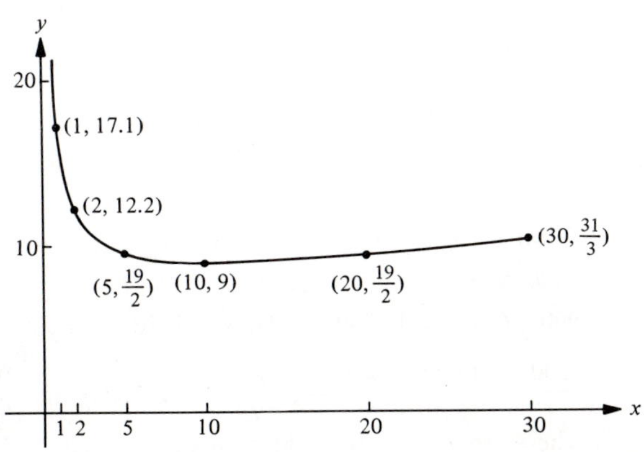
y
20
10
(1, 17.1)
(2, 12.2)
$(5, \frac{19}{2})$
(10, 9)
$(20, \frac{19}{2})$
$(30, \frac{31}{3})$
x
1 2 5 10 20 30

3.7 Page 165

1. (a) 0; (b) $-\frac{2}{3}$; (c) $-2, 0, 2$. **3.** (a) 1; (b) 1.

5. Assume that $f'(x) < 0$ at each point of (a, b). Let x_1 and x_2 be numbers such that $a < x_1 < x_2 < b$ and show that $f(x_1) > f(x_2)$. As in the proof of Theorem 1(*i*),

$$\frac{f(x_2) - f(x_1)}{x_2 - x_1} = f'(c), \qquad x_1 < c < x_2.$$

Since $f'(c) < 0$ and $x_2 - x_1 > 0$, $f(x_2) - f(x_1) < 0$ and hence $f(x_1) > f(x_2)$.

7. By Theorem 10, $G(x) - F(x) = C$, where C is a constant, since two functions with equal derivatives on (a, b) differ by a constant on (a, b). Therefore, $G(x) = F(x) + C$. If $F(x) = x^2 + 3$ and $G(x) = x^2 + 7$, then $C = 4$.

4.1 Page 170

1. $\frac{x^4}{4} + C$. **3.** $\frac{x^3}{3} - \frac{3x^2}{2} + C$. **5.** $-\frac{1}{t} + C$. **7.** $\frac{x^{\pi+1}}{\pi + 1} - \frac{x^3}{3} + \frac{3x^2}{2} + 8x + C$. **9.** $\frac{2x^3}{3} - \frac{11x^2}{2} + 9x + C$. **11.** $\frac{5x^7}{7} - \frac{2x^3}{3} + C$; to check, show that $D_x\left(\frac{5x^7}{7} - \frac{2x^3}{3} + C\right) = 5x^6 - 2x^2$. **13.** $\frac{x^3}{3} + 2x - \frac{1}{x} + C$; to check, show that $D_x\left(\frac{x^3}{3} + 2x - \frac{1}{x} + C\right) = \frac{(x^2 + 1)^2}{x^2}$. **15.** Yes, because $g'(x) = f(x)$. **17.** No, because $F'(x) \neq f(x)$. **19.** $-\frac{2}{\sqrt{x}} - \frac{3}{2x^{2/3}} + C$.

21. $4x - \frac{8}{\sqrt{x}} - \frac{1}{2x^2} + C$. **23.** $-\frac{2}{x^2} + \frac{4}{3x^3} - \frac{1}{4x^4} + C$.

25. $2y^4 - 12y^3 + 27y^2 - 27y + C$.

27. (a) $D_x(x + C) = 1 + 0 = 1$. Therefore, $\int 1\, dx = \int dx = x + C$.
(b) $D_x[k \int f(x)\, dx] = k\, D_x[\int f(x)\, dx] = kf(x)$, from which Theorem 2(*ii*) follows.

4.2 Page 175

1. $\frac{1}{8}(3x^2 + 1)^8 + C$. **3.** $\frac{1}{6}(4x + 3)^{3/2} + C$. **5.** $\frac{1}{3}(x^2 + 3)^{3/2} + C$.
7. $\frac{1}{2}\sqrt{2x^2 + 5} + C$. **9.** $\frac{1}{3}(1 + \sqrt{x})^6 + C$. **11.** $\frac{1}{3}x^3 + \frac{2}{9}(3x)^{3/2} + C$.
13. $\frac{2}{5}x^{5/2} + C$. **15.** $-\frac{1}{9(3x + 1)^3} + C$. **17.** $D_x[\frac{1}{2}(2x^2 + 5)^{1/2} + C] = (\frac{1}{2})(\frac{1}{2})(2x^2 + 5)^{-1/2}(4x) = \frac{x}{\sqrt{2x^2 + 5}}$. **19.** (*b*); $\frac{1}{3}(9 + x^2)^{3/2} + C$.

21. $y = \frac{50}{3(3x + 1)^2} + C$, provided that $0 \leq x \leq 5$.

4.3 Page 182

1. (a) 1; (b) 2; (c) 6; (d) 2. **3.** (a) $y = 2x^2 + 1$; (b) $y = x^3 + 4x - 2$.
5. $y = \frac{1}{3}x^3 + x + \frac{8}{3}$. **7.** $s = \frac{3}{8}(t^2 + 4)^{4/3} + 2$. **9.** $C(x) = 1000x - x^2 + 2000$; $C(10) = 11{,}900$. **11.** $s = -16t^2 + v_0t$. **13.** $\frac{dP}{dy} = \frac{(200 - 10y)(200 + 10y)}{y^2}$; $\frac{dP}{dy} > 0$ on (8, 20) and $\frac{dP}{dy} < 0$ on (20, 25).
Therefore, maximum profit occurs at $y = \$20$.

5.1 Page 189

1. $S_3 = \frac{137}{16}$. **3.** $S_4 = \frac{239}{64}$. **5.** $S_5 = \frac{153}{32}$. **7.** $\frac{1}{2}, \frac{1}{2}, \frac{1}{4}, \frac{1}{2}, \frac{3}{4}, \frac{1}{2}; \frac{3}{4}$. **9.** $\frac{1}{2}$.
11. **(a)** $\frac{71}{8}$; **(b)** $T_n = \frac{1}{2}\Delta x[f(x_0) + 2(f(x_1) + f(x_2) + \cdots + f(x_{n-1})) + f(x_n)]$.
13. **(a)** $\frac{1}{2} + \frac{2}{3} + \frac{3}{4} + \frac{4}{5} + \frac{5}{6}$; **(b)** $\Sigma_{k=1}^{8} 5k = 5 + 10 + 15 + 20 + 25 + 30 + 35 + 40 = 5(1 + 2 + 3 + 4 + 5 + 6 + 7 + 8) = 5\Sigma_{k=1}^{8} k$;
(c) $\Sigma_{k=1}^{n} k^2$; **(d)** two possible answers are $\Sigma_{k=3}^{8} 5k$ and $\Sigma_{k=1}^{6} (10 + 5k)$;
(e) $\Sigma_{k=1}^{3} f(c_k)\,\Delta x_k$.

5.2 Page 198

1. $\frac{26}{3}$. **3.** $-\frac{4}{3}$. **5.** 10. **7.** $\frac{1}{9}$.
9. The hypothesis of Theorem 2 is not satisfied. The function f given by $f(x) = 1/x^2$ is not continuous at $x = 0$; therefore, it is not continuous on $[-1, 1]$.
11. Integrand: x^2; upper limit: 4; lower limit: 3; variable of integration: x.
13. $\frac{37}{3}$. **15.** $\frac{49}{3}$. **17.** $\frac{2}{27}$. **19.** $\$(-\frac{1}{9})$.

5.3 Page 206

1. 3. **3.** 4. **5.** -3. **7.** -4. **9.** $\dfrac{\sqrt{39}}{3}$. **11.** $\frac{1}{2}$. **13.** $\frac{86}{3}$.

15. $\frac{44}{3}$. **17.** $\dfrac{P\pi R^4}{8\eta l}$. **19.** $\frac{23}{3}$.

5.4 Page 213

1. $\frac{7}{3}$. **3.** $\frac{9}{2}$. **5.** $\frac{9}{2}$. **7.** 12. **9.** $\frac{32}{3}$. **11.** 4. **13.** **(a)** $\frac{686}{15}$; **(b)** $\frac{80}{3}$.
15. **(a)** \$533.33; **(b)** \$133.33.

5.5 Page 219

1. Each side of the equation equals 28. **3.** $\dfrac{5}{7}; \dfrac{5}{n}$. **5.** 6. **7.** $\frac{45}{2}$.

9. $\frac{88}{3}$. **11.** $\frac{26}{3}$.

5.6 Page 227

1. w; [1, 5]. **3.** x^2. **5.** $x\sqrt{1 + x^2}$. **7.** Left side is zero by Definition 2(*i*); right side is $F(b) - F(b) = 0$. **9.** $\frac{20}{27}$; probability of death by age 80 is $\frac{20}{27}$. **11.** .6.

6.1 Page 235

1. Domain: set of all real numbers; range: set of all positive numbers; increasing; 0. (See figure on p. 322.)

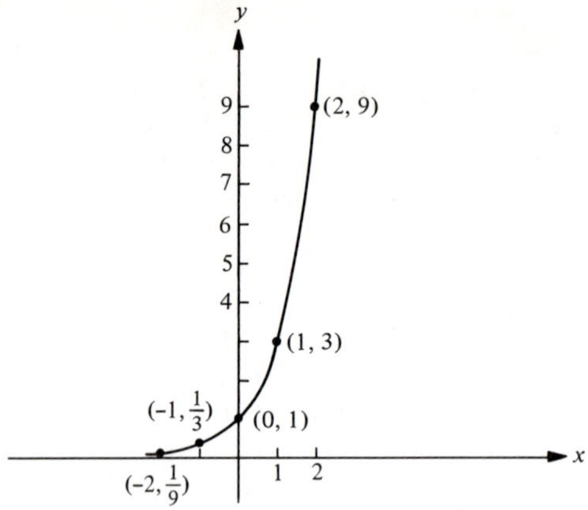

3.

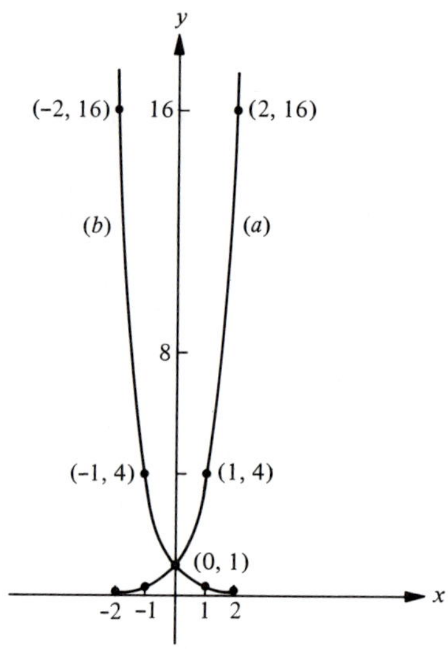

5. **(a)** 4; **(b)** -4; **(c)** not possible. **7.** **(a)** 32; **(b)** 16; **(c)** 9; **(d)** 125.
9. 2.488. **11.** e^2. **13.** **(a)** \$104.06; **(b)** \$104.08.

15.

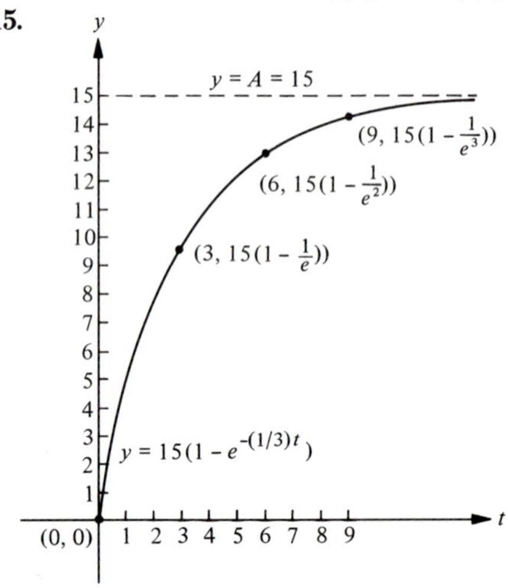

6.2 Page 241

1. (**a**) 3; (**b**) 3; (**c**) -2.

3. Yes, f and g are inverse functions; domain of f: set of all real numbers; range of f: set of all positive real numbers; domain of g: set of all positive real numbers; range of g: set of all real numbers.

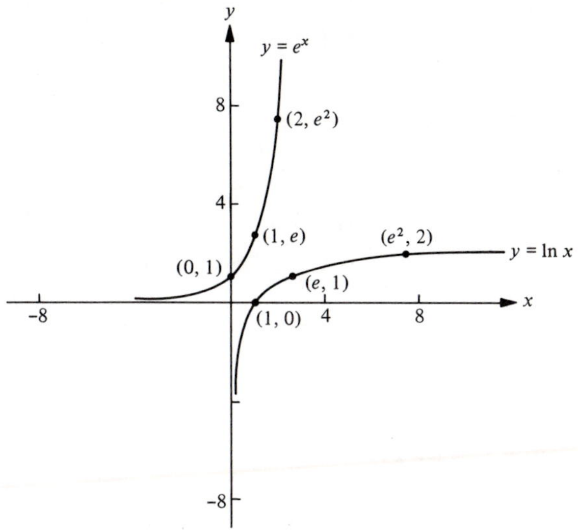

5.

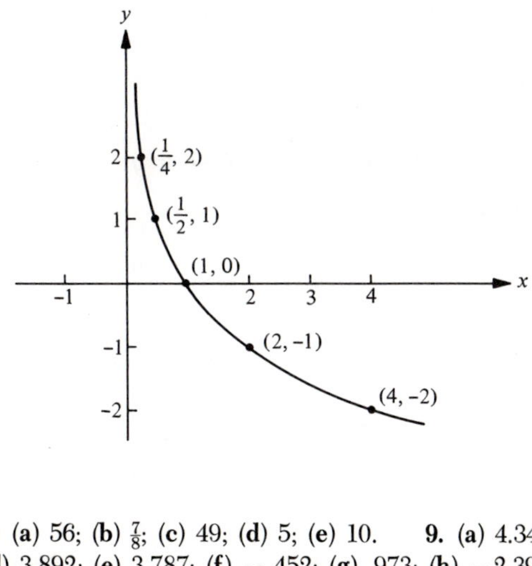

7. (**a**) 56; (**b**) $\frac{7}{8}$; (**c**) 49; (**d**) 5; (**e**) 10. **9.** (**a**) 4.344; (**b**) .452; (**c**) 1.494; (**d**) 3.892; (**e**) 3.787; (**f**) $-.452$; (**g**) .973; (**h**) -2.398; (**i**) 1.946; (**j**) .417; (**k**) 7; (**l**) 49; (**m**) 1.232.

11. Let $x = \log_b M$; then $b^x = M$. Let $y = \log_b N$; then $b^y = N$.

$$MN = b^x b^y = b^{x+y},$$
$$\log_b MN = x + y = \log_b M + \log_b N.$$

13. (**a**) *Hint:* in part (*viii*) let $M = b$; then use part (*ii*). (**b**) *Hint:* in part (*x*) let $M = b$; then use part (*ii*). **15.** The combination is symmetric about $y = x$. (See figure on p. 324.)

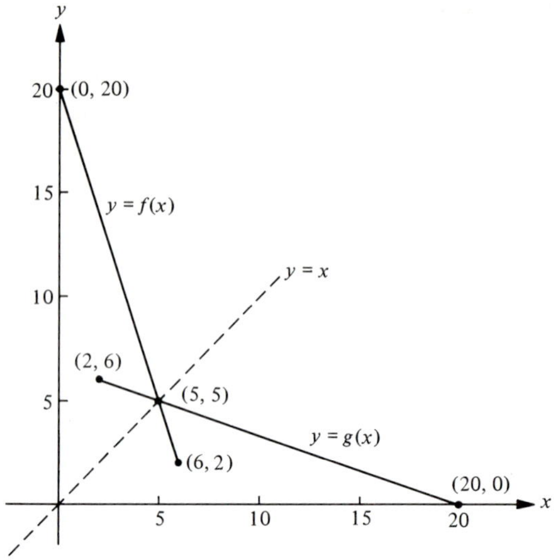

6.3 Page 248

1. **(a)** $\dfrac{2\log_3 e}{2x+1}$; **(b)** $\dfrac{3(\log_{10} e)x^2}{x^3+2}$; **(c)** $\dfrac{2}{2x+5}$; **(d)** $\dfrac{1}{2x}$; **(e)** $\dfrac{8(1+\ln x)^7}{x}$; **(f)** $x^3(1+4\ln x)$; **(g)** $2(\ln 3)3^{2x}$; **(h)** $2^x(1+(\ln 2)x)$; **(i)** $\dfrac{e^x+e^{-x}}{2}$; **(j)** $\dfrac{e^{\sqrt{2x+1}}}{\sqrt{2x+1}}$; **(k)** $\dfrac{27e^{3x}}{(9+e^{3x})^2}$; **(l)** $\dfrac{e^{2x}}{\sqrt{1+e^{2x}}}$.

3. Let $y = u^n$, where $u > 0$. Then $\ln y = n \ln u$ and $\dfrac{1}{y} D_x y = \dfrac{n}{u} D_x u$. Therefore,

$$D_x y = \frac{ny}{u} D_x u = \frac{nu^n}{u} D_x u = nu^{n-1} D_x u.$$

5. **(a)** $\left(\dfrac{1}{\sqrt{e}}, -\dfrac{1}{2e}\right)$ minimum point; **(b)** $\left(\dfrac{1}{e}, -\dfrac{1}{e}\right)$ minimum point; **(c)** $(0, 1)$ maximum point; **(d)** $\left(-\dfrac{1}{\sqrt{2}}, -\sqrt{\dfrac{2}{e}}\right)$ minimum point, $\left(\dfrac{1}{\sqrt{2}}, \sqrt{\dfrac{2}{e}}\right)$ maximum point.

7. **(a)** 327 ounces; **(b)** The rate of growth is 39 ounces per month at the instant the child is 4 months old, but this rate is always changing. In fact, it can be shown that the rate of growth gradually decreases throughout the fifth month. **9.** 1.39.

6.4 Page 256

1. **(a)** $\dfrac{1}{4}\dfrac{7^{4x}}{\ln 7} + C$; **(b)** $\dfrac{1}{2}\dfrac{2^{1+x^2}}{\ln 2} + C = \dfrac{(2)^{x^2}}{\ln 2} + C$; **(c)** $\dfrac{e^x+e^{-x}}{2} + C$; **(d)** $\dfrac{1}{2}\ln|2x+1| + C$; **(e)** $\dfrac{1}{3}\ln|x^3+2| + C$; **(f)** $\dfrac{(\ln x)^2}{2} + C$. **3.** **(a)** $3(e-1)$;

(**b**) 2; (**c**) $\frac{1}{11}$. **5.** $x - \ln|x+1| + C$. **7.** (**a**) $e^2 - 1$; (**b**) $e - 1$; (**c**) $\dfrac{2}{\ln 3} - \dfrac{3}{2}$; (**d**) $\frac{1}{2}\ln 2$; (**e**) $\ln 5$.

6.5 Page 263

1. $y = 900(\frac{5}{3})^{t/5}$. **3.** 130 milligrams. **5.** (**a**) 632 bacteria; (**b**) 2.95 hours.
7. \$633. **9.** 19 days.

6.6 Page 267

1. $\dfrac{1}{12}\ln\left|\dfrac{3+2x}{3-2x}\right| + C$. **3.** Formula 24; $\frac{1}{4}(2x-1)\sqrt{4x^2-4x+10} + \frac{9}{4}\ln|2x-1+\sqrt{4x^2-4x+10}| + C$.
5. Formula 16; $\dfrac{1}{4}\left(\dfrac{3}{3+2x} + \ln|3+2x|\right) + C$.
7. Formulas 18 and 6; $\frac{1}{2}xe^{2x} - \frac{1}{4}e^{2x} + C$.
9. Formula 21; $\frac{1}{3}x^3\ln 2x - \frac{1}{9}x^3 + C$. **11.** $\frac{1}{2}(e^2+1)$.

6.7 Page 277

1. Converges; 1.

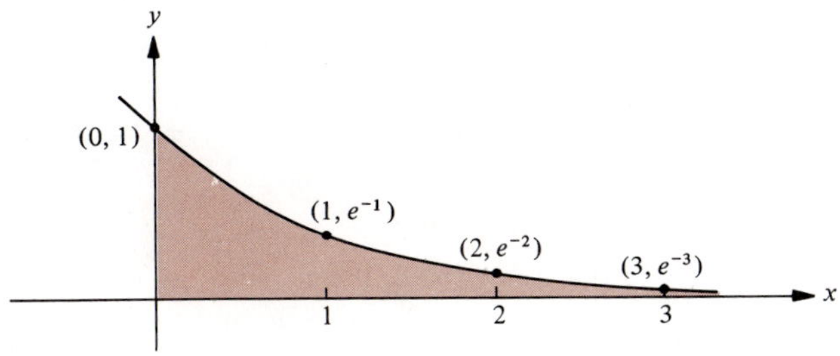

3. Converges; $\frac{1}{2}$.

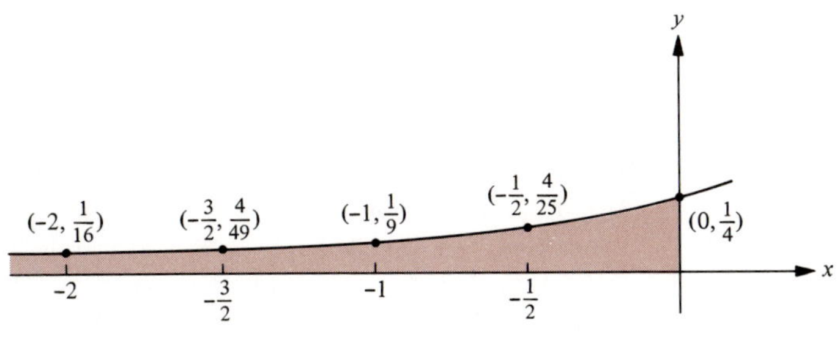

5. Diverges. **7.** Converges; $\frac{1}{18}$. **9.** Diverges. **11.** Converges; $\frac{1}{2}$.
13. Converges; 0. **15.** 0; the mean of the standard normal distribution.
17. (**a**) .3085; (**b**) 2.27 per cent; (**c**) 68.26 per cent.

7.1 Page 283

1. (**a**) 10; (**b**) -2. **3.** (**a**) 8; (**b**) 10.

5.

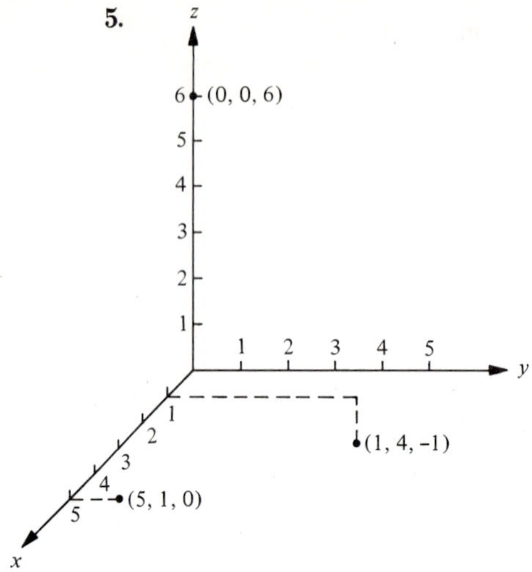

7.

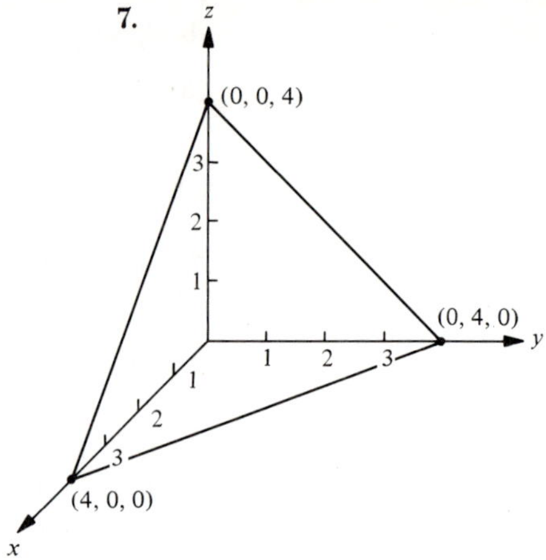

9. w, x, and y are independent variables; C is the dependent variable.
11. **(a)** $\sqrt{29}$; **(b)** 7.

7.2 Page 289

1. $\dfrac{\partial f}{\partial x} = 3x^2 + 3y^2$; $\dfrac{\partial f}{\partial y} = 6xy + 3y^2$. **3.** $\dfrac{\partial z}{\partial x} = 2x + ye^{xy}$; $\dfrac{\partial z}{\partial y} = xe^{xy}$.

5. $f_x(2, 1) = 17$; $f_y(2, 1) = 18$. **7.** $\dfrac{\partial z}{\partial x} = yw^2e^{xyw^2} + \dfrac{1}{w + y}$; $\dfrac{\partial z}{\partial y} = xw^2e^{xyw^2} - \dfrac{x}{(w + y)^2}$.

9. In the plane $x = 2$, the slope of the tangent line to the curve of intersection of $z = 9 - x^2 - y^2$ and the plane $x = 2$ at $(2, 1, 4)$ is $f_y(2, 1) = -2$.
11. Each side of the equation equals $9x^2y^2 + 4x$. **13.** 4; 0.

15. **(a)** $224(w + 2x + 3y)^6$; **(b)** $168(w + 2x + 3y)^6$; **(c)** $112(w + 2x + 3y)^6$.

17. $\dfrac{\partial P}{\partial w} = 2 + x$; $\dfrac{\partial P}{\partial x} = 3 + w$; $\dfrac{\partial P}{\partial y} = 4$.

7.3 Page 296

1. $(10, 8, 600)$ neither. **3.** $(1, 2, 5)$ minimum point. **5.** $(0, 0, \frac{31}{16})$ minimum point; $(-\frac{1}{2}, -\frac{1}{4}, 2)$ neither. **7.** $x = 4$, $y = 1$.
9. $(w, x, y, z) = (-2, 4, 6, 50)$ is the only critical point and it is a maximum point.
11. $y = -\frac{42}{37}x - \frac{51}{37}$.

7.4 Page 303

1. Minimum $f(x, y)$ is 20 at $x = 2$, $y = 4$. **3.** Each number is 12.
5. $w = 3$, $x = 3$, $y = 6$. **7.** Maximum $f(x, y)$ is $10{,}000\sqrt{10}$ at $x = 1000$, $y = 1000$. **9.** Length: 100 feet; width: 50 feet. **11.** $w = 6$, $x = 2$, $y = 2$.

Index